Learn Java 12 Programming

A step-by-step guide to learning essential concepts in Java
SE 10, 11, and 12

Nick Samoylov

BIRMINGHAM - MUMBAI

Learn Java 12 Programming

Copyright © 2019 Packt Publishing

Commissioning Editor: Richa Tripathi
Acquisition Editor: Denim Pinto
Content Development Editor: Digvijay Bagul
Technical Editor: Neha Pande
Copy Editor: Safis Editing
Project Coordinator: Prajakta Naik
Proofreader: Safis Editing
Indexer: Tejal Daruwale Soni
Graphics: Tom Scaria
Production Coordinator: Tom Scaria

First published: April 2019
Production reference: 2250619

Published by Packt Publishing Ltd.
Livery Place
35 Livery Street
Birmingham
B3 2PB, UK.

ISBN 978-1-78995-705-1

www.packtpub.com

To my wife Luda, a software developer too, who was the source of my inspiration
and the sound technical advice for this book

– Nick Samoylov

`mapt.io`

Mapt is an online digital library that gives you full access to over 5,000 books and videos, as well as industry leading tools to help you plan your personal development and advance your career. For more information, please visit our website.

Why subscribe?

- Spend less time learning and more time coding with practical eBooks and Videos from over 4,000 industry professionals

- Improve your learning with Skill Plans built especially for you

- Get a free eBook or video every month

- Mapt is fully searchable

- Copy and paste, print, and bookmark content

Packt.com

Did you know that Packt offers eBook versions of every book published, with PDF and ePub files available? You can upgrade to the eBook version at `www.packt.com` and as a print book customer, you are entitled to a discount on the eBook copy. Get in touch with us at `customercare@packtpub.com` for more details.

At `www.packt.com`, you can also read a collection of free technical articles, sign up for a range of free newsletters, and receive exclusive discounts and offers on Packt books and eBooks.

Contributors

About the author

Nick Samoylov graduated from Moscow Institute of Physics and Technology, working as a theoretical physicist and learning to program as a tool for testing his mathematical models. After the demise of the USSR, Nick created and successfully ran a software company, but was forced to close it under the pressure of governmental and criminal rackets. In 1999, with his wife Luda and two daughters, he emigrated to the USA and has been living in Colorado since then, working as a Java programmer. In his free time, Nick likes to write and hike in the Rocky Mountains.

Many thanks to Kaushlendra Kumar Singh for support and Venkat Subramaniam for reading parts of the draft and providing the valuable feedback.

About the reviewer

Aristides Villarreal Bravo is a Java developer, a member of the NetBeans Dream Team, and a Java User Groups leader. He lives in Panama. He has organized and participated in various conferences and seminars related to Java, JavaEE, NetBeans, the NetBeans platform, free software, and mobile devices. He is the author of *jmoordb*, and writes tutorials and blogs about Java, NetBeans, and web development. He has participated in several interviews on sites about topics such as NetBeans, NetBeans DZone, and JavaHispano. He is a developer of plugins for NetBeans.

Packt is searching for authors like you

If you're interested in becoming an author for Packt, please visit `authors.packtpub.com` and apply today. We have worked with thousands of developers and tech professionals, just like you, to help them share their insight with the global tech community. You can make a general application, apply for a specific hot topic that we are recruiting an author for, or submit your own idea.

Table of Contents

Preface

The purpose of this book is to equip the readers with a solid understanding of Java fundamentals and to lead them through a series of practical steps from the basics to the actual real programming. The discussion and examples aim to stimulate the growth of the reader's professional intuition by using proven programming principles and practices. The book starts with the basics and brings the readers up to the latest programming technologies, considered on a professional level.

After finishing this book, you will be able to do the following:

- Install and configure your Java development environment.
- Install and configure your **Integrated Development Environment (IDE)**–essentially, your editor.
- Write, compile, and execute Java programs and tests.
- Understand and use Java language fundamentals.
- Understand and apply object-oriented design principles.
- Master the most frequently used Java constructs.
- Learn how to access and manage data in the database from Java application.
- Enhance your understanding of network programming.
- Learn how to add graphical user interface for better interaction with your application.
- Become familiar with the functional programming.
- Understand the most advanced data processing technologies—streams, including parallel and reactive streams.
- Learn and practice creating microservices and building a reactive system.
- Learn the best design and programming practices.
- Envision Java's future and learn how you can become part of it.

Who this book is for

This book is for those who would like to start a new career in the modern Java programming profession, as well as those who do it professionally already and would like to refresh their knowledge of the latest Java and related technologies and ideas.

What this book covers

Chapter 1, *Getting Started with Java 12*, begins with the basics, first explaining what "Java" is and defining its main terms, then going on to how to install the necessary tools to write and run (execute) a program. This chapter also describes the basic Java language constructs, illustrating them with examples that can be executed immediately.

Chapter 2, *Object-Oriented Programming (OOP)*, presents the concepts of object-oriented programming and how they are implemented in Java. Each concept is demonstrated with specific code examples. The Java language constructs of class and interface are discussed in detail, as well as overloading, overriding, hiding, and use of the `final` keyword. The last section of the chapter is dedicated to presenting the power of polymorphism.

Chapter 3, *Java Fundamentals*, presents to the reader a more detailed view of Java as a language. It starts with the code organization in packages and a description of the accessibility levels of classes (interfaces) and their methods and properties (fields). The reference types as the main types of Java's object-oriented nature are presented in much detail, followed by a list of reserved and restricted keywords and a discussion of their usage. The chapter ends with the methods of conversion between primitive types, and from a primitive type to the corresponding reference type and back.

Chapter 4, *Exception Handling*, tells the reader about the syntax of the Java constructs related to exception handling and the best practices to address (handle) exceptions. The chapter ends with the related topic of the assertion statement that can be used to debug the application code in production.

Chapter 5, *Strings, Input/Output, and Files*, discusses the String class methods, as well as popular string utilities from standard libraries and the Apache Commons project. An overview of Java input/output streams and related classes of the `java.io` package follow along with some classes of the `org.apache.commons.io` package. The file-managing classes and their methods are described in a dedicated section.

Chapter 6, *Data Structures, Generics, and Popular Utilities*, presents the Java collections framework and its three main interfaces, List, Set, and Map, including discussion and demonstration of generics. The equals() and hashCode() methods are also discussed in the context of Java collections. Utility classes for managing arrays, objects, and time/date values have corresponding dedicated sections, too.

Chapter 7, *Java Standard and External Libraries*, provides an overview of the functionality of the most popular packages of **Java Class Library**
(**JCL**): java.lang, java.util, java.time, java.io and java.nio, java.sql and javax.sql, java.net, java.lang.math, java.math, java.awt, javax.swing, and javafx. The most popular external libraries are represented by the org.junit, org.mockito, org.apache.log4j, org.slf4j,
and org.apache.commons packages. This chapter helps the reader to avoid writing custom code in cases where such functionality already exists and can be just imported and used out of the box.

Chapter 8, *Multithreading and Concurrent Processing*, presents the ways to increase Java application performance by using workers (threads) that process data concurrently. It explains the concept of Java threads and demonstrates their usage. It also talks about the difference between parallel and concurrent processing, and how to avoid unpredictable results caused by the concurrent modification of a shared resource.

Chapter 9, *JVM Structure and Garbage Collection*, provides the readers with an overview of JVM's structure and behavior, which are more complex than we usually expect. One of the service threads, called *garbage collection*, performs an important mission of releasing the memory from unused objects. After reading this chapter, the readers will understand better what constitutes Java application execution, Java processes inside JVM, garbage collection, and how JVM works in general.

Chapter 10, *Managing Data in a Database*, explains and demonstrates how to manage—that is, insert, read, update, and delete–data in a database from a Java application. It also provides a short introduction to the SQL language and basic database operations: how to connect to a database, how to create a database structure, how to write a database expression using SQL, and how to execute them.

Chapter 11, *Network Programming*, describes and discusses the most popular network protocols, **User Datagram Protocol (UDP)**, **Transmission Control Protocol (TCP)**, **HyperText Transfer Protocol (HTTP)**, and WebSocket, and their support for JCL. It demonstrates how to use these protocols and how to implement client-server communication in Java code. The APIs reviewed include URL-based communication and the latest Java HTTP Client API.

Chapter 12, *Java GUI Programming*, provides an overview of Java GUI technologies and demonstrates how the JavaFX kit can be used to create a GUI application. The latest versions of JavaFX not only provide many helpful features, but also allow preserving and embedding legacy implementations and styles.

Chapter 13, *Functional Programming*, explains what a functional interface is, provides an overview of functional interfaces that come with JDK, and defines and demonstrates lambda expressions and how to use them with functional interfaces, including using method reference.

Chapter 14, *Java Standard Streams*, talks about the processing of data streams, which are different from the I/O streams reviewed in Chapter 5, *Strings, Input/Output, and Files*. It defines what data streams are, how to process their elements using methods (operations) of the `java.util.stream.Stream` object, and how to chain (connect) stream operations in a pipeline. It also discusses the stream's initialization and how to process the stream in parallel.

Chapter 15, *Reactive Programming*, introduces the Reactive Manifesto and the world of reactive programming. It starts with defining and discussing the main related concepts – "asynchronous", "non-blocking", "responsive", and so on. Using them, it then defines and discusses reactive programming, the main reactive frameworks, and talks about RxJava in more details.

Chapter 16, *Microservices*, explains how to build microservices – the foundational component for creating a reactive system. It discusses what a microservice is, how big or small they can be, and how existing microservices frameworks support message-driven architecture. The discussion is illustrated with a detailed code demonstration of a small reactive system built using the Vert.x toolkit.

Chapter 17, *Java Microbenchmark Harness*, presents the **Java Microbenchmark Harness (JMH)** project that allows us to measure various code performance characteristics. It defines what JMH is, how to create and run a benchmark, what the benchmark parameters are, and outlines supported IDE plugins. The chapter ends with some practical demo examples and recommendations.

Chapter 18, *Best Practices for Writing High-Quality Code*, introduces Java idioms and the most popular and useful practices for designing and writing application code.

Chapter 19, *Java Gets New Features*, talks about current the most significant projects that will add new features to Java and enhance it in other aspects. After reading this chapter, the reader will understand how to follow Java development and will be able to envision the roadmap of future Java releases. If so desired, the reader can become the JDK source contributor too.

To get the most out of this book

Read the chapters systematically and answer the quiz questions at the end of each chapter. Clone or just download the source code repository (see the following sections) and run all the code samples that demonstrate the discussed topics. For getting up to speed in programming, there is nothing better than executing the provided examples, modifying them, and trying your own ideas. Code is truth.

Download the example code files

You can download the example code files for this book from your account at www.packt.com. If you purchased this book elsewhere, you can visit www.packt.com/support and register to have the files emailed directly to you.

You can download the code files by following these steps:

1. Log in or register at www.packt.com
2. Select the **SUPPORT** tab
3. Click on **Code Downloads & Errata**
4. Enter the name of the book in the **Search** box and follow the onscreen instructions

Once the file is downloaded, please make sure that you unzip or extract the folder using the latest version of:

- WinRAR/7-Zip for Windows
- Zipeg/iZip/UnRarX for Mac
- 7-Zip/PeaZip for Linux

The code bundle for the book is also hosted on GitHub at https://github.com/PacktPublishing/Learn-Java-12-Programming. In case there's an update to the code, it will be updated on the existing GitHub repository.

We also have other code bundles from our rich catalog of books and videos available at https://github.com/PacktPublishing/. Check them out!

Download the color images

We also provide a PDF file that has color images of the screenshots/diagrams used in this book. You can download it here: https://www.packtpub.com/sites/default/files/downloads/9781789957051_ColorImages.pdf.

Conventions used

There are a number of text conventions used throughout this book.

`CodeInText`: Indicates code words in text, database table names, folder names, filenames, file extensions, pathnames, dummy URLs, user input, and Twitter handles. Here is an example: "When an exception is thrown inside a `try` block, it redirects control flow to the first `catch` clause."

A block of code is set as follows:

```
void someMethod(String s){
    try {
        method(s);
    } catch (NullPointerException ex){
        //do something
    } catch (Exception ex){
        //do something else
    }
}
```

When we wish to draw your attention to a particular part of a code block, the relevant lines or items are set in bold:

```
class TheParentClass {
    private int prop;
    public TheParentClass(int prop){
        this.prop = prop;
    }
    // methods follow
}
```

Any command-line input or output is written as follows:

```
--module-path /path/JavaFX/lib \
        :-add-modules=javafx.controls,javafx.fxml
```

Bold: Indicates a new term, an important word, or words that you see onscreen. For example, words in menus or dialog boxes appear in the text like this. Here is an example: "Select a value for **Project SDK** (java version 12, if you have installed JDK 12 already) and click **Next**."

 Warnings or important notes appear like this.

Tips and tricks appear like this.

Get in touch

Feedback from our readers is always welcome.

General feedback: If you have questions about any aspect of this book, mention the book title in the subject of your message and email us at customercare@packtpub.com.

Errata: Although we have taken every care to ensure the accuracy of our content, mistakes do happen. If you have found a mistake in this book, we would be grateful if you would report this to us. Please visit www.packt.com/submit-errata, selecting your book, clicking on the Errata Submission Form link, and entering the details.

Piracy: If you come across any illegal copies of our works in any form on the Internet, we would be grateful if you would provide us with the location address or website name. Please contact us at copyright@packt.com with a link to the material.

If you are interested in becoming an author: If there is a topic that you have expertise in and you are interested in either writing or contributing to a book, please visit authors.packtpub.com.

Reviews

Please leave a review. Once you have read and used this book, why not leave a review on the site that you purchased it from? Potential readers can then see and use your unbiased opinion to make purchase decisions, we at Packt can understand what you think about our products, and our authors can see your feedback on their book. Thank you!

For more information about Packt, please visit packt.com.

Section 1: Overview of Java Programming

The first part of the book brings the reader into the world of Java programming. It starts with basic Java-related definitions and the main terms, walks the reader through the installation of the necessary tools and Java itself, and explains how to run (execute) a Java program and the examples of the code provided with this book.

With the basics in place, we then explain and discuss **Object-Oriented Programming (OOP)** principles, how Java implements them, and how a programmer can take advantage of them to write high-quality code that is easy to maintain.

The book proceeds by presenting a more detailed view of Java as a language. It explains how the code is organized in packages, defines all the main types, and the list of reserved and restricted keywords. All the discussions are illustrated with specific code examples.

This section contains the following chapters:

Chapter 1, *Getting Started with Java 12*

Chapter 2, *Java Object-Oriented Programming (OOP)*

Chapter 3, *Java Fundamentals*

Getting Started with Java 12

This chapter is about how to start learning Java 12 and Java in general. We will begin with the basics, first explaining what Java is and its main terms, followed by how to install the necessary tools to write and run (execute) a program. In this respect, Java 12 is not much different to the previous Java versions, so this chapter's content is applies to the older versions too.

We will describe and demonstrate all the necessary steps for building and configuring a Java programming environment. This is the bare minimum that you have to have on the computer in order to start programming. We also describe the basic Java language constructs and illustrate them with examples that can be executed immediately.

The best way to learn a programming language, or any language for that matter, is to use it, and this chapter guides the reader on how they can do this with Java. The topics covered in this chapter include the following:

- How to install and run Java
- How to install and run an **Integrated Development Environment (IDE)**
- Java primitive types and operators
- String types and literals
- Identifiers and variables
- Java statements

How to install and run Java

When somebody says "*Java*," they may mean quite different things:

- **Java programming language**: A high-level programming language that allows an intent (a program) to be expressed in a human-readable format that can be translated in the binary code executable by a computer
- **Java compiler**: A program that can read a text written in the Java programming language and translate it into a bytecode that can be interpreted by **Java Virtual Machine (JVM)** in the binary code executable by a computer
- **Java Virtual Machine (JVM)**: A program that reads a compiled Java program and interprets it into the binary code that is executable by a computer
- **Java Development Kit (JDK)**: The collection of programs (tools and utilities), including Java compiler, JVM, and supporting libraries, which allow the compilation and execution of a program written in the Java language

The following section walks the reader through the installation of the JDK of Java 12 and the basic related terms and commands.

What is JDK and why do we need it?

As we have mentioned already, JDK includes a Java compiler and JVM. The task of the compiler is to read a .java file that contains the text of the program written in Java (called **source code**) and transform (compile) it into a bytecode stored in a .class file. The JVM can then read the .class file, interpret the bytecode in a binary code, and send it to the operating system for execution. Both the compiler and JVM have to be invoked explicitly from the command line.

To support the .java file compilation and its bytecode execution, JDK installation also includes standard Java libraries called **Java Class Library (JCL)**. If the program uses a third-party library, it has to be present during compilation and execution. It has to be referred from the same command line that invokes the compiler and later when the bytecode is executed by JVM. JCL, on the other hand, does not need to be referred to explicitly. It is assumed that the standard Java libraries reside in the default location of the JDK installation, so the compiler and JVM know where to find them.

If you do not need to compile a Java program and would like to run only the already compiled `.class` files, you can download and install **Java Runtime Environment** (JRE). For example, it consists of a subset of the JDK and does not include a compiler.

Sometimes, JDK is referred to as a **Software Development Kit (SDK)**, which is a general name for a collection of software tools and supporting libraries that allow the creation of an executable version of a source code written using a certain programming language. So, JDK is an SDK for Java. This means it is possible to call JDK an SDK.

You may also hear the terms *Java platform* and *Java edition* in relation to a JDK. A typical platform is an operating system that allows a software program to be developed and executed. Since JDK provides its own operating environment, it is called a platform too. An **edition** is a variation of a Java platform (JDK) assembled for a specific purpose. There are five Java platform editions, as listed here:

- **Java Platform Standard Edition (Java SE)**: This includes JVM, JCL, and other tools and utilities.
- **Java Platform Enterprise Edition (Java EE)**: This includes Java SE, servers (computer programs that provide services to the applications), JCL, other libraries, code samples, and tutorials, and other documentation for developing and deploying large-scale, multi-tiered, and secure network applications.
- **Java Platform Micro Edition (Java ME)**: This is a subset of Java SE with some specialized libraries for developing and deploying Java applications for embedded and mobile devices, such as phones, personal digital assistants, TV set-top boxes, printers, and sensors. A variation of Java ME (with its own JVM implementation) is called **Android SDK**. It was developed by Google for Android programming.
- **Java Card**: This is the smallest of the Java editions and is intended for developing and deploying Java applications onto small embedded devices such as smart cards. It has two editions: **Java Card Classic Edition**, for smart cards, based on ISO7816 and ISO14443 communication, and **Java Card Connected Edition**, which supports a web application model and TCP/IP as basic protocol and runs on high-end secure microcontrollers.

So, to install Java means to install JDK, which also means to install Java platform on one of the listed editions. In this book, we are going to talk about and use only Java SE.

Installing Java SE

All the recently released JDKs are listed on the official Oracle
page: www.oracle.com/technetwork/java/javase/overview/index.html (we will call it
an **Installation Home Page** for the further references).

Here are the steps that need to be followed to install Java SE:

1. Find the link to the Java SE version you are looking for (**Java SE 12** in this case) and click on it.
2. You will be presented with the various links, one of which is **Installation Instructions**. Alternatively, you could get to this page by clicking the **Downloads** tab.
3. Click the **DOWNLOAD** link under the title **Oracle JDK**.
4. A new screen will give you the option to accept or decline a **License Agreement** using a radio button and a list of links to various JDK installers.
5. Read the **License Agreement** and make your decision. If you do not accept it, you cannot download the JDK. If you accept the **License Agreement**, you can select the JDK installer from the available list.
6. You need to choose the installer that fits your operating system and the format (extension) you are familiar with.
7. If in doubt, go back to the **Installation Home Page**, select the **Downloads** tab, and click the **Installation Instructions** link.
8. Follow the steps that correspond to your operating system.
9. The JDK is installed successfully when the `java -version` command on your computer displays the correct Java version, as demonstrated in the following screenshot, for example:

```
demo>
demo> java -version
java version "12" 2019-03-19
Java(TM) SE Runtime Environment (build 12+33)
Java HotSpot(TM) 64-Bit Server VM (build 12+33, mixed mode, sharing)
demo>
```

Commands, tools, and utilities

If you follow the installation instructions, you may have noticed a link (**Installed Directory Structure of JDK**) given under the **Table of Contents**. It brings you to a page that describes the location of the installed JDK on your computer and the content of each directory of the JDK root directory. The `bin` directory contains all the executables that constitute the Java commands, tools, and utilities. If the directory `bin` is not added to the `PATH` environment variable automatically, consider doing it manually so that you can launch a Java executable from any directory.

In the previous section, we have already demonstrated the `Java` command, `java -version`. A list of the other Java executables available (commands, tools, and utilities) can be found in the **Java SE documentation** (`https://www.oracle.com/technetwork/java/javase/documentation/index.html`) by clicking the **Java Platform Standard Edition Technical Documentation site** link, and then the **Tools Reference** link on the next page. You can learn more about each executable tool by clicking its link.

You can also run each of the listed executables on your computer using one of the following options: `-?`, `-h`, `--help`, or `-help`. It will display a brief description of the executable and all its options.

The most important Java commands are the following:

- `javac`: This reads a `.java` file, compiles it, and creates one or more corresponding `.class` files, depending on how many Java classes are defined in the `.java` file.
- `java`: This executes a `.class` file.

These are the commands that make programming possible. Each Java programmer must have a good understanding of their structure and capabilities. But if you are new to Java programming and use an IDE (see the *How to install and run an IDE* section), you do not need to master these commands immediately. A good IDE hides them from you by compiling a `.java` file automatically every time you make a change to it. It also provides a graphical element that runs the program every time you click it.

Another very useful Java tool is `jcmd`. This facilitates communication with, and diagnosis of, any of the currently running Java processes (JVM) and has many options. But in its simplest form, without any option, it lists all the currently running Java processes and their **Process IDs (PIDs)**. You can use it to see whether you have run-away Java processes. If you have, you can then kill such a process using the PID provided.

How to install and run an IDE

What used to be just a specialized editor that allowed checking the syntax of the written program the same way a Word editor checks the syntax of an English sentence gradually evolved into an **Integrated Development Environment** (IDE). This bears its main function in the name. It integrates all the tools necessary for writing, compiling, and then executing a program under one **Graphical User Interface** (GUI). Using the power of Java Compiler, the IDE identifies syntax errors immediately and then helps to improve code quality by providing context-dependent help and suggestions.

Selecting an IDE

There are several IDEs available for a Java programmer, such as **NetBeans**, **Eclipse**, **IntelliJ IDEA**, **BlueJ**, **DrJava**, **JDeveloper**, **JCreator**, **jEdit**, **JSource**, **jCRASP**, and **jEdit**, to name a few. The most popular ones are NetBeans, Eclipse, and IntelliJ IDEA.

NetBeans development started in 1996 as a Java IDE student project at Charles University in Prague. In 1999, the project and the company created around the project were acquired by Sun Microsystems. After Oracle acquired Sun Microsystems, NetBeans became open source, and many Java developers have since contributed to the project. It was bundled with JDK 8 and became an official IDE for Java development. In 2016, Oracle donated it to the Apache Software Foundation.

There is a NetBeans IDE for Windows, Linux, Mac, and Oracle Solaris. It supports multiple programming languages and can be extended with plugins. As of the time of writing, NetBeans is bundled only with JDK 8, but NetBeans 8.2 can work with JDK 9 too and uses features introduced with JDK 9, such as Jigsaw, for example. On `netbeans.apache.org`, you can read more about NetBeans IDE and download the latest version, which is 11.0 as of the time of this writing.

Eclipse is the most widely used Java IDE. The list of plugins that add new features to the IDE is constantly growing, so it is not possible to enumerate all the IDE's capabilities. The Eclipse IDE project has been developed since 2001 as open source software. A non-profit, member-supported corporation Eclipse foundation was created in 2004 with the goal of providing the infrastructure (version control systems, code review systems, build servers, the download sites, and so on) and a structured process. None of the thirty something employees of the Foundation is working on any of 150 Eclipse-supported projects.

The sheer number and variety of the Eclipse IDE plugins create a certain challenge for a beginner because you have to find your way around different implementations of the same, or similar features, that can, on occasion, be incompatible and may require deep investigation, as well as a clear understanding of all the dependencies. Nevertheless, Eclipse IDE is very popular and has solid community support. You can read about the Eclipse IDE and download the latest release from `www.eclipse.org/ide`.

The IntelliJ IDEA has two versions: a paid one and a free community edition. The paid version is consistently ranked as the best Java IDE, but the community edition is listed among the three leading Java IDEs too. The JetBrains software company that develops the IDE has offices in Prague, Saint Petersburg, Moscow, Munich, Boston, and Novosibirsk. The IDE is known for its deep Intelligence that is *"giving relevant suggestions in every context: instant and clever code completion, on-the-fly code analysis, and reliable refactoring tools,"* as stated by the authors while describing the product on their website (`www.jetbrains.com/idea`). In the *Installing and configuring IntelliJ IDEA* section, we will walk you through the installation and configuration of IntelliJ IDEA community edition.

Installing and configuring IntelliJ IDEA

These are the steps you need to follow in order to download and install IntelliJ IDEA:

1. Download an installer of IntelliJ community edition from `www.jetbrains.com/idea/download`.
2. Launch the installer and accept all the default values.
3. Select **.java** on the **Installation Options** screen. We assume you have installed JDK already, so you do not check the **Download and install JRE** option.
4. The last installation screen has a checkbox, **Run IntelliJ IDEA**, that you can check to start the IDE automatically. Alternatively, you can leave the checkbox unchecked and launch the IDE manually once the installation is complete.
5. When the IDE starts the first time, it asks whether you would like to **Import IntelliJ IDEA settings**. Check the **Do not import settings** checkbox if you have not used the IntelliJ IDEA before and would like to reuse the settings.
6. The following screen or two asks whether you accept the **JetBrains Privacy Policy** and whether you would like to pay for the license or prefer to continue to use the free community edition or free trial (this depends on the particular download you get).

7. Answer the questions whichever way you prefer, and if you accept the privacy policy, the **Customize IntelliJ IDEA** screen asks you to choose a theme, white (**IntelliJ**) or dark (**Darcula**).

8. When offered the buttons **Skip All and Set Defaults** and **Next: Default plugins**, select **Next: Default plugins**, as it will give you the option to configure the IDE beforehand.

9. When presented with the **Tune IDEA to your tasks** screen, select the **Customize...** link for the following three, one at a time:

   - **Build Tools**: Select **Maven** and click the **Save Changes and Go Back** button.
   - **Version Controls**: Select the version control system you prefer (optional) and click **Save Changes and Go Back**.
   - **Test Tools**: Select **JUnit** or any other test framework you prefer (optional) and click **Save Changes and Go Back**.

10. If you decide to change the set values, you can do this later by selecting from the topmost menu, **File**, **Settings**, on Windows, or **Preferences** on Linux and macOS.

Creating a project

Before you start writing your program, you need to create a project. There are several ways to create a project in IntelliJ IDEA, which is the same for any IDE, as follows:

1. **Create New Project**: This creates a new project from scratch.
2. **Import Project**: This facilitates reading of the existing source code from the filesystem.
3. **Open**: This facilitates reading of the existing project from the filesystem.
4. **Check out from Version Control**: This facilitates reading of the existing project from the version control system.

In this book, we will walk you through the first option only—using the sequence of guided steps provided by the IDE. The other two options are much simpler and do not require additional explanations. Once you have learned how to create a new project from scratch, the other ways to bring up a project in the IDE will be very easy for you.

Start by clicking the **Create New Project** link and proceed further as follows:

1. Select a value for **Project SDK** (Java Version 12, if you have installed JDK 12 already) and click **Next**.

2. Do not check **Create project from template** (if checked, the IDE generates a canned program, Hello world, and similar, which we do not need) and click **Next**.

3. Select the desired project location in the **Project location** field (this is where your new code will reside).

4. Enter anything you like in the **Project name** field (for example, the project for the code in this book is called learnjava) and click the Finish button.

5. You will see the following project structure:

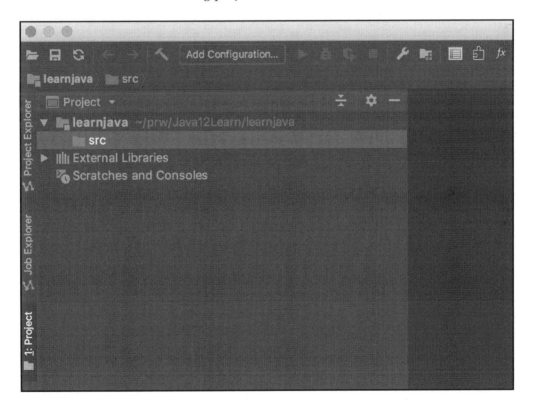

6. Right-click on the project name (`learnjava`) and select **Add Framework Support** from the drop-down menu. On the following pop-up window, select **Maven**:

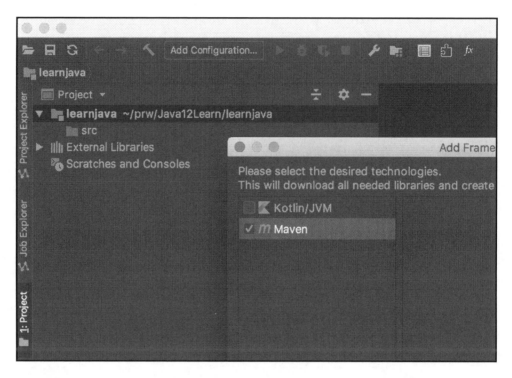

7. **Maven** is a project configuration tool. Its primary function is to manage project dependencies. We will talk about it shortly. For now, we will use its other responsibility, to define and hold the project code identity using three properties:

- `groupId`: To identify a group of projects within an organization or an open source community
- `artifactId`: To identify a particular project within the group
- `version`: To identify the version of the project

The main goal is to make the identity of a project unique among all the projects of the world. To help avoid a `groupId` clash, the convention requires that you start building it from the organization domain name in reverse. For example, if a company has the domain name `company.com`, the group IDs of its projects should start with `com.company`. That is why, for the code in this book, we use the `groupId` value `com.packt.learnjava`.

Let's set it. Click **OK** on the **Add Framework Support** pop-up window and you will see a newly generated `pom.xml` file as follows:

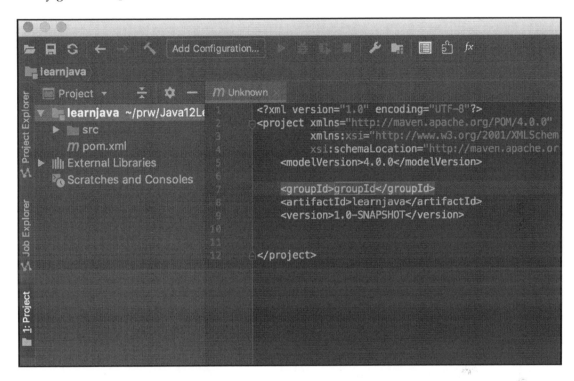

At the same time, in the lower-right corner of the screen, another small window will pop up:

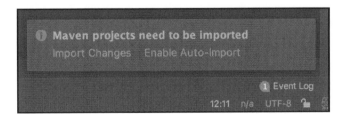

Click the **Enable Auto-Import** link. This will make writing code easier: all the new classes you will start using will be imported automatically. We will talk about class importing in due time.

Now, let's enter `groupId`, `artifactId`, and `version` values:

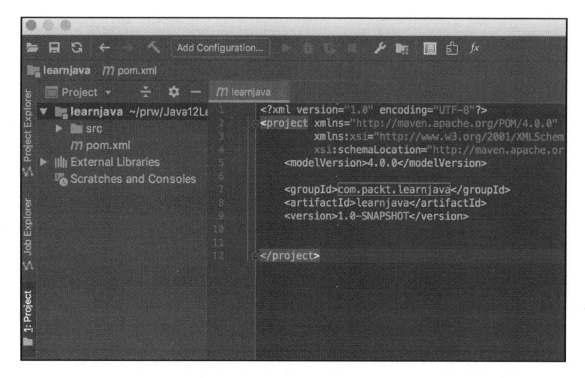

Now, if somebody would like to use the code of your project in their application, they would refer to it by the three values shown and Maven (if they use it) will bring it in (if you upload your project in the publicly shared Maven repository, of course). Read more about Maven at `https://maven.apache.org/guides`.

Another function of the `groupId` value is to define the root directory of the folders tree that holds your project code. Let's open the `src` folder; you will see the following directory structure beneath it:

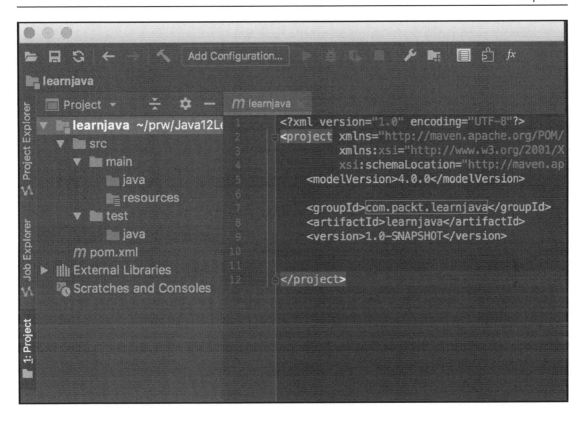

The `java` folder under `main` will hold the application code, while the `java` folder under `test` will hold the test code.

Let's create our first program using the following steps:

1. Right-click on `java`, select **New**, and then click **Package**:

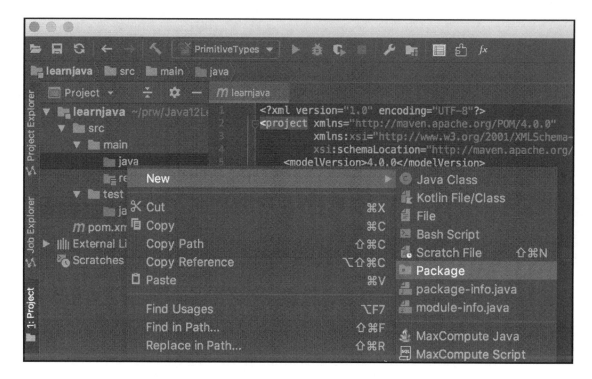

2. In the **New Package** window provided, type
   `com.packt.learnjava.ch01_start` as follows:

3. Click **OK** and you should see in the left panel a set of new folders, the last of them being `com.packt.learnjava.ch01_start`:

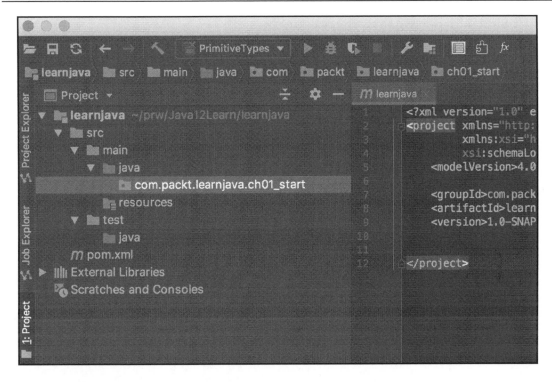

4. Right-click on it, select **New**, and then click **Java Class**:

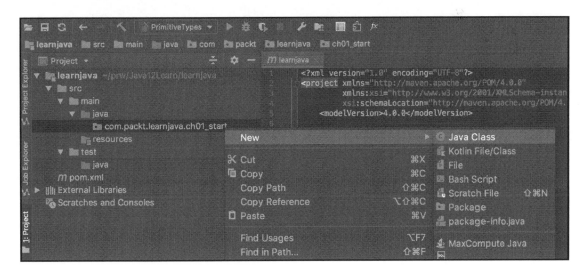

5. In the input window provided, type `PrimitiveTypes`:

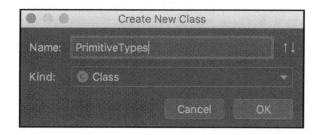

6. Click **OK** and you will see the first Java class, `PrimitiveTypes`, created in the `com.packt.learnjava.ch01_start` package:

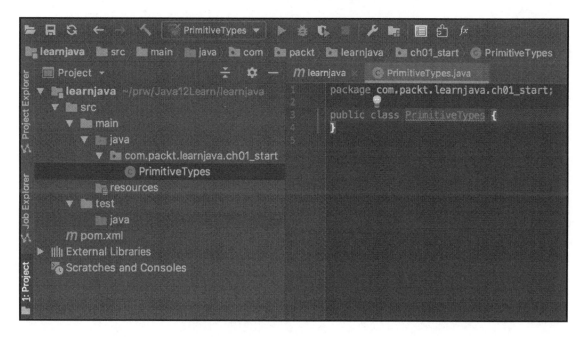

The package reflects the Java class location in the filesystem. We will talk about it in `Chapter 2`, *Java Object-Oriented Programming (OOP)*. Now, in order to run a program, we create a `main()` method. If present, this method can be executed and serve as an entry point into the application. It has a certain format, as follows:

```
package com.packt.learnjava.ch01_start;

public class PrimitiveTypes {

    public static void main(String... args){

    }
}
```

This has to have the following attributes:

- `public`: Freely accessible from outside the package
- `static`: Should be able to be called without creating an object of the class it belongs to

It should also be the following:

- Return `void` (nothing).
- Accept a `String` array as an input or `varargs` as we have done. We will talk about `varargs` in Chapter 2, *Java Object-Oriented Programming (OOP)*. For now, suffice to say that `String[] args` and `String... args` define essentially the same input format.

We explained how to run the main class using a command line in the *Executing examples from the command line* section. You can read more about Java command-line arguments in the official Oracle documentation: `https://docs.oracle.com/javase/tutorial/essential/environment/cmdLineArgs.html`. It is also possible to run the examples from IntelliJ IDEA.

Notice the two green triangles to the left in the following screenshot. By clicking any of them, you can execute the `main()` method. For example, let's display `Hello, world!`.

In order to do this, type the following line inside the `main()` method:

```
System.out.println("Hello, world!");
```

Then, click one of the green triangles:

You should get the following output in the terminal area as follows:

From now on, every time we are going to discuss code examples, we will run them the same way, by using the `main()` method. While doing this, we will not capture a screenshot but put the result in comments, because such a style is easier to follow. For example, the following code displays how the previous code demonstration would look in this style:

```
System.out.println("Hello, world!");       //prints: Hello, world!
```

It is possible to add a comment (any text) to the right of the code line separated by the double slash `//`. The compiler does not read this text and just keeps it as it is. The presence of a comment does not affect performance and is used to explain the programmer's intent to humans.

Importing a project

We are going to demonstrate project importing using the source code for this book. We assume that you have Maven installed (`https://maven.apache.org/install.html`) on your computer and that you have Git (`https://gist.github.com/derhuerst/1b15ff4652a867391f03`) installed too, and can use it. We also assume that you have installed JDK 12, as was described in the *Installation of Java SE* section.

To import the project with the code examples for this book, follow these steps:

1. Go to the source repository (`https://github.com/PacktPublishing/Learn-Java-12-Programming`) and click the **Clone or download** link, as shown in the following screenshot:

2. Click the **Clone or download** link and then copy the provided URL:

3. Select a directory on your computer where you would like the source code to be placed and then run the following Git command:

4. A new `Learn-Java-12-Programming` folder is created, as shown in the
   following screenshot:

```
demo>
demo> git clone https://github.com/PacktPublishing/Learn-Java-12-Programming.git
Cloning into 'Learn-Java-12-Programming'...
remote: Enumerating objects: 13, done.
remote: Counting objects: 100% (13/13), done.
remote: Compressing objects: 100% (10/10), done.
remote: Total 735 (delta 0), reused 6 (delta 0), pack-reused 722
Receiving objects: 100% (735/735), 3.36 MiB | 3.06 MiB/s, done.
Resolving deltas: 100% (268/268), done.
demo> ls
Learn-Java-12-Programming
demo>
```

Alternatively, instead of cloning, you can download the source as a `.zip` file
using the link **Download ZIP** shown on the preceding screenshot. Unarchive the
downloaded source in a directory on your computer where you would like the
source code to be placed, and then rename the newly created folder by removing
the suffix "-master" from its name, making sure that the folder's name is `Learn-
Java-12-Programming`.

5. The new `Learn-Java-12-Programming` folder contains the Maven project with
   all the source code from this book. Now run the IntelliJ IDEA and click **File** in the
   topmost menu, then **New** and **Project from Existing Sources...**:

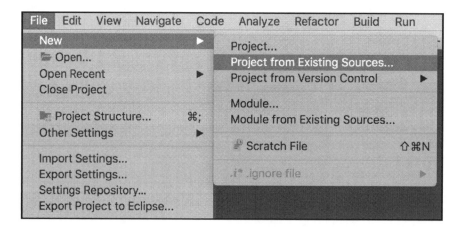

6. Select the `Learn-Java-12-Programming` folder created in step 4 and click the **Open** button:

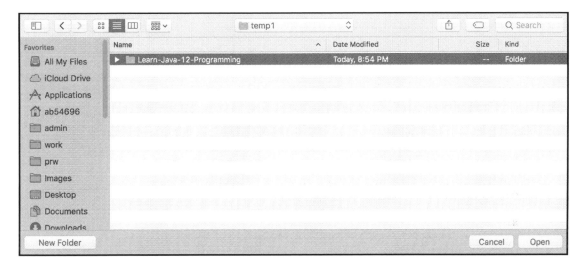

7. Accept the default settings and click the **Next** button on each of the following screens until you reach a screen that shows a list of the JDKs installed and the **Finish** button:

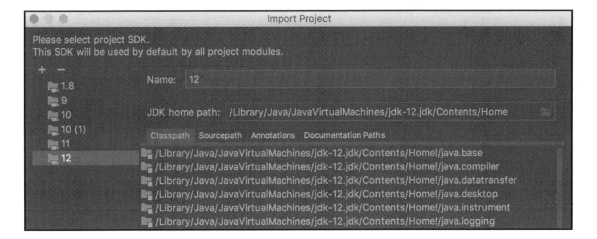

8. Select 12 and click **Finish**. You will see the project imported into your IntelliJ IDEA:

9. Wait until the following small window shows up in the bottom-right corner:

You may not want to wait and continue with step 12. Just do steps 10 and 11 when the window pops up later. If you miss this window, you may click the **Event Log** link any time later, and you will be presented with the same options.

10. Click on it; then click the **Add as Maven Project** link:

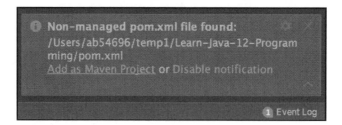

11. Any time the following window shows up, click **Enable Auto-Import**:

You may not want to wait and continue with step 12. Just do step 11 when the window pops up later. If you miss this window, you may click the **Event Log** link any time later, and you will be presented with the same options.

12. Select the Project structure symbol, which is the third from the right on the following screenshot:

13. If you have the **main** and **test** modules listed, remove them by highlighting them and clicking the minus symbol (−) as shown on the following screen:

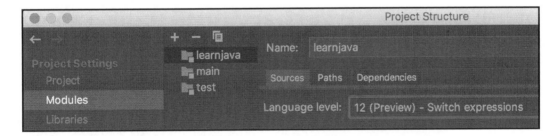

14. Here's how the final list of modules should look:

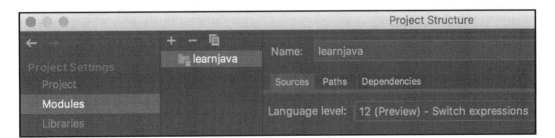

15. Click **OK** in the bottom-right corner and get back to your project. Click **Learn-Java-12-Programming** in the left pane and continue going down in the source tree until you see the following list of classes:

16. Click on the green arrow in the right pane and execute any class you want. The result you will be able to see in the **Run** window is similar to the following:

Executing examples from the command line

To execute the examples from the command line, follow these steps:

1. Go to the `Learn-Java-12-Programming` folder created in *step 4* in the *Importing a project* section, where the `pom.xml` file is located, and run the `mvn clean package` command:

```
demo>
demo> cd Learn-Java-12-Programming
demo> ls
LICENSE          README.md        learnjava.iml   pom.xml          src              target
demo> mvn clean package
[INFO] Scanning for projects...
[INFO]
[INFO] -------------------< com.packt.learnjava:learnjava >--------------------
[INFO] Building learnjava 1.0-SNAPSHOT
[INFO] --------------------------------[ jar ]---------------------------------
```

2. Select the example you would like to run. For example, assuming you would like to run `ControlFlow.java`, run the following command:

```
java -cp target/learnjava-1.0-SNAPSHOT.jar:target/libs/* \
com.packt.learnjava.ch01_start.ControlFlow
```

You will see the following results:

```
demo>
demo> java -cp target/learnjava-1.0-SNAPSHOT.jar:target/libs/* com.packt.learnjava.ch01_start.ControlFlow

Selection statements:

Iteration statements:
0 1 2 3 4
0 1 2 3 4
0 1 2
```

3. If you would like to run example files from the `ch05_stringsIoStreams` package, run the same command with a different package and class name:

```
java -cp target/learnjava-1.0-SNAPSHOT.jar:target/libs/* \
com.packt.learnjava.ch05_stringsIoStreams.Files
```

If your computer has a Windows system, use the following command as one line:

```
java -cp target\learnjava-1.0-SNAPSHOT.jar;target\libs\*
com.packt.learnjava.ch05_stringsIoStreams.Files
```

Note that a Windows command has a different slash and semicolon (;) as the classpath separator.

4. The results will be as follows:

```
demo>
demo> java -cp target/learnjava-1.0-SNAPSHOT.jar:target/libs/* com.packt.learnjava.ch05_stringsIoStreams.Files

dir1.list(): demo2
dir1.listFiles(): demo1/demo2
dir.list(): file1.txt file2.txt
dir.listFiles(): demo1/demo2/file1.txt demo1/demo2/file2.txt
File.listRoots(): /
```

5. This way you can run any class that has the `main()` method in it. The content of the `main()` method will be executed.

Java primitive types and operators

With all the main programming tools in place, we can start talking about Java as a language. The language syntax is defined by Java Language Specification, which you can find on `https://docs.oracle.com/javase/specs`. Don't hesitate to refer to it every time you need some clarification. It is not as daunting as many people assume.

All the values in Java are divided into two categories: `reference` types and `primitive` types. We start with primitive types and operators as the natural entry point to any programming language. In this chapter, we will also discuss one reference type called `String` (see the *String type and literals* section).

All primitive types can be divided into two groups: the `boolean` type and the `numeric` types.

Boolean type

There are only two `boolean` type values in Java: `true` and `false`. Such a value can only be assigned to a variable of a `boolean` type, for example:

```
boolean b = true;
```

A `boolean` variable is typically used in control flow statements, which we are going to discuss in the *Java statements* section. Here is one example:

```
boolean b = x > 2;
if(b){
    //do something
}
```

In the code, we assign to the b variable the result of the evaluation of the x > 2 expression. If the value of x is greater than 2, the b variable gets the assigned value, `true`. Then, the code inside the braces, `{}`, is executed.

Numeric types

Java numeric types form two groups: integral types (`byte`, `char`, `short`, `int`, and `long`) and floating-point types (`float` and `double`).

Integral types

Integral types consume the following amount of memory:

- `byte`: 8 bit
- `char`: 16 bit
- `short`: 16 bit
- `int`: 32 bit
- `long`: 64 bit

The char type is an unsigned integer that can hold a value (called a **code point**) from 0 to 65,535 inclusive. It represents a Unicode character, which means there are 65,536 Unicode characters. Here are three records form the basic Latin list of Unicode characters:

Code point	Unicode escape	Printable symbol	Description
33	\u0021	!	Exclamation mark
50	\u0032	2	Digit two
65	\u0041	A	Latin capital letter "A"

The following code demonstrates the properties of the char type:

```java
char x1 = '\u0032';
System.out.println(x1);    //prints: 2

char x2 = '2';
System.out.println(x2);    //prints: 2
x2 = 65;
System.out.println(x2);    //prints: A

char y1 = '\u0041';
System.out.println(y1);    //prints: A

char y2 = 'A';
System.out.println(y2);    //prints: A
y2 = 50;
System.out.println(y2);    //prints: 2

System.out.println(x1 + x2);   //prints: 115
System.out.println(x1 + y1);   //prints: 115
```

The last two lines from the code example explain why the char type is considered an integral type because the char values can be used in arithmetic operations. In such a case, each char value is represented by its code point.

The range of values of other integral types is as follows:

- byte: from -128 to 127 inclusive
- short: from $-32,768$ to $32,767$ inclusive
- int: from $-2.147.483.648$ to $2.147.483.647$ inclusive
- long: from $-9,223,372,036,854,775,808$ to $9,223,372,036,854,775,807$ inclusive

You can always retrieve the maximum and minimum value of each primitive type from a corresponding Java constant as follows:

```
System.out.println(Byte.MIN_VALUE);        //prints: -128
System.out.println(Byte.MAX_VALUE);        //prints:  127
System.out.println(Short.MIN_VALUE);       //prints: -32768
System.out.println(Short.MAX_VALUE);       //prints:  32767
System.out.println(Integer.MIN_VALUE);     //prints: -2147483648
System.out.println(Integer.MAX_VALUE);     //prints:  2147483647
System.out.println(Long.MIN_VALUE);        //prints: -9223372036854775808
System.out.println(Long.MAX_VALUE);        //prints:  9223372036854775807
System.out.println((int)Character.MIN_VALUE); //prints: 0
System.out.println((int)Character.MAX_VALUE); //prints: 65535
```

The construct `(int)` in the last two lines, is an example of **cast operator** usage. It forces the conversion of a value from one type to another in cases where such a conversion is not always guaranteed to be successful. As you can see from our examples, some types allow bigger values than other types. But the programmer may know that the value of a certain variable can never exceed the maximum value of the target type, and the cast operator is the way a programmer can force their opinion on the compiler. Otherwise, without a cast operator, a compiler would raise an error and would not allow the assignment. However, a programmer may be mistaken and the value may become bigger. In such a case, a runtime error will be raised during the execution time.

There are types that, in principle, cannot be cast to other types though, or not to all types at least. For example, a `boolean` type value cannot be cast to an integral type value.

Floating-point types

There are two types in this group of primitive types—`float` and `double`:

- `float`: 32 bit
- `doubele`: 64 bit

Their positive maximum and minimum possible values are as follows:

```
System.out.println(Float.MIN_VALUE);   //prints: 1.4E-45
System.out.println(Float.MAX_VALUE);   //prints: 3.4028235E38
System.out.println(Double.MIN_VALUE);  //prints: 4.9E-324
System.out.println(Double.MAX_VALUE);  //prints: 1.7976931348623157E308
```

The maximum and minimum negative values are the same as those just shown, only with a minus sign (–) in front of them. So, effectively, the values `Float.MIN_VALUE` and `Double.MIN_VALUE` are not the minimal values, but the precision of the corresponding type. A zero value can be either `0.0` or `-0.0` for each of the floating-point types.

The special feature of the floating-point type is the presence of a dot (`.`), which separates integer and fractional parts of the number. By default, in Java, a number with a dot is assumed to be a `double` type. For example, the following is assumed to be a double value:

```
42.3
```

This means that the following assignment causes a compilation error:

```
float f = 42.3;
```

To indicate that you would like it to be treated as a `float` type, you need to add either f or F. For example, the following assignments do not cause an error:

```
float f = 42.3f;
float d = 42.3F;

double a = 42.3f;
double b = 42.3F;

float x = (float)42.3d;
float y = (float)42.3D;
```

As you may have noticed from the example, d and D indicate a `double` type. But we were able to cast them to the `float` type because we are confident that `42.3` is well inside the range of possible `float` type values.

Default values of primitive types

In some cases, a variable has to be assigned a value even when a programmer did not want to do that. We will talk about such cases in `Chapter 2`, *Java Object-Oriented Programming (OOP)*. The default primitive type value in such cases is as follows:

- The `byte`, `short`, `int`, and `long` types have the default value 0.
- The `char` type has the default value \u0000, with the code point 0.
- The `float` and `double` types have the default value 0.0.
- The `boolean` type has the default value `false`.

Literals of primitive types

The representation of a value is called a **literal**. The `boolean` type has two literals: `true` and `false`. Literals of the `byte`, `short`, `int`, and `long` integral types have the `int` type by default:

```
byte b = 42;
short s = 42;
int i = 42;
long l = 42;
```

In addition, to indicate a literal of a `long` type, you can append the letter `l` or `L` to the end:

```
long l1 = 42l;
long l2 = 42L;
```

The letter `l` can be easily confused with the number 1, so using `L` (instead of `l`) for this purpose is a good practice.

So far, we have expressed integral literals in a decimal number system. Meanwhile, the literals of the `byte`, `short`, `int`, and `long` types can also be expressed in the binary (base 2, digits 0-1), octal (base 8, digits 0-7), and hexadecimal (base 16, digits 0-9 and a-f) number systems. A binary literal starts with `0b` (or `0B`), followed by the value expressed in a binary system. For example, the decimal `42` is expressed as `101010` $= 2\hat{\ }0*0 + 2\hat{\ }1*1 + 2\hat{\ }2*0 + 2\hat{\ }3 *1 + 2\hat{\ }4 *0 + 2\hat{\ }5 *1$ (we start from the right 0). An octal literal starts with `0`, followed by the value expressed in an octal system, so `42` is expressed as `52` $= 8\hat{\ }0*2+ 8\hat{\ }1*5$. The hexadecimal literal starts with `0x` (or with `0X`), followed by a value expressed in a hexadecimal system. So, the `42` is expressed as `2a` $= 16\hat{\ }0*a + 16\hat{\ }1*2$ because, in the hexadecimal system, the symbols `a` to `f` (or `A` to `F`) map to the decimal values 10 to 15. Here is the demonstration code:

```
int i = 42;
System.out.println(Integer.toString(i, 2));        // 101010
System.out.println(Integer.toBinaryString(i));     // 101010
System.out.println(0b101010);                       // 42

System.out.println(Integer.toString(i, 8));        // 52
System.out.println(Integer.toOctalString(i));      // 52
System.out.println(052);                            // 42

System.out.println(Integer.toString(i, 10));       // 42
System.out.println(Integer.toString(i));           // 42
System.out.println(42);                             // 42

System.out.println(Integer.toString(i, 16));       // 2a
```

```
System.out.println(Integer.toHexString(i));          // 2a
System.out.println(0x2a);                            // 42
```

As you can see, Java provides methods that convert decimal system values to the systems with different bases. All these expressions of numeric values are called literals.

One feature of numeric literals makes them human-friendly. If the number is large, it is possible to break it into triples separated by an underscore (_) sign. Observe the following, for example:

```
int i = 354_263_654;
System.out.println(i);    //prints: 354263654

float f = 54_436.98f;
System.out.println(f);    //prints: 54436.98

long l = 55_763_948L;
System.out.println(l);    //prints: 55763948
```

The compiler ignores an embedded underscore sign.

The `char` type has two kinds of literals: a **single character** or an **escape sequence**. We have seen examples of `char` type literals when discussing the numeric types:

```
char x1 = '\u0032';
char x2 = '2';
char y1 = '\u0041';
char y2 = 'A';
```

As you can see, the character has to be enclosed in single quotes.

An escape sequence starts with a backslash (\) followed by a letter or another character. Here is the full list of escape sequences:

- \b: backspace BS, Unicode escape \u0008
- \t: horizontal tab HT, Unicode escape \u0009
- \n: line feed LF, Unicode escape \u000a
- \f: form feed FF, Unicode escape \u000c
- \r: carriage return CR, Unicode escape \u000d
- \": double quote ", Unicode escape \u0022
- \': single quote ', Unicode escape \u0027
- \\: backslash \, Unicode escape \u005c

From the eight escape sequences, only the last three are represented by a symbol. They are used when this symbol cannot be otherwise displayed. Observe the following, for example:

```
System.out.println("\"");    //prints: "
System.out.println('\'');    //prints: '
System.out.println('\\');    //prints: \
```

The rest are used more as control codes that direct the output device to do something:

```
System.out.println("The back\bspace");      //prints: The bacspace
System.out.println("The horizontal\ttab");  //prints: The horizontal    tab
System.out.println("The line\nfeed");       //prints: The line
                                            //        feed
System.out.println("The form\ffeed");       //prints: The form feed
System.out.println("The carriage\rreturn"); //prints: return
```

As you can see, \b deletes a previous symbol, \t inserts a tab space, \n breaks the line and begins the new one, \f forces the printer to eject the current page and to continue printing at the top of another, and /r starts the current line anew.

New compact number format

The java.text.NumberFormat class presents numbers in various formats. It also allows formats to be adjusted to those provided, including locales. We mention it here only because of a new feature added to this class in Java 12. It is called a **compact** or **short number format**.

It represents a number in the locale-specific, human-readable form. Observe the following, for example:

```
NumberFormat fmt = NumberFormat.getCompactNumberInstance(Locale.US,
                                      NumberFormat.Style.SHORT);
System.out.println(fmt.format(42_000));         //prints: 42K
System.out.println(fmt.format(42_000_000));     //prints: 42M

NumberFormat fmtP = NumberFormat.getPercentInstance();
System.out.println(fmtP.format(0.42));          //prints: 42%
```

As you can see, to access this capability, you have to acquire a particular instance of the NumberFormat class, sometimes based on the locale and style provided.

Operators

There are 44 operators in Java. These are listed in the following table:

Operators	Description
`+` `-` `*` `/` `%`	Arithmetic unary and binary operators
`++` `--`	Increment and decrement unary operators
`==` `!=`	Equality operators
`<` `>` `<=` `>=`	Relational operators
`!` `&` `|`	Logical operators
`&&` `||` `?:`	Conditional operators
`=` `+=` `-=` `*=` `/=` `%=`	Assignment operators
`&=` `|=` `^=` `<<=` `>>=` `>>>=`	Assignment operators
`&` `|` `~` `^` `<<` `>>` `>>>`	Bitwise operators
`->` `::`	Arrow and method reference operators
`new`	Instance creation operator
`.`	Field access/method invocation operator
`instanceof`	Type comparison operator
`(target type)`	Cast operator

We will not describe the not-often-used assignment operators `&=`, `|=`, `^=`, `<<=`, `>>=`, `>>>=` and bitwise operators. You can read about them in the Java specification (`https://docs.oracle.com/javase/specs`). Arrow `->` and method reference `::` operators will be described in Chapter 14, *Functional Programming*. The instance creation operator `new`, the field access/method invocation operator `.`, and the type comparison operator `instanceof` will be discussed in Chapter 2, *Java Object-Oriented Programming (OOP)*. As for the `cast` operator, we have described it already in the *Integral types* section.

Arithmetic unary (+ and -) and binary operators (+, -, *, /, and %)

Most of the arithmetic operators and positive and negative signs (**unary** operators) are quite familiar to us. The modulus operator, %, divides the left-hand operand by the right-hand operand and returns the remainder, as follows:

```
int x = 5;
System.out.println(x % 2);    //prints: 1
```

It is also worth mentioning that the division of two integer numbers in Java loses the fractional part because Java assumes the result should be an integer number two, as follows:

```
int x = 5;
System.out.println(x / 2);    //prints: 2
```

If you need the fractional part of the result to be preserved, convert one of the operands into a floating-point type. Here are a few, among many ways, in which to do this:

```
int x = 5;
System.out.println(x / 2.);              //prints: 2.5
System.out.println((1. * x) / 2);        //prints: 2.5
System.out.println(((float)x) / 2);      //prints: 2.5
System.out.println(((double) x) / 2);    //prints: 2.5
```

Increment and decrement unary operators (++ and --)

The ++ operator increases the value of an integral type by 1, while the -- operator decreases it by 1. If placed before (prefix) the variable, it changes its value by 1 before the variable value is returned. But when placed after (postfix) the variable, it changes its value by 1 after the variable value is returned. Here are a few examples:

```
int i = 2;
System.out.println(++i);    //prints: 3
System.out.println(i);      //prints: 3
System.out.println(--i);    //prints: 2
System.out.println(i);      //prints: 2
System.out.println(i++);    //prints: 2
System.out.println(i);      //prints: 3
System.out.println(i--);    //prints: 3
System.out.println(i);      //prints: 2
```

Equality operators (== and !=)

The == operator means **equals**, while the != operator means **not equals**. They are used to compare values of the same type and return the `boolean` value `true` if the operand's values are equal, or `false` otherwise. Observe the following, for example:

```
int i1 = 1;
int i2 = 2;
System.out.println(i1 == i2);        //prints: false
System.out.println(i1 != i2);        //prints: true
System.out.println(i1 == (i2 - 1));  //prints: true
System.out.println(i1 != (i2 - 1));  //prints: false
```

Exercise caution, though, while comparing values of floating-point types, especially when you compare the results of calculations. Using relational operators (<, >, <=, and >=) in such cases is much more reliable, because calculations such as 1/3, for example, results in a never-ending fractional part 0.33333333... and ultimately depends on the precision implementation (a complex topic that is beyond the scope of this book).

Relational operators (<, >, <=, and >=)

Relational operators compare values and return a `boolean` value. Observe the following, for example:

```
int i1 = 1;
int i2 = 2;
System.out.println(i1 > i2);         //prints: false
System.out.println(i1 >= i2);        //prints: false
System.out.println(i1 >= (i2 - 1));  //prints: true
System.out.println(i1 < i2);         //prints: true
System.out.println(i1 <= i2);        //prints: true
System.out.println(i1 <= (i2 - 1));  //prints: true
float f = 1.2f;
System.out.println(i1 < f);          //prints: true
```

Logical operators (!, &, and |)

The logical operators can be defined as follows:

- The ! binary operator returns `true` if the operand is `false`, otherwise `false`.
- The & binary operator returns `true` if both of the operands are `true`.
- The | binary operator returns `true` if at least one of the operands is `true`.

Here is an example:

```
boolean b = true;
System.out.println(!b);     //prints: false
System.out.println(!!b);    //prints: true
boolean c = true;
System.out.println(c & b); //prints: true
System.out.println(c | b); //prints: true
boolean d = false;
System.out.println(c & d); //prints: false
System.out.println(c | d); //prints: true
```

Conditional operators (&&, ||, and ? :)

The && and || operators produce the same results as the & and | logical operators we have just demonstrated:

```
boolean b = true;
boolean c = true;
System.out.println(c && b); //prints: true
System.out.println(c || b); //prints: true
boolean d = false;
System.out.println(c && d); //prints: false
System.out.println(c || d); //prints: true
```

The difference is that the && and || operators do not always evaluate the second operand. For example, in the case of the && operator, if the first operand is false, the second operand is not evaluated because the result of the whole expression will be false anyway. Similarly, in the case of the || operator, if the first operand is true, the whole expression will be clearly evaluated to true without evaluating the second operand. We can demonstrate this in the following code:

```
int h = 1;
System.out.println(h > 3 & h++ < 3);   //prints: false
System.out.println(h);                 //prints: 2
System.out.println(h > 3 && h++ < 3);  //prints: false
System.out.println(h);                 //prints: 2
```

The ? : operator is called a **ternary operator**. It evaluates a condition (before the sign ?) and, if it results in true, assigns to a variable the value calculated by the first expression (between the ? and : signs); otherwise, it assigns the value calculated by the second expression (after the : sign):

```
int n = 1, m = 2;
float k = n > m ? (n * m + 3) : ((float)n / m);
System.out.println(k);            //prints: 0.5
```

Assignment operators (=, +=, -=, *=, /=, and %=)

The = operator just assigns the specified value to a variable:

```
x = 3;
```

Other assignment operators calculate a new value before assigning it:

- x += 42 assigns to x the result of the addition operation, x = x + 42.
- x -= 42 assigns to x the result of the subtraction operation, x = x - 42.
- x *= 42 assigns to x the result of the multiplication operation, x = x * 42.
- x /= 42 assigns to x the result of the division operation, x = x / 42.
- x %= 42 assigns the remainder of the division operation, x = x + x % 42.

Here is how these operators work:

```
float a = 1f;
a += 2;
System.out.println(a); //prints: 3.0
a -= 1;
System.out.println(a); //prints: 2.0
a *= 2;
System.out.println(a); //prints: 4.0
a /= 2;
System.out.println(a); //prints: 2.0
a %= 2;
System.out.println(a); //prints: 0.0
```

String types and literals

We have just described the primitive value types of the Java language. All the other value types in Java belong to a category of **reference types**. Each reference type is a more complex construct than just a value. It is described by a **class**, which serves as a template for creating an **object**, a memory area that contains the values and methods (the processing code) defined in the class. An object is created by the `new` operator. We will talk about classes and objects in more detail in `Chapter 2`, *Java Object-Oriented Programming (OOP)*.

In this chapter, we will talk about one of the reference types called `String`. It is represented by the `java.lang.String` class, which belongs, as you can see, to the most foundational package of JDK, `java.lang`. The reason we introduce the `String` class that early is that it behaves in some respects very similar to primitive types, despite being a reference type.

A reference type is so called because, in the code, we do not deal with the values of this type directly. The value of a reference type is more complex than the primitive type value. It is called an object and requires more complex memory allocation, so a reference type variable contains a memory reference. It points *(refers)* to the memory area where the object resides, hence the name.

This nature of the reference type requires particular attention when a reference-type variable is passed into a method as a parameter. We will discuss this in more detail in `Chapter 3`, *Java Fundamentals*. For now, in relation to `String`, we will see how `String`, being a reference type, helps to optimize the memory usage by storing each `String` value only once.

String literals

The `String` class represents character strings in Java programs. We have seen several such strings. We have seen `Hello, world!`, for example. That is a `String` literal.

Another example of a literal is `null`. Any reference class can refer to a `null` literal. It represents a reference value that does not point to any object. In the case of a `String` type, it appears as follows:

```
String s = null;
```

But a literal that consists of characters enclosed in double quotes (`"abc"`, `"123"`, and `"a42%$#"`, for example) can only be of a `String` type. In this respect, the `String` class, being a reference type, has something in common with primitive types. All `String` literals are stored in a dedicated section of memory called the **string pool**, and two literals equally spelled to represent the same value from the pool:

```
String s1 = "abc";
String s2 = "abc";
System.out.println(s1 == s2);    //prints: true
System.out.println("abc" == s1); //prints: true
```

The JVM authors have chosen such an implementation to avoid duplication and improve memory usage. The previous code examples look very much like operations involving primitive types, don't they? But when a `String` object is created using a `new` operator, the memory for the new object is allocated outside the string pool, so references of two `String` objects, or any other objects for that matter, are always different:

```
String o1 = new String("abc");
String o2 = new String("abc");
System.out.println(o1 == o2);    //prints: false
System.out.println("abc" == o1); //prints: false
```

If necessary, it is possible to move the string value created with the `new` operator to the string pool using the `intern()` method:

```
String o1 = new String("abc");
System.out.println("abc" == o1);            //prints: false
System.out.println("abc" == o1.intern());   //prints: true
```

In the previous code, the `intern()` method attempted to move the newly created `"abc"` value into the string pool, but discovered that such a literal exists there already, so it reused the literal from the string pool. That is why the references in the last line in the preceding example are equal.

The good news is that you probably will not need to create `String` objects using the `new` operator, and most Java programmers never do it. But when a `String` object is passed into your code as an input and you have no control over its origin, comparison by reference only may cause an incorrect result (if the strings have the same spelling but were created by the `new operator`). That is why, when the equality of two strings by spelling (and case) is necessary, in order to compare two literals or `String` objects, the `equals()` method is a better choice:

```
String o1 = new String("abc");
String o2 = new String("abc");
System.out.println(o1.equals(o2));          //prints: true
```

```
System.out.println(o2.equals(o1));      //prints: true
System.out.println(o1.equals("abc"));   //prints: true
System.out.println("abc".equals(o1));   //prints: true
System.out.println("abc".equals("abc")); //prints: true
```

We will talk about the `equals()` and other methods of the `String` class shortly.

Another feature that makes `String` literals and objects look like primitive values is that they can be added using the arithmetic operator, +:

```
String s1 = "abc";
String s2 = "abc";
String s = s1 + s2;
System.out.println(s);             //prints: abcabc
System.out.println(s1 + "abc");    //prints: abcabc
System.out.println("abc" + "abc"); //prints: abcabc

String o1 = new String("abc");
String o2 = new String("abc");
String o = o1 + o2;
System.out.println(o);             //prints: abcabc
System.out.println(o1 + "abc");    //prints: abcabc
```

No other arithmetic operator can be applied to a `String` literal or an object.

A new `String` literal, called a **raw string literal**, was introduced with Java 12. It facilitates the preservation of indents and multiple lines without adding white spaces in quotes. For example, here is how a programmer would add the indentation before Java 12 and use \n to break the line:

```
String html = "<html>\n" +
       "    <body>\n" +
       "             <p>Hello World.</p>\n" +
       "    </body>\n" +
       "</html>\n";
```

And here is how the same result is achieved with Java 12:

```
String html = `<html>
               <body>
                   <p>Hello World.</p>
               </body>
           </html>
           `;
```

As you can see, a raw string literal consists of one or more characters enclosed in a backtick (`` ` ``) (\u0060), also called a **backquote** or an **accent grave**.

String immutability

Since all String literals can be shared, JVM authors make sure that, once stored, a String variable cannot be changed. It helps not only avoid the problem of concurrent modification of the same value from different places of the code, but also prevents unauthorized modification of a String value, which often represents a username or password.

The following code looks like a String value modification:

```
String str = "abc";
str = str + "def";
System.out.println(str);        //prints: abcdef
str = str + new String("123");
System.out.println(str);        //prints: abcdef123
```

But, behind the scenes, the original "abc" literal remains intact. Instead, a few new literals were created: "def", "abcdef", "123", and "abcdef123". To prove it, we have executed the following code:

```
String str1 = "abc";
String r1 = str1;
str1 = str1 + "def";
String r2 = str1;
System.out.println(r1 == r2);       //prints: false
System.out.println(r1.equals(r2)); //prints: false
```

As you can see, the r1 and r2 variables refer to different memories, and the objects they refer to are spelled differently too.

We will talk more about strings in Chapter 5, *Strings, Input/Output, and Files*.

Identifiers and variables

From our school days, we have an intuitive understanding of what a variable is. We think of it as a name that represents a value. We solve problems using such variables as x gallons of water or n miles of a distance, and similar. In Java, the name of a variable is called an **identifier** and can be constructed by certain rules. Using an identifier, a variable can be declared (defined) and initialized.

Identifier

According to Java Language Specification (`https://docs.oracle.com/javase/specs`), an identifier (a variable name) can be a sequence of Unicode characters that represent letters, digits 0-9, a dollar sign (`$`), or an underscore (`_`).

Other limitations are as follows:

- The first symbol of an identifier cannot be a digit.
- An identifier cannot have the same spelling as a keyword (see Java keywords in `Chapter 3`, *Java Fundamentals*).
- It cannot be spelled as the `boolean` literal `true` or `false`, and or as the literal `null`.
- And, since Java 9, an identifier cannot be just an underscore (`_`).

Here are a few unusual but legal examples of identifiers:

```
$
_42
αρετη
String
```

Variable declaration (definition) and initialization

A variable has a name (an identifier) and a type. Typically, it refers to the memory where a value is stored, but may refer to nothing (`null`) or not refer to anything at all (then, it is not initialized). It can represent a class property, an array element, a method parameter, and a local variable. The last one is the most frequently used kind of variable.

Before a variable can be used, it has to be declared and initialized. In some other programming languages, a variable can also be *defined*, so Java programmers sometimes use the word *definition* as a synonym of *declaration*, which is not exactly correct.

Here is the terminology review with examples:

```
int x;      //declartion of variable x
x = 1;      //initialization of variable x
x = 2;      //assignment of variable x
```

Initialization and assignment look the same. The difference is in their sequence: the first assignment is called an **initialization**. Without an initialization, a variable cannot be used.

Declaration and initialization can be combined in a single statement. Observe the following, for example:

```
float $ = 42.42f;
String _42 = "abc";
int αρετη = 42;
double String = 42.;
```

Type holder var

In Java 10, a sort of type holder, `var`, was introduced. The Java Language Specification defines it as follows: "`var` *is not a keyword, but an identifier with a special meaning as the type of a local variable declaration.*"

In practical terms, it lets a compiler figure out the nature of the declared variable as follows:

```
var x = 1;
```

In the preceding example, the compiler can reasonably assume that x has the primitive type, `int`.

As you could guess, to accomplish that, a declaration on its own would not suffice:

```
var x;     //compilation error
```

That is, without initialization, the compiler cannot figure out the type of the variable when `var` is used.

Java statements

A **Java statement** is a minimal construct that can be executed. It describes an action and ends with a semicolon (;). We have seen many statements already. For example, here are three statements:

```
float f = 23.42f;
String sf = String.valueOf(f);
System.out.println(sf);
```

The first line is a declaration statement combined with an assignment statement. The second line is also a declaration statement combined with an assignment statement and method invocation statement. The third line is just a method invocation statement.

Here is the list of Java statement types:

- An empty statement that consists of only one symbol, ; (semicolon)
- A class or interface declaration statement (we will talk about this in `Chapter 2,` *Java Object-Oriented Programming (OOP)*)
- A local variable declaration statement: `int x;`
- A synchronized statement: this is beyond the scope of this book
- An expression statement
- A control flow statement

An expression statement can be one of the following:

- A method invocation statement: `someMethod();`
- An assignment statement: `n = 23.42f;`
- An object creation statement: `new String("abc");`
- A unary increment or decrement statement: `++x ;` or `--x;` or `x++;` or `x--;`

We will talk more about expression statements in the *Expression statements* section.

A control flow statement can be one of the following:

- A selection statement: `if-else` or `switch-case`
- An iteration statement: `for`, or `while`, or `do-while`
- An exception-handling statement: `throw`, or `try-catch`, or `try-catch-finally`
- A branching statement: `break`, or `continue`, or `return`

We will talk more about control statements in the *Control flow statements* section.

Expression statements

An **expression statement** consists of one or more expressions. An expression typically includes one or more operators. It can be evaluated, which means it can produce a result of one of the following types:

- A variable: `x = 1`, for example.
- A value: `2*2`, for example.

- Nothing, when the expression is an invocation of a method that returns `void`. Such a method is said to produce only a side effect: `void someMethod()`, for example

Consider the following expression:

```
x = y++;
```

The preceding expression assigns a value to an `x` variable and has a side effect of adding 1 to the value of the `y` variable.

Another example would be a method that prints a line:

```
System.out.println(x);
```

The `println()` method returns nothing and has a side effect of printing something. By its form, an expression can be one of the following:

- A primary expression: a literal, a new object creation, a field or method access (invocation)
- A unary operator expression: `x++`, for example
- A binary operator expression: `x*y`, for example
- A ternary operator expression: `x > y ? true : false`, for example
- A lambda expression: `x -> x + 1` (see `Chapter 14`, *Functional Programming*)

If an expression consists of other expressions, the parenthesis is often used to identify each of the expressions clearly. This way, it is easier to understand and to set the expressions precedence.

Control flow statements

When a Java program is executed, it is executed statement by statement. Some statements have to be executed conditionally, based on the result of an expression evaluation. Such statements are called **control flow statements** because, in computer science, a control flow (or flow of control) is the order in which individual statements are executed or evaluated.

A control flow statement can be one of the following:

- A selection statement: `if-else` or `switch-case`
- An iteration statement: `for,` or `while,` or `do-while`
- An exception-handling statement: `throw,` or `try-catch,` or `try-catch-finally`
- A branching statement: `break,` or `continue,` or `return`

Selection statements

The selection statements are based on an expression evaluation and have four variations:

- `if` (expression) {do something}
- `if` (expression) {do something} `else` {do something else}
- `if` (expression) {do something} `else` `if` {do something else} `else` {do something else}
- `switch` case statement

Here are examples of `if` statements:

```
if(x > y){
    //do something
}

if(x > y){
    //do something
} else {
    //do something else
}

if(x > y){
    //do something
} else if (x == y){
    //do something else
} else {
    //do something different
}
```

A `switch...case` statement is a variation of the `if...else` statement:

```
switch(x){
    case 5:                 //means: if(x = 5)
        //do something
        break;
    case 7:
        //do something else
        break;
    case 12:
        //do something different
        break;
    default:
        //do something completely different
        //if x is not 5, 7, or 12
}
```

As you can see, the `switch...case` statement forks the execution flow based on the value of the variable. The `break` statement allows the `switch...case` statement to be executed. Otherwise, all the following cases would be executed.

In Java 12, a new feature is introduced in a preview mode; a `switch...case` statement in a less verbose form:

```
void switchDemo1(int x){
    switch (x) {
        case 1, 3 -> System.out.print("1 or 3");
        case 4    -> System.out.print("4");
        case 5, 6 -> System.out.print("5 or 6");
        default   -> System.out.print("Not 1,3,4,5,6");
    }
    System.out.println(": " + x);
}
```

As you can see, it uses an arrow, `->`, and does not use a `break` statement. To take advantage of this feature, you have to add an `--enable-preview` option to the `javac` and `java` commands. If you run the examples from IDE, you need to add this option to the configuration. In IntelliJ IDEA, the option should be added to two configuration screens: for the compiler and for the runtime:

1. Open the **Preferences** screen and put it as the **Compilation** option of the **learnjava** module, as shown in the following screenshot:

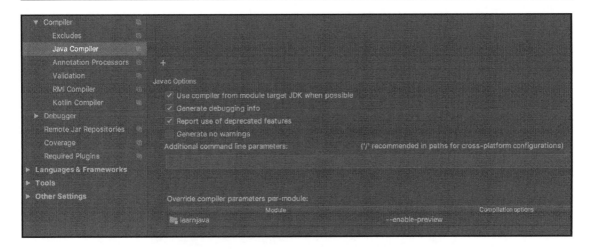

2. Select **Run** on the topmost horizontal menu:

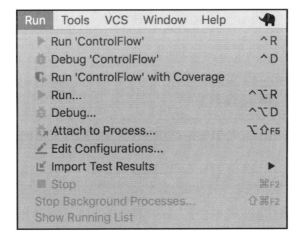

3. Click **Edit Configurations...** and add the **VM option** to the **ControlFlow** application that will be used at runtime:

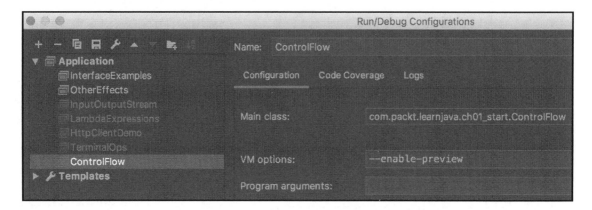

We have added the `--enable-preview` option, as we have just described, and executed the `switchDemo1()` method with different parameters:

```
switchDemo1(1);     //prints: 1 or 3: 1
switchDemo1(2);     //prints: Not 1,3,4,5,6: 2
switchDemo1(5);     //prints: 5 or 6: 5
```

You can see the results from the comments.

If several lines of code have to be executed in each case, you can just put braces, `{}`, around the block of code, as follows:

```
switch (x) {
    case 1, 3 -> {
                    //do something
                }
    case 4     -> {
                    //do something else
                }
    case 5, 6 -> System.out.println("5 or 6");
    default    -> System.out.println("Not 1,3,4,5,6");
}
```

The Java 12 `switch...case` statement can even return a value. For example, here is the case when another variable has to be assigned based on the `switch...case` statement result:

```
void switchDemo2(int i){
    boolean b = switch(i) {
```

```
                case 0 -> false;
                case 1 -> true;
                default -> false;
        };
        System.out.println(b);
    }
```

If we execute the `switchDemo2()` method, the results are going to be as follows:

```
switchDemo2(0);      //prints: false
switchDemo2(1);      //prints: true
switchDemo2(2);      //prints: false
```

It looks like a nice improvement. If this feature will prove to be useful, it will be included in the future Java releases as a permanent feature.

Iteration statements

An **iteration statement** can be one of the following three forms:

- A `while` statement
- A `do...while` statement
- A `for` statement, also called a `loop` statement

A `while` statement appears as follows:

```
while (boolean expression){
        //do something
}
```

Here is a specific example:

```
int n = 0;
while(n < 5){
  System.out.print(n + " "); //prints: 0 1 2 3 4
  n++;
}
```

In some examples, instead of the `println()` method, we use the `print()` method, which does not feed another line (does not add a line feed control at the end of its output). The `print()` method displays the output in one line.

The do...while statement has a very similar form:

```
do {
    //do something
} while (boolean expression)
```

It differs from the while statement by always executing the block of statements at least once, before evaluating the expression:

```
int n = 0;
do {
    System.out.print(n + " ");    //prints: 0 1 2 3 4
    n++;
} while (n < 5);
```

As you can see, it behaves the same way in the case when the expression is true at the first iteration. But if the expression evaluates to false, the results are different:

```
int n = 6;
while (n < 5) {
    System.out.print(n + " ");    //prints:
    n++;
}

n = 6;
do {
    System.out.print(n + " ");    //prints: 6
    n++;
} while (n < 5);
```

The for statement syntax is as follows:

```
for(init statements; boolean expression; update statements) {
  //do what has to be done here
}
```

Here is how the for statement works:

- init statements initialize a variable.
- boolean expression is evaluated using the current variable value: if true, the block of statements is executed, otherwise, the for statement exits.
- update statements update, the variable, and the boolean expression is evaluated again with this new value: if true, the block of statements is executed, otherwise the for statement exits.
- Unless exited, the final step is repeated.

As you can see, if you aren't careful, you can get into an infinite loop:

```
for (int x = 0; x > -1; x++){
    System.out.print(x + " ");   //prints: 0 1 2 3 4 5 6 ...
}
```

So, you have to make sure that the `boolean` expression guarantees eventual exit from the loop:

```
for (int x = 0; x < 3; x++){
    System.out.print(x + " ");   //prints: 0 1 2
}
```

The following example demonstrates multiple initialization and update statements:

```
for (int x = 0, y = 0; x < 3 && y < 3; ++x, ++y){
    System.out.println(x + " " + y);
}
```

And here is the variation of the preceding `for` statements for demo purposes:

```
for (int x = getInitialValue(), i = x == -2 ? x + 2 : 0, j = 0;
    i < 3 || j < 3 ; ++i, j = i) {
    System.out.println(i + " " + j);
}
```

If the `getInitialValue()` method is implemented like `int getInitialValue(){ return -2; }`, then the preceding two `for` statements produce exactly the same results.

To iterate over an array of values, you can use an array index:

```
int[] arr = {24, 42, 0};
for (int i = 0; i < arr.length; i++){
    System.out.print(arr[i] + " ");   //prints: 24 42 0
}
```

Alternatively, you can use a more compact form of a `for` statement that produces the same result, as follows:

```
int[] arr = {24, 42, 0};
for (int a: arr){
    System.out.print(a + " ");   //prints: 24 42 0
}
```

This last form is especially useful with a collection as shown here:

```
List<String> list = List.of("24", "42", "0");
for (String s: list){
    System.out.print(s + " ");  //prints: 24 42 0
}
```

We will talk about collections in `Chapter 6`, *Data Structures, Generics, and Popular Utilities.*

Exception-handling statements

In Java, there are classes called **exceptions** that represent the events that disrupt the normal execution flow. They typically have names that end with *Exception*: `NullPointerException`, `ClassCastException`, `ArrayIndexOutOfBoundsException`, to name a few.

All the exception classes extend the `java.lang.Exception` class, which, in turn, extends the `java.lang.Throwable` class (we will explain what this means in `Chapter 2`, *Java Object-Oriented Programming (OOP)*). That's why all exception objects have a common behavior. They contain information about the cause of the exceptional condition and the location of its origination (line number of the source code).

Each exception object can be generated (thrown) either automatically by JVM or by the application code, using the keyword `throw`. If a block of code throws an exception, you can use a `try-catch` or `try-catch-finally` construct to capture the thrown exception object and redirect the execution flow to another branch of code. If the surrounding code does not catch the exception object, it propagates all the way out of the application into the JVM and forces it to exit (and abort the application execution). So, it is good practice to use `try-catch` or `try-catch-finally` in all the places where an exception can be raised and you do not want your application to abort execution.

Here is a typical example of exception handling:

```
try {
    //x = someMethodReturningValue();
    if(x > 10){
        throw new RuntimeException("The x value is out of range: " + x);
    }
    //normal processing flow of x here
} catch (RuntimeException ex) {
    //do what has to be done to address the problem
}
```

In the preceding code snippet, `normal processing flow` will be not executed in the case of `x > 10`. Instead, the `do what has to be done` block will be executed. But, in the case `x <= 10`, the `normal processing flow` block will be run and the `do what has to be done` block will be ignored.

Sometimes, it is necessary to execute a block of code anyway, whether an exception was thrown/caught or not. Instead of repeating the same code block in two places, you can put it in a `finally` block, as follows:

```
try {
    //x = someMethodReturningValue();
    if(x > 10){
        throw new RuntimeException("The x value is out of range: " + x);
    }
    //normal processing flow of x here
} catch (RuntimeException ex) {
    System.out.println(ex.getMessage());
                        //prints: The x value is out of range: ...
    //do what has to be done to address the problem
} finally {
    //the code placed here is always executed
}
```

We will talk about exception handling in more detail in `Chapter 4`, *Exception Handling*.

Branching statements

Branching statements allow breaking of the current execution flow and continuation of execution form the first line after the current block or from a certain (labeled) point of the control flow.

A branching statement can be one of the following:

- `break`
- `continue`
- `return`

We have seen how `break` was used in `switch-case` statements. Here is another example:

```
String found = null;
List<String> list = List.of("24", "42", "31", "2", "1");
for (String s: list){
    System.out.print(s + " ");              //prints: 24 42 31
    if(s.contains("3")){
```

```
                found = s;
                break;
            }
        }
    }
    System.out.println("Found " + found);   //prints: Found 31
```

If we need to find the first list element that contains "3", we can stop executing as soon as condition s.contains("3") is evaluated to true. The remainder list elements are ignored.

In a more complicated scenario, with nested for statements, it is possible to set a label (with a : column) that indicates which for statement has to be exited:

```
    String found = null;
    List<List<String>> listOfLists = List.of(
            List.of("24", "16", "1", "2", "1"),
            List.of("43", "42", "31", "3", "3"),
            List.of("24", "22", "31", "2", "1")
    );
    exit: for(List<String> l: listOfLists){
        for (String s: l){
            System.out.print(s + " ");       //prints: 24 16 1 2 1 43
            if(s.contains("3")){
                found = s;
                break exit;
            }
        }
    }
    System.out.println("Found " + found);   //prints: Found 43
```

We have chosen the label name exit, but we could call it any other name too.

The continue statement works similarly, as follows:

```
    String found = null;
    List<List<String>> listOfLists = List.of(
                List.of("24", "16", "1", "2", "1"),
                List.of("43", "42", "31", "3", "3"),
                List.of("24", "22", "31", "2", "1")
    );
    String checked = "";
    cont: for(List<String> l: listOfLists){
            for (String s: l){
                System.out.print(s + " "); //prints: 24 16 1 2 1 43 24 22 31
                if(s.contains("3")){
                    continue cont;
```

```
            }
        checked += s + " ";
    }
}
System.out.println("Found " + found);    //prints: Found 43
System.out.println("Checked " + checked);
                            //prints: Checked 24 16 1 2 1 24 22
```

It differs from `break` by telling which of the `for` statements to continue, and not to exit. A `return` statement is used to return a result from a method:

```
String returnDemo(int i){
    if(i < 10){
        return "Not enough";
    } else if (i == 10){
        return "Exactly right";
    } else {
        return "More than enough";
    }
}
```

As you can see, there can be several `return` statements in a method, each returning a different value under different circumstances. If the method returns nothing (`void`), the return statement is not required, although it is frequently used for better readability, as follows:

```
void returnDemo(int i){
    if(i < 10){
        System.out.println("Not enough");
        return;
    } else if (i == 10){
        System.out.println("Exactly right");
        return;
    } else {
        System.out.println("More than enough");
        return;
    }
}
```

Statements are the building blocks of Java programming. They are like sentences in English, the complete expressions of intent that can be acted upon. They can be compiled and executed. Programming is expressing an action plan in statements.

With this, the explanation of the basics of Java is concluded.

Congratulations for getting through it!

Summary

This chapter introduced you to the exciting world of Java programming. We started with explaining the main terms, and then explained how to install the necessary tools, JDK and IDE, and how to configure and use them.

With a development environment in place, we provided the reader with the basics of Java as a programming language. We have described Java primitive types, the `String` type, and their literals. We have also defined what an identifier is, and what a variable is, and finished with a description of the main types of Java statements. All the points of the discussion were illustrated by the specific code examples.

In the next chapter, we are going to talk about the object-oriented aspects of Java. We will introduce the main concepts, explain what a class is, what an interface is, and the relationship between them. The terms *overloading*, *overriding*, and *hiding* will also be defined and demonstrated in code examples, as well as the usage of the `final` keyword.

Quiz

1. What does JDK stand for?

    a. Java Document Kronos
    b. June Development Karate
    c. Java Development Kit
    d. Java Developer Kit

2. What does JCL stand for?

    a. Java Classical Library
    b. Java Class Library
    c. Junior Classical Liberty
    d. Java Class Libras

3. What does Java SE stand for?

    a. Java Senior Edition
    b. Java Star Edition
    c. Java Structural Elections
    d. Java Standard Edition

4. What does IDE stand for?

    a. Initial Development Edition
    b. Integrated Development Environment
    c. International Development Edition
    d. Integrated Development Edition

5. What are Maven's functions?

    a. Project building
    b. Project configuration
    c. Project documentation
    d. Project cancellation

6. What are Java primitive types?

    a. `boolean`
    b. `numeric`
    c. `integer`
    d. `string`

7. What are Java primitive types?

    a. `long`
    b. `bit`
    c. `short`
    d. `byte`

8. What is a *literal*?

    a. A letter-based string
    b. A number-based string
    c. A variable representation
    d. A value representation

9. Which of the following are literals?

    a. `\\`

    b. `2_0`

    c. `2__0f`

    d. `\f`

10. Which of the following are Java operators?

    a. `%`

    b. `$`

    c. `&`

    d. `->`

11. What does the following code snippet print?

```
int i = 0; System.out.println(i++);
```

    a. 0

    b. 1

    c. 2

    d. 3

12. What does the following code snippet print?

```
boolean b1 = true;
boolean b2 = false;
System.out.println((b1 & b2) + " " + (b1 && b2));
```

    a. false true

    b. false false

    c. true false

    d. true true

13. What does the following code snippet print?

```
int x = 10;
x %= 6;
System.out.println(x);
```

    a. 1
    b. 2
    c. 3
    d. 4

14. What is the result of the following code snippet?

```
System.out.println("abc" - "bc");
```

    a. a
    b. abc-bc
    c. Compilation error
    d. Execution error

15. What does the following code snippet print?

```
System.out.println("A".repeat(3).lastIndexOf("A"));
```

    a. 1
    b. 2
    c. 3
    d. 4

16. What are the correct identifiers?

    a. `int __` (two underscores)
    b. `2a`
    c. `a2`
    d. `$`

17. What does the following code snippet print?

```
for (int i=20, j=-1; i < 23 && j < 0; ++i, ++j){
        System.out.println(i + " " + j + " ");
}
```

       a. 20 -1 21 0

       b. Endless loop

       c. 21 0

       d. 20 -1

18. What does the following code snippet print?

```
int x = 10;
try {
    if(x++ > 10){
        throw new RuntimeException("The x value is out of the
range: " + x);
    }
    System.out.println("The x value is within the range: " + x);
} catch (RuntimeException ex) {
    System.out.println(ex.getMessage());
}
```

       a. Compilation error

       b. The x value is out of the range: 11

       c. The x value is within the range: 11

       d. Execution time error

19. What does the following code snippet print?

```
int result = 0;
List<List<Integer>> source = List.of(
        List.of(1, 2, 3, 4, 6),
        List.of(22, 23, 24, 25),
        List.of(32, 33)
);
cont: for(List<Integer> l: source){
    for (int i: l){
        if(i > 7){
            result = i;
            continue cont;
        }
    }
}
System.out.println("result=" + result);
```

a. result = 22
b. result = 23
c. result = 32
d. result = 33

20. Select all the following statements that are correct:

    a. A variable can be declared
    b. A variable can be assigned
    c. A variable can be defined
    d. A variable can be determined

21. Select all the correct Java statement types from the following:

    a. An executable statement
    b. A selection statement
    c. A method end statement
    d. An increment statement

2
Java Object-Oriented Programming (OOP)

Object-Oriented Programming (OOP) was born out of the necessity for better control over concurrent modification of the shared data, which was the curse of pre-OOP programming. The core of the idea was not to allow direct access to the data but to do it only through the dedicated layer of code. Since the data needs to be passed around and modified in the process, the concept of an object was conceived. In the most general sense, an *object* is a set of data that can be passed around and accessed only through the set of methods passed along too. This data is said to compose an **object state**, while the methods constitute the **object behavior**. The object state is hidden (**encapsulated**) from direct access.

Each object is constructed based on a certain template called a **class**. In other words, a class defines a class of objects. Each object has a certain **interface**, a formal definition of the way other objects can interact with it. Originally, it was said that one object sends a message to another object by calling its method. But this terminology did not hold, especially after actual message-based protocols and systems were introduced.

To avoid code duplication, a parent-child relationship between objects was introduced: it was said that one class can inherit behavior from another class. In such a relationship, the first class is called a **child class**, or a **subclass**, while the second is called a **parent** or **base class** or **superclass**.

Another form of relationship was defined between classes and interfaces: it is said that a class can *implement* an interface. Since an interface describes how you can interact with an object, but not how an object responds to the interaction, different objects can behave differently while implementing the same interface.

In Java, a class can have only one direct parent but can implement many interfaces. The ability to behave as any of its ancestors and adhere to multiple interfaces is called **polymorphism**.

In this chapter, we will look at these OOP concepts and how they are implemented in Java. The topics discussed include the following:

- OOP concepts
- Class
- Interface
- Overloading, overriding, and hiding
- Final variable, method, and class
- Polymorphism in action

OOP concepts

As we have already stated in the introduction, the main OOP concepts are as follows:

- **Object/Class**: It defines a state (data) and behavior (methods) and holds them together
- **Inheritance**: It propagates behavior down the chain of classes connected via parent-child relationships
- **Abstraction/Interface**: It describes how the object data and behavior can be accessed. It isolates (abstracts) an object's appearance from its implementations (behavior)
- **Encapsulation**: It hides the state and details of the implementation
- **Polymorphism**: It allows an object to assume an appearance of implemented interfaces and behave as any of the ancestor classes

Object/class

In principle, you can create a very powerful application with minimal usage of classes and objects. It became even easier to do this after functional programming was added to Java 8, to JDK, which allowed you to pass around behavior as a function. Yet passing data (state) still requires classes/objects. This means that the position of Java as an OOP language remains intact.

A class defines the types of all internal object properties that hold the object state. A class also defines object behavior expressed by the code of the methods. It is possible to have a class/object without a state or without a behavior. Java also has a provision for making the behavior accessible statically—without creating an object. But these possibilities are no more than just additions to the object/class concept that was introduced for keeping the state and behavior together.

To illustrate this concept, a class `Vehicle`, for example, defines the properties and behavior of a vehicle in principle. Let's make the model simple and assume that a vehicle has only two properties: weight and engine of a certain power. It also can have a certain behavior: it can reach a certain speed in a certain period of time, depending on the values of its two properties. This behavior can be expressed in a method that calculates the speed the vehicle can reach in a certain period of time. Every object of the `Vehicle` class will have a specific state (values of its properties) and the speed calculation will result in a different speed in the same time period.

All Java code is contained inside methods. A **method** is a group of statements that have (optional) input parameters and a return a value (also optional). In addition, each method can have side effects: it can display a message or write data into the database, for example. Class/object behavior is implemented in the methods.

To follow our example, speed calculations could reside in a `double calculateSpeed(float seconds)` method, for instance. As you can guess, the name of the method is `calculateSpeed`. It accepts a number of seconds (with a fractional part) as a parameter and returns the speed value as `double`.

Inheritance

As we have mentioned already, objects can establish a parent-child relationship and share properties and behavior this way. For example, we can create a `Car` class that inherits properties (weight, for example) and behavior (speed calculation) of the `Vehicle` class. In addition, the child class can have its own properties (number of passengers, for example) and car-specific behavior (soft shock absorption, for example). But if we create a `Truck` class as the vehicle's child, its additional truck-specific property (payload, for example) and behavior (hard shock absorption) will be different.

It is said that each object of the `Car` class or of the `Truck` class has a parent object of the `Vehicle` class. But objects of the `Car` and `Truck` class do not share the specific `Vehicle` object (every time a child object is created, a new parent object is created first). They share only the parent's behavior. That is why all the child objects can have the same behavior but different states. That is one way to achieve code reusability. It may not be flexible enough when the object behavior has to change dynamically. In such cases, object composition (bringing behavior from other classes) or functional programming would be more appropriate (see `Chapter 13`, *Functional Programming*).

It is possible to make a child behave differently than the inherited behavior would do. To achieve it, the method that captures the behavior can be re-implemented in the `child` class. It is said that a child can *override* the inherited behavior. We will explain how to do it shortly (see the *Overloading, overriding, and hiding* section). If, for example, the `Car` class has its own method for speed calculation, the corresponding method of the parent class `Vehicle` is not inherited, but the new speed calculation, implemented in the child class, is used instead.

Properties of a parent class can be inherited (but not overridden) too. However, class properties are typically declared private; they cannot be inherited—that's the point of encapsulation. See the description of various access levels—`public`, `protected`, and `private`—in the *Access modifiers* section.

If the parent class inherits some behavior from another class, the child class acquires (inherits) this behavior too, unless, of course, the parent class overrides it. There is no limit to how long the chain of inheritance can be.

The parent-child relationship in Java is expressed using the `extends` keyword:

```
class A { }
class B extends A { }
class C extends B { }
class D extends C { }
```

In this code, the classes A, B, C, and D have the following relationships:

- Class D inherits from classes C, B, and A
- Class C inherits from classes B and A
- Class B inherits from class A

All non-private methods of class A are inherited (if not overridden) by classes B, C, and D.

Abstraction/interface

The name of a method and the list of its parameter types is called a **method signature**. It describes how the behavior of an object (of `Car` or `Truck`, in our example) can be accessed. Such a description together with a `return` type is presented as an interface. It does not say anything about the code that does calculations—only about the method name, parameters' types, their position in the parameter list, and the result type. All the implementation details are hidden (encapsulated) within the class that *implements* this interface.

As we have mentioned already, a class can implement many different interfaces. But two different classes (and their objects) can behave differently even when they implement the same interface.

Similarly to classes, interfaces can have a parent-child relationship using the `extends` keyword too:

```
interface A { }
interface B extends A {}
interface C extends B {}
interface D extends C {}
```

In this code, the interfaces A, B, C, and D have the following relationships:

- An Interface D inherits from interfaces C, B, and A
- An Interface C inherits from interfaces B and A
- An Interface B inherits from interface A

All non-private methods of interface A are inherited by interfaces B, C, and D.

Abstraction/interface also reduces dependency between different sections of the code, thus increasing its maintainability. Each class can be changed without the need to coordinate it with its clients, as long as the interface stays the same.

Encapsulation

Encapsulation is often defined either as a data hiding or a bundling together of publicly accessible methods and privately accessible data. In a broad sense, encapsulation is a controlled access to the object properties.

The snapshot of values of object properties is called an **object state.** The object state is the data that is encapsulated. So, encapsulation addresses the main issue that motivated the creation of object-oriented programming: better management of concurrent access to the shared data. For example:

```
class A {
    private String prop = "init value";
    public void setProp(String value){
        prop = value;
    }
    public String getProp(){
        return prop;
    }
}
```

As you can see, to read or to modify the value of the prop property, we cannot access it directly because of the access modifier private. Instead, we can do it only via the methods setProp(String value) and getProp().

Polymorphism

Polymorphism is the ability of an object to behave as an object of a different class or as an implementation of a different interface. It owes its existence to all the concepts that have been mentioned previously: inheritance, interface, and encapsulation. Without them, polymorphism would not be possible.

Inheritance allows an object to acquire or override the behaviors of all its ancestors. An interface hides from the client code the name of the class that implemented it. The encapsulation prevents exposing the object state.

In the following sections, we will demonstrate all these concepts in action and look at the specific usage of polymorphism in the *Polymorphism in action* section.

Class

Java program is a sequence of statements that express an executable action. The statements are organized in methods, and methods are organized in classes. One or more classes are stored in .java files. They can be compiled (transformed from the Java language into a bytecode) by the Java compiler javac and stored in .class files. Each .class file contains one compiled class only and can be executed by JVM.

A `java` command starts JVM and tells it which class is the `main` one, the class that has the method called `main()`. The `main` method has a particular declaration: it has to be `public static`, must return `void`, has the name `main`, and accepts a single parameter of an array of a `String` type.

JVM *loads* the main class into memory, finds the `main()` method, and starts executing it, statement by statement. The `java` command can also pass parameters (arguments) that the `main()` method receives as an array of `String` values. If JVM encounters a statement that requires the execution of a method from another class, that class (its `.class` file) is loaded into the memory too and the corresponding method is executed. So, a Java program flow is all about loading classes and executing their methods.

Here is an example of the main class:

```
public class MyApp {
   public static void main(String[] args){
      AnotherClass an = new AnotherClass();
      for(String s: args){
         an.display(s);
      }
   }
}
```

It represents a very simple application that receives any number of parameters and passes them, one by one, into the `display()` method of `AnotherClass` class. As JVM starts, it loads the `MyApp` class from the `MyApp.class` file first. Then it loads the `AnotherClass` class from the `AnotherClass.class` file, creates an object of this class using the `new` operator (we will talk about it shortly), and calls on it the `display()` method.

And here is the `AnotherClass` class:

```
public class AnotherClass {
   private int result;
   public void display(String s){
      System.out.println(s);
   }
   public int process(int i){
      result = i *2;
      return result;
   }
   public int getResult(){
      return result;
   }
}
```

As you can see, the display() method is used for its side effect only—it prints out the passed-in value and returns nothing (void). The AnotherClass class has other two methods:

- The process() method doubles the input integer, stores it in its result property, and returns the value to the caller
- The getResult() method allows getting the result from the object any time later

These two methods are not used in our demo application. We have shown them just for the demonstration that a class can have properties (result, in this case) and many other methods.

The private keyword makes the value accessible only from inside the class, from its methods. The public keyword makes a property or a method accessible by any other class.

Method

As we have stated already, Java statements are organized as methods:

```
<return type> <method name>(<list of parameter types>){
    <method body that is a sequence of statements>
}
```

We have seen a few examples already. A method has a name, a set of input parameters or no parameters at all, a body inside { } brackets, and a return type or void keyword that indicates that the method does not return any value.

The method name and the list of parameter types together are called the **method signature**. The number of input parameters is called an **arity**.

 Two methods have the same *signature* if they have the same name, the same arity, and the same sequence of types in the list of input parameters.

The following two methods have the same signature:

```
double doSomething(String s, int i){
    //some code goes here
}

double doSomething(String i, int s){
    //some code other code goes here
}
```

The code inside the methods may be different even if the signature is the same.

The following two methods have different signatures:

```
double doSomething(String s, int i){
    //some code goes here
}

double doSomething(int s, String i){
    //some code other code goes here
}
```

Just a change in the sequence of parameters makes the signature different, even if the method name remains the same.

Varargs

One particular type of parameter requires mentioning because it is quite different than all others. It is declared a type followed by three dots. It is called **varargs**, which stands for **variable arguments**. But, first, let's briefly define what an array is in Java.

An **array** is a data structure that holds elements of the same type. The elements are referenced by a numerical index. That's enough, for now. We talk about an array more in Chapter 6, *Data Structures, Generics, and Popular Utilities*.

Let's start with an example. Let's declare method parameters using varargs:

```
String someMethod(String s, int i, double... arr){
 //statements that compose method body
}
```

When the someMethod method is called, the java compiler matches the arguments from left to right. Once it gets to the last varargs parameter, it creates an array of the remaining arguments and passes it to the method. Here is a demo code:

```
public static void main(String... args){
    someMethod("str", 42, 10, 17.23, 4);

}

private static String someMethod(String s, int i, double... arr){
    System.out.println(arr[0] + ", " + arr[1] + ", " + arr[2]);
                                        //prints: 10.0, 17.23, 4.0
    return s;
}
```

As you can see, the varargs parameter acts like an array of the specified type. It can be listed as the last or the only parameter of a method. That is why, sometimes, you can see the main method declared as in the preceding example.

Constructor

When an object is created, JVM uses a **constructor**. The purpose of a constructor is to initialize the object state to assign values to all the declared properties. If there is no constructor declared in the class, JVM just assigns to the properties default values. We have talked about the default values for primitive types: it is 0 for integral types, 0.0 for floating-point types, and `false` for boolean types. For other Java reference types (see Chapter 3, *Java Fundamentals*), the default value is `null`, which means that the property of a reference type is not assigned any value.

> When there is no constructor declared in a class, it is said that the class has a default constructor without parameters provided by the JVM.

If necessary, it is possible to declare any number of constructors explicitly, each taking a different set of parameters to set the initial state. Here is an example:

```
class SomeClass {
    private int prop1;
    private String prop2;
    public SomeClass(int prop1){
        this.prop1 = prop1;
    }
    public SomeClass(String prop2){
        this.prop2 = prop2;
    }
    public SomeClass(int prop1, String prop2){
        this.prop1 = prop1;
        this.prop2 = prop2;
    }
    // methods follow
}
```

If a property is not set by a constructor, the default value of the corresponding type is going to be assigned to it automatically.

When several classes are related along the same line of succession, the parent object is created first. If the parent object requires the setting of non-default initial values to its properties, its constructor must be called as the first line of the child constructor using the super keyword as follows:

```
class TheParentClass {
    private int prop;
    public TheParentClass(int prop){
        this.prop = prop;
    }
    // methods follow
}

class TheChildClass extends TheParentClass{
  private int x;
  private String prop;
  private String anotherProp = "abc";
  public TheChildClass(String prop){
  super(42);
  this.prop = prop;
  }
  public TheChildClass(int arg1, String arg2){
  super(arg1);
  this.prop = arg2;
  }
  // methods follow
}
```

In the preceding code example, we added two constructors to TheChildClass: one that always passes 42 to the constructor of TheParentClass and another that accepts two parameters. Note the x property that is declared but not initialized explicitly. It is going to be set to value 0, the default value of the int type, when an object of TheChildClass is created. Also, note the anotherProp property that is initialized explicitly to the value of "abc". Otherwise, it would be initialized to the value null, the default value of any reference type, including String.

Logically, there are three cases when an explicit definition of a constructor in the class is not required:

- When neither the object nor any of its parents do not have properties that need to be initialized
- When each property is initialized along with the type declaration (int x = 42, for example)
- When default values for the properties initialization are good enough

Nevertheless, it is possible that a constructor is still implemented even when all three conditions (mentioned in the list) are met. For example, you may want to execute some statements that initialize some external resource—a file or another database—the object will need as soon as it is created.

As soon as an explicit constructor is added, the default constructor is not provided and the following code generates an error:

```java
class TheParentClass {
    private int prop;
    public TheParentClass(int prop){
        this.prop = prop;
    }
    // methods follow
}

class TheChildClass extends TheParentClass{
    private String prop;
    public TheChildClass(String prop){
        //super(42);          //No call to the parent's contuctor
        this.prop = prop;
    }
    // methods follow
}
```

To avoid the error, either add a constructor without parameters to `TheParentClass` or call an explicit constructor of the parent class as the first statement of the child's constructor. The following code does not generate an error:

```java
class TheParentClass {
    private int prop;
    public TheParentClass() {}
    public TheParentClass(int prop){
        this.prop = prop;
    }
    // methods follow
}

class TheChildClass extends TheParentClass{
    private String prop;
    public TheChildClass(String prop){
        this.prop = prop;
    }
    // methods follow
}
```

One important aspect to note is that constructors, although they look like methods, are not methods or even members of the class. A constructor doesn't have a return type and always has the same name as the class. Its only purpose is to be called when a new instance of the class is created.

The new operator

The new operator creates an object of a class (it also can be said it **instantiates a class** or **creates an instance of a class**) by allocating memory for the properties of the new object and returning a reference to that memory. This memory reference is assigned to a variable of the same type as the class used to create the object or the type of its parent:

```
TheChildClass ref1 = new TheChildClass("something");
TheParentClass ref2 = new TheChildClass("something");
```

Here is an interesting observation. In the code, both object references `ref1` and `ref2`, provide access to the methods of `TheChildClass` and `TheParentClass`. For example, we can add methods to these classes as follows:

```
class TheParentClass {
    private int prop;
    public TheParentClass(int prop){
        this.prop = prop;
    }
    public void someParentMethod(){}
}

class TheChildClass extends TheParentClass{
    private String prop;
    public TheChildClass(int arg1, String arg2){
        super(arg1);
        this.prop = arg2;
    }
    public void someChildMethod(){}
}
```

Then we can call them using any of the following references:

```
TheChildClass ref1 = new TheChildClass("something");
TheParentClass ref2 = new TheChildClass("something");
ref1.someChildMethod();
ref1.someParentMethod();
((TheChildClass) ref2).someChildMethod();
ref2.someParentMethod();
```

Note that, to access the child's methods using the parent's type reference, we had to cast it to the child's type. Otherwise, the compiler generates an error. That is possible because we have assigned to the parent's type reference the reference to the child's object. That is the power of polymorphism. We will talk more about it in the *Polymorphism in action* section.

Naturally, if we had assigned the parent's object to the variable of the parent's type, we would not be able to access the child's method even with casting, as the following example shows:

```
TheParentClass ref2 = new TheParentClass(42);
((TheChildClass) ref2).someChildMethod();  //compiler's error
ref2.someParentMethod();
```

The area where memory for the new object is allocated is called **heap**. The JVM has a process called **garbage collection** that watches for the usage of this area and releases memory for usage as soon as an object is not needed anymore. For example, look at the following method:

```
void someMethod(){
    SomeClass ref = new SomeClass();
    ref.someClassMethod();
    //other statements follow
}
```

As soon as the execution of the `someMethod()` method is completed, the object of `SomeClass` is not accessible anymore. That's what the garbage collector notices and releases the memory occupied by this object. We will talk about the garbage collection process in Chapter 9, *JVM Structure and Garbage Collection*.

Class java.lang.Object

In Java, all classes are children of the `Object` class by default, even if you do not specify it implicitly. The `Object` class is declared in the `java.lang` package of the standard JDK library. We will define what *package* is in the *Packages, importing, and access* section and describe libraries in Chapter 7, *Java Standard and External Libraries*.

Let's look back at the example we have provided in the *Inheritance* section:

```
class A { }
class B extends A {}
class C extends B {}
class D extends C {}
```

All classes, A, B, C, D, are children of the `Object` class, which has ten methods that every class inherits:

- `public String toString()`
- `public int hashCode()`
- `public boolean equals (Object obj)`
- `public Class getClass()`
- `protected Object clone()`
- `public void notify()`
- `public void notifyAll()`
- `public void wait()`
- `public void wait(long timeout)`
- `public void wait(long timeout, int nanos)`

The first three, `toString()`, `hashCode()`, and `equals()`, are the most often used methods and often re-implemented (overridden). The `toString()` method is typically used to print the state of the object. Its default implementation in JDK looks like this:

```
public String toString() {
    return getClass().getName()+"@"+Integer.toHexString(hashCode());
}
```

If we use it on the object of the `TheChildClass` class, the result will be as follows:

```
TheChildClass ref1 = new TheChildClass("something");
System.out.println(ref1.toString());
//prints: com.packt.learnjava.ch02_oop.Constructor$TheChildClass@72ea2f77
```

By the way, there is no need to call `toString()` explicitly while passing an object into the `System.out.println()` method and similar output methods, because they do it inside the method anyway and `System.out.println(ref1)`, in our case, produces the same result.

So, as you can see, such an output is not human-friendly, so it is a good idea to override the `toString()` method. The easiest way to do it is by using IDE. For example, in IntelliJ IDEA, right-click inside `TheChildClass` code, as shown in the following screenshot:

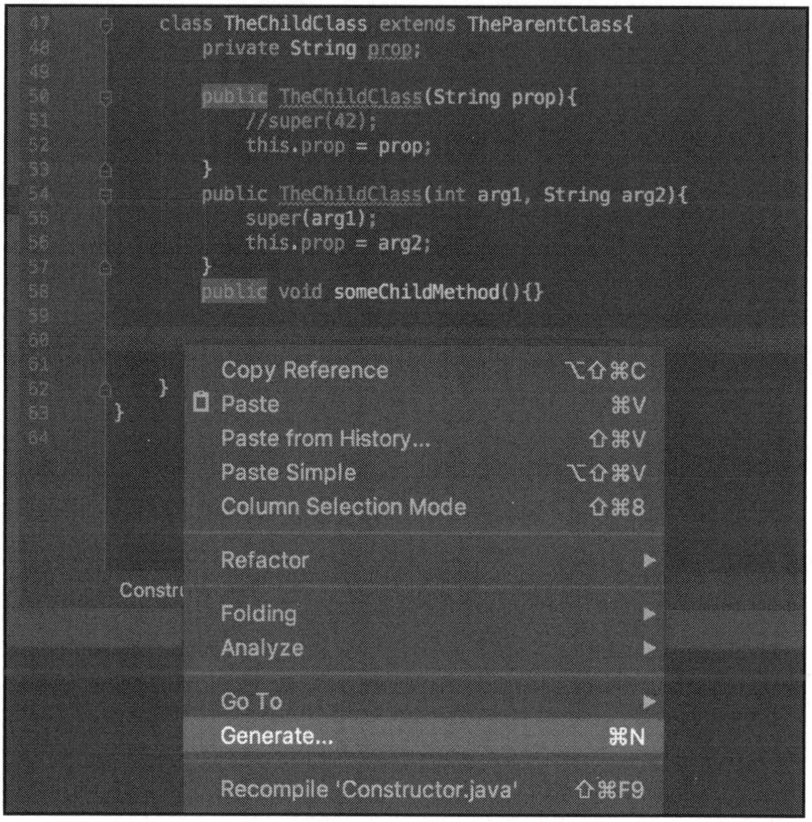

Select and click **Generate...** and then select and click **toString()**, as shown in the following screenshot:

```
46
47        class TheChildClass extends TheParentClass{
48            private String prop;
49
50            public TheChildClass(String prop){
51                //super(42);
52                this.prop = prop;
53            }
54            public TheChildClass(int arg1, String arg2){
55                super(arg1);
56                this.prop = arg2;
57            }
58            public void someChildMethod(){}
59
60
61                    Generate
62
63        Constructor
64        Getter
           Setter
           Getter and Setter
           equals() and hashCode()
           toString()
           Override Methods...        ^O
           Delegate Methods...
           Copyright

:tor$TheChildClass@72ea2f77
```

The new pop-up window will enable you to select which properties to include in the
`toString()` method. Select only properties of `TheChildClass` as follows:

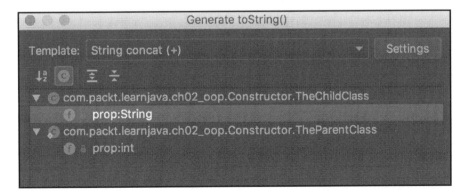

After you click the **OK** button, the following code will be generated:

```
@Override
public String toString() {
    return "TheChildClass{" +
            "prop='" + prop + '\'' +
            '}';
}
```

If there were more properties in the class and you had selected them, more properties and their values would be included in the method output. If we print the object now, the result will be this:

```
TheChildClass ref1 = new TheChildClass("something");
System.out.println(ref1.toString());
                        //prints: TheChildClass{prop='something'}
```

That is why the `toString()` method is often overridden and even included in the services of an IDE.

We will talk about the `hashCode()` and `equals()` methods in more detail in Chapter 6, *Data Structures, Generics, and Popular Utilities*.

The `getClass()` and `clone()` methods are used not as often. The `getClass()` method returns an object of the `Class` class that has many methods which provide various system information. The most used method is the one that returns the name of the class of the current object. The `clone()` method can be used to copy the current object. It works just fine as long as all the properties of the current object are of primitive types. But, if there is a reference type property, the `clone()` method has to be re-implemented so that the copy of the reference type can be done correctly. Otherwise, only the reference will be copied, not the object itself. Such a copy is called a **shallow copy**, which may be good enough in some cases. The `protected` keyword indicates that only children of the class can access it. See the *Packages, importing, and access* section.

The last five of the class `Object` methods are used for communication between threads—the lightweight processes for concurrent processing. They are typically not re-implemented.

Instance and static properties and methods

So far, we have seen mostly methods that can be invoked only on an object (instance) of a class. Such methods are called **instance methods**. They typically use the values of the object properties (the object state). Otherwise, if they do not use the object state, they can be made static and invoked without creating an object. The example of such a method is the main() method. Here is another example:

```
class SomeClass{
    public static void someMethod(int i){
        //do something
    }
}
```

This method can be called as follows:

```
SomeClass.someMethod(42);
```

Static methods can be called on an object too, but it is considered bad practice as it hides the static nature of the method from a human who tries to understand the code. Besides, it raises a compiler warning and, depending on the compiler implementation, may even generate a compiler error.

Similarly, a property can be declared static and thus accessible without creating an object. For example:

```
class SomeClass{
    public static String SOME_PROPERTY = "abc";
}
```

This property can be accessed directly via class too, as follows:

```
System.out.println(SomeClass.SOME_PROPERTY);   //prints: abc
```

Having such a static property works against the idea of the state encapsulation and may cause all the problems of concurrent data modification because it exists as a single copy in JVM memory and all the methods that use it, share the same value. That is why a static property is typically used for two purposes:

- To store a constant—a value that can be read but not modified (also called a **read-only value**)
- To store a stateless object that is expensive to create or that keeps read-only values

A typical example of a constant is a name of a resource:

```
class SomeClass{
    public static final String INPUT_FILE_NAME = "myFile.csv";
}
```

Note the `final` keyword in front of the static property. It tells the compiler and JVM that this value, once assigned, cannot change. An attempt to do it generates an error. It helps to protect the value and express clearly the intent to have this value as a constant. When a human tries to understand how the code works, such seemingly small details make the code easier to understand.

That said, consider using interfaces for such a purpose. Since Java 1.8, all the fields declared in an interface are implicitly static and final, so there is less chance you'll forget to declare a value to be final. We will talk about interfaces shortly.

When an object is declared a static final class property, it does not mean all its properties become final automatically. It only protects the property from assigning another object of the same type. We will discuss the complicated procedure of concurrent access of an object property in Chapter 8, *Multithreading and Concurrent Processing*. Nevertheless, programmers often use static final objects to store the values that are read-only just by the way they are used in the application. A typical example would be an application configuration information. Once created after reading from a disk, it is not changed, even if it could be. Also, caching of data is obtained from an external resource.

Again, before using such a class property for this purpose, consider using an interface that provides more default behavior that supports a read-only functionality.

Similar to static properties, static methods can be invoked without creating an instance of the class. Consider, for example, the following class:

```
class SomeClass{
    public static String someMethod() {
        return "abc";
    }
}
```

We can call the preceding method by using just a class name:

```
System.out.println(SomeClass.someMethod()); //prints: abc
```

Interface

In the *Abstraction/Interface* section, we talked about an interface in general terms. In this section, we are going to describe a Java language construct that expresses it.

An **interface** presents what can be expected of an object. It hides the implementation and exposes only method signatures with return values. For example, here is an interface that declares two abstract methods:

```
interface SomeInterface {
    void method1();
    String method2(int i);
}
```

And here is a class that implements it:

```
class SomeClass implements SomeInterface{
    public void method1(){
        //method body
    }
    public String method2(int i) {
        //method body
        return "abc";
    }
}
```

An interface cannot be instantiated. An object of an interface type can be created only by creating an object of a class that *implements* this interface:

```
SomeInterface si = new SomeClass();
```

If not all of the abstract methods of the interface have been implemented, the class must be declared abstract and cannot be instantiated. See the *Interface versus abstract class* section.

An interface does not describe how the object of the class can be created. To discover that, you must look at the class and see what constructors it has. An interface also does not describe the static class methods. So, an interface is the public face of a class instance (object) only.

With Java 8, interface acquired an ability to have not just abstract methods (without a body), but really implemented ones. According to The Java Language Specification, "*The body of an interface may declare members of the interface, that is, fields, methods, classes, and interfaces.*" Such a broad statement brings up the question, what is the difference between an interface and a class? One principal difference we have pointed out already is this: an interface cannot be instantiated; only a class can be instantiated.

Another difference is that a non-static method implemented inside an interface is declared `default` or `private`. By contrast, a `default` declaration is not available for the class methods.

Also, fields in an interface are implicitly public, static, and final. By contrast, class properties and methods are not static or final by default. The implicit (default) access modifier of a class itself, its fields, methods, and constructors is package-private, which means it is visible only within its own package.

Default methods

To get an idea about the function of default methods in an interface, let's look at an example of an interface and a class that implements it, as follows:

```java
interface SomeInterface {
    void method1();
    String method2(int i);
    default int method3(){
        return 42;
    }
}

class SomeClass implements SomeInterface{
    public void method1(){
        //method body
    }
    public String method2(int i) {
        //method body
        return "abc";
    }
}
```

We can now create an object of `SomeClass` class and make the following call:

```java
SomeClass sc = new SomeClass();
sc.method1();
sc.method2(22);   //returns: "abc"
sc.method3();     //returns: 42
```

As you can see, `method3()` is not implemented in the `SomeClass` class, but it looks as if the class has it. That is one way to add a new method to an existing class without changing it—by adding the default method to the interface the class implements.

Let's now add `method3()` implementation to the class too, as follows:

```
class SomeClass implements SomeInterface{
    public void method1(){
        //method body
    }
    public String method2(int i) {
        //method body
        return "abc";
    }
    public int method3(){
        return 15;
    }
}
```

Now the interface implementation of `method3()` will be ignored:

```
SomeClass sc = new SomeClass();
sc.method1();
sc.method2(22);   //returns: "abc"
sc.method3();     //returns: 15
```

The purpose of the default method in an interface is to provide a new method to the classes (that implement this interface) without changing them. But the interface implementation is ignored as soon as a class implements the new method too.

Private methods

If there are several default methods in an interface, it is possible to create private methods accessible only by the default methods of the interface. They can be used to contain common functionality, instead of repeating it in every default method:

```
interface SomeInterface {
    void method1();
    String method2(int i);
    default int method3(){
        return getNumber();
    }
    default int method4(){
        return getNumber() + 22;
    }
    private int getNumber(){
        return 42;
    }
}
```

This concept of private methods is not different to private methods in classes (see the *Packages, importing, and access* section). The private methods cannot be accessed from outside the interface.

Static fields and methods

Since Java 8, all the fields declared in an interface are implicitly public, static, and final constants. That is why an interface is a preferred location for the constants. You do not need to add `public static final` to their declarations.

As for the static methods, they function in an interface in the same way as in a class:

```
interface SomeInterface{
    static String someMethod() {
        return "abc";
    }
}
```

Note, there is no need to mark the interface method as `public`. All non-private interface methods are public by default.

We can call the preceding method by using just an interface name:

```
System.out.println(SomeInetrface.someMethod()); //prints: abc
```

Interface versus abstract class

We have mentioned already that a class can be declared `abstract`. It may be a regular class that we do not want to be instantiated, or it may be a class that contains (or inherits) abstract methods. In the latter case, we must declare such a class as `abstract` to avoid a compilation error.

In many respects, an abstract class is very similar to an interface. It forces every child class that extends it to implement the abstract methods. Otherwise, the child cannot be instantiated and has to be declared abstract itself.

However, a few principal differences between an interface and abstract class make each of them useful in different situations:

- An abstract class can have a constructor, while an interface cannot.
- An abstract class can have a state, while an interface cannot.

- The fields of an abstract class can be `public`, `private` or `protected`, `static` or not, `final` or not, while, in an interface, fields are always `public`, `static`, and `final`.
- The methods in an abstract class can be `public`, `private`, or `protected`, while the interface methods can be `public`, or `private` only.
- If the class you would like to amend extends another class already, you cannot use an abstract class, but you can implement an interface because a class can extend only one other class, but can implement multiple interfaces

See an example of an abstract usage in the *Polymorphism in action* section.

Overloading, overriding, and hiding

We have already mentioned overriding in the *Inheritance* and *Abstraction/Interface* sections. It is a replacing of a non-static method implemented in a parent class with the method of the same signatures in the child class. A default method of an interface also can be overridden in the interface that extends it.

Hiding is similar to overriding but applies only to static methods and static, as well as properties of the instance .

Overloading is creating several methods with the same name and different parameters (thus, different signatures) in the same class or interface.

In this section, we will discuss all these concepts and demonstrate how they work for classes and interfaces.

Overloading

It is not possible to have two methods in the same interface or a class with the same signature. To have a different signature, the new method has to have either a new name or a different list of parameter types (and the sequence of the type does matter). Having two methods with the same name but a different list of parameter types constitutes overloading. Here are a few examples of a legitimate method of overloading in an interface:

```
interface A {
    int m(String s);
    int m(String s, double d);
    default int m(String s, int i) { return 1; }
    static int m(String s, int i, double d) { return 1; }
}
```

Note that no two of the preceding methods have the same signature, including the default and static methods. Otherwise, a compiler's error would be generated. Neither designation as default, nor static, plays any role in the overloading. A return type does not affect the overloading either. We use `int` as a return type everywhere just to make the examples less cluttered.

Method overloading is done similarly in a class:

```
class C {
    int m(String s){ return 42; }
    int m(String s, double d){ return 42; }
    static int m(String s, double d, int i) { return 1; }
}
```

And, it does not matter where the methods with the same name are declared. The following method overloading is not different to the previous example, as follows:

```
interface A {
    int m(String s);
    int m(String s, double d);
}
interface B extends A {
    default int m(String s, int i) { return 1; }
    static int m(String s, int i, double d) { return 1; }
}
class C {
    int m(String s){ return 42; }
}
class D extends C {
    int m(String s, double d){ return 42; }
    static int m(String s, double d, int i) { return 1; }
}
```

A private non-static method can be overloaded only by a non-static method of the same class.

Overloading happens when methods have the same name but a different list of parameter types and belong to the same interface (or class) or to different interfaces (or classes), one of which is an ancestor to another. A private method can be overloaded only by a method in the same class.

Overriding

In contrast to overloading, which happens with the static and non-static methods, method overriding happens only with non-static methods and only when they have *exactly the same signature* and *belong to different interfaces (or classes)*, one of which is an ancestor to another.

 The overriding method resides in the child interface (or class), while the overridden method has the same signature and belongs to one of the ancestor interfaces (or classes). A private method cannot be overridden.

The following are examples of a method overriding in an interface:

```
interface A {
    default void method(){
        System.out.println("interface A");
    }
}
interface B extends A{
    @Override
    default void method(){
        System.out.println("interface B");
    }
}
class C implements B { }
```

If we call the method() using the C class instance, the result will be as follows:

```
C c = new C();
c.method();        //prints: interface B
```

Please notice the usage of the annotation @Override. It tells the compiler that the programmer thinks that the annotated method overrides a method of one of the ancestor interfaces. This way, the compiler can make sure that the overriding does happen and generates an error if not. For example, a programmer may misspell the name of the method as follows:

```
interface B extends A{
    @Override
    default void metod(){
        System.out.println("interface B");
    }
}
```

If that happens, the compiler generates an error because there is no method `metod()` to override. Without the annotation `@Overrride`, this mistake may go unnoticed by the programmer and the result would be quite different:

```
C c = new C();
c.method();        //prints: interface A
```

The same rules of overriding apply to the class instance methods. In the following example, the C2 class overrides a method of the C1 class:

```
class C1{
    public void method(){
        System.out.println("class C1");
    }
}
class C2 extends C1{
    @Override
    public void method(){
        System.out.println("class C2");
    }
}
```

The result is as follows:

```
C2 c2 = new C2();
c2.method();        //prints: class C2
```

And, it does not matter how many ancestors are between the class or interface with the overridden method and the class or interface with the overriding method:

```
class C1{
    public void method(){
        System.out.println("class C1");
    }
}
class C3 extends C1{
    public void someOtherMethod(){
        System.out.println("class C3");
    }
}
class C2 extends C3{
    @Override
    public void method(){
        System.out.println("class C2");
    }
}
```

The result of the preceding method's overriding will still be the same.

Hiding

Hiding is considered by many to be a complicated topic, but it should not be, and we will try to make it look simple.

The name *hiding* came from the behavior of static properties and methods of classes and interfaces. Each static property or method exists as a single copy in the JVM's memory because they are associated with the interface or class, not with an object. And interface or class exists as a single copy. That is why we cannot say that the child's static property or method overrides the parent's static property or method with the same name. All static properties and methods are loaded into the memory only once when the class or interface is loaded and stay there, not copied anywhere. Let's look at the example.

Let's create two interfaces that have a parent-child relationship and static fields and methods with the same name:

```
interface A {
    String NAME = "interface A";
    static void method() {
        System.out.println("interface A");
    }
}
interface B extends A {
    String NAME = "interface B";
    static void method() {
        System.out.println("interface B");
    }
}
```

Please note the capital case for an identifier of an interface field. That is the convention often used to denote a constant, whether it is declared in an interface or in a class. Just to remind you, a constant in Java is a variable that, once initialized, cannot be re-assigned another value. An interface field is a constant by default because any field in an interface is *final* (see the *Final properties, methods, and classes* section).

If we print NAME from the B interface and execute its method(), we get the following result:

```
System.out.println(B.NAME); //prints: interface B
B.method();                  //prints: interface B
```

It looks very much like overriding, but, in fact, it is just that we call a particular property or a method associated with this particular interface.

Similarly, consider the following classes:

```
public class C {
    public static String NAME = "class C";
    public static void method(){
        System.out.println("class C");
    }
    public String name1 = "class C";
}
public class D extends C {
    public static String NAME = "class D";
    public static void method(){
        System.out.println("class D");
    }
    public String name1 = "class D";
}
```

If we try to access the static members of D class using the class itself, we will get what we asked for:

```
System.out.println(D.NAME);    //prints: class D
D.method();                    //prints: class D
```

The confusion appears only when a property or a static method is accessed using an object:

```
C obj = new D();

System.out.println(obj.NAME);          //prints: class C
System.out.println(((D) obj).NAME);    //prints: class D

obj.method();                          //prints: class C
((D)obj).method();                     //prints: class D

System.out.println(obj.name1);          //prints: class C
System.out.println(((D) obj).name1);   //prints: class D
```

The `obj` variable refers to the object of the D class, and the casting proves it, as you can see in the preceding example. But, even if we use an object, trying to access a static property or method brings us the members of the class that was used as the declared variable type. As for the instance property in the last two lines of the example, the properties in Java do not conform to polymorphic behavior and we get the `name1` property of the parent C class, instead of the expected property of the child D class.

To avoid confusion with static members of a class, always access them using the class, not an object. To avoid confusion with instance properties, always declare them private and access via methods.

To illustrate the last tip, consider the following classes:

```
class X {
    private String name = "class X";
    public String getName() {
        return name;
    }
    public void setName(String name) {
        this.name = name;
    }
}
class Y extends X {
    private String name = "class Y";
    public String getName() {
        return name;
    }
    public void setName(String name) {
        this.name = name;
    }
}
```

If we run the same test for the instance properties as we did for classes C and D, the result will be this:

```
X x = new Y();
System.out.println(x.getName());        //prints: class Y
System.out.println(((Y)x).getName());   //prints: class Y
```

Now we access instance properties using methods, which are subjects for an overriding effect and do not have unexpected results anymore.

To close the discussion of hiding in Java, we would like to mention another type of hiding, namely, when a local variable hides the instance or static property with the same name. Here is a class that does it:

```
public class HidingProperty {
    private static String name1 = "static property";
    private String name2 = "instance property";

    public void method() {
        var name1 = "local variable";
        System.out.println(name1);      //prints: local variable
```

```
        var name2 = "local variable";   //prints: local variable
        System.out.println(name2);

        System.out.println(HidingProperty.name1); //prints: static property
        System.out.println(this.name2);          //prints: instance property
    }
}
```

As you can see, the local variable `name1` hides the static property with the same name, while the local variable `name2` hides the instance property. It is possible still to access the static property using the class name (see `HidingProperty.name1`). Please note that, despite being declared `private`, it is accessible from inside the class.

The instance property can always be accessed by using the `this` keyword that means **current object**.

Final variable, method, and classes

We have mentioned a `final` property several times in relation to the notion of a constant in Java. But that is only one case of using the `final` keyword. It can be applied to any variable in general. Also, a similar constraint can be applied to a method and even a class too, thus preventing the method from being overridden and the class being extended.

Final variable

The `final` keyword placed in front of a variable declaration makes this variable immutable after the initialization. For example:

```
final String s = "abc";
```

The initialization can even be delayed:

```
final String s;
s = "abc";
```

In the case of an object property, this delay can last only until the object is created. This means that the property can be initialized in the constructor. For example:

```
private static class A {
    private final String s1 = "abc";
    private final String s2;
    private final String s3;   //error
    private final int x;       //error
```

```
        public A() {
            this.s1 = "xyz";        //error
            this.s2 = "xyz";
        }
}
```

Notice that, even during the object construction, it is not possible to initialize the property twice—during declaration and in the constructor. It is also interesting to note that a final property has to be initialized explicitly. As you can see from the preceding example, the compiler does not allow the initialization of the final property to a default value.

It is also possible to initialize a `final` property in an initialization block:

```
class B {
    private final String s1 = "abc";
    private final String s2;
    {
        s1 = "xyz"; //error
        s2 = "abc";
    }
}
```

In the case of a static property, it is not possible to initialize it in a constructor, so it has to be initialized either during its declaration or in a static initialization block:

```
class C {
    private final static String s1 = "abc";
    private final static String s2;
    static {
        s1 = "xyz"; //error
        s2 = "abc";
    }
}
```

In an interface, all fields are always final, even if they are not declared as such. Since neither constructor nor initialization block is not allowed in an interface, the only way to initialize an interface field is during declaration. Failing to do it results in a compilation error:

```
interface I {
    String s1;   //error
    String s2 = "abc";
}
```

Final method

A method declared `final` cannot be overridden in a child class or hidden in the case of a static method. For example, the `java.lang.Object` class, which is the ancestor of all classes in Java, has some of its methods declared `final`:

```
public final Class getClass()
public final void notify()
public final void notifyAll()
public final void wait() throws InterruptedException
public final void wait(long timeout) throws InterruptedException
public final void wait(long timeout, int nanos)
                                    throws InterruptedException
```

All private methods and uninherited methods of a `final` class are effectively final because you cannot override them.

Final class

A `final` class cannot be extended. It cannot have children, which makes all the methods of the class effectively `final` too. This feature is used for security or when a programmer would like to make sure the class functionality cannot be overridden, overloaded, or hidden because of some other design considerations.

Polymorphism in action

Polymorphism is the most powerful and useful feature of OOP. It uses all other OOP concepts and features we have presented so far. It is the highest conceptual point on the way to mastering Java programming. After it, the rest of the book will be mostly about Java language syntax and JVM functionality.

As we have stated in the *OOP concepts* section, **polymorphism** is the ability of an object to behave as an object of different classes or as an implementation of different interfaces. If you search the word polymorphism on the Internet, you will find that it is "*the condition of occurring in several different forms.*" Metamorphosis is "*a change of the form or nature of a thing or person into a completely different one, by natural or supernatural means.*" So, **Java polymorphism** is the ability of an object to behave as if going through a metamorphosis and to exhibit completely different behaviors under different conditions.

We will present this concept in a practical hands-on way, using an **object factory**—a specific programming implementation of a factory, which is a *method that returns objects of a varying prototype or class.* See `https://en.wikipedia.org/wiki/Factory_(object-oriented_programming)`.

Object factory

The idea behind the object factory is to create a method that returns a new object of a certain type under certain conditions. For example, look at the `CalcUsingAlg1` and `CalcUsingAlg2` classes:

```
interface CalcSomething{ double calculate(); }
class CalcUsingAlg1 implements CalcSomething{
    public double calculate(){ return 42.1; }
}
class CalcUsingAlg2 implements CalcSomething{
    private int prop1;
    private double prop2;
    public CalcUsingAlg2(int prop1, double prop2) {
        this.prop1 = prop1;
        this.prop2 = prop2;
    }
    public double calculate(){ return prop1 * prop2; }
}
```

As you can see, they both implement the same interface, `CalcSomething`, but use different algorithms. Now, let's say that we decided that the selection of the algorithm used will be done in a property file. Then we can create the following object factory:

```
class CalcFactory{
    public static CalcSomething getCalculator(){
        String alg = getAlgValueFromPropertyFile();
        switch(alg){
            case "1":
                return new CalcUsingAlg1();
            case "2":
                int p1 = getAlg2Prop1FromPropertyFile();
                double p2 = getAlg2Prop2FromPropertyFile();
                return new CalcUsingAlg2(p1, p2);
            default:
                System.out.println("Unknown value " + alg);
                return new CalcUsingAlg1();
        }
    }
}
```

The factory selects which algorithm to use based on the value returned by the `getAlgValueFromPropertyFile()` method. In the case of the second algorithm, it also uses the `getAlg2Prop1FromPropertyFile()` methods and `getAlg2Prop2FromPropertyFile()` to get the input parameters for the algorithm. But this complexity is hidden from a client:

```
CalcSomething calc = CalcFactory.getCalculator();
double result = calc.calculate();
```

We can add new algorithm variations, change the source for the algorithm parameters or the process of the algorithm selection, but the client will not need to change the code. And that is the power of polymorphism.

Alternatively, we could use inheritance to implement polymorphic behavior. Consider the following classes:

```
class CalcSomething{
    public double calculate(){ return 42.1; }
}
class CalcUsingAlg2 extends CalcSomething{
    private int prop1;
    private double prop2;
    public CalcUsingAlg2(int prop1, double prop2) {
        this.prop1 = prop1;
        this.prop2 = prop2;
    }
    public double calculate(){ return prop1 * prop2; }
}
```

Then our factory may look as follows:

```
class CalcFactory{
    public static CalcSomething getCalculator(){
        String alg = getAlgValueFromPropertyFile();
        switch(alg){
            case "1":
                return new CalcSomething();
            case "2":
                int p1 = getAlg2Prop1FromPropertyFile();
                double p2 = getAlg2Prop2FromPropertyFile();
                return new CalcUsingAlg2(p1, p2);
            default:
                System.out.println("Unknown value " + alg);
                return new CalcSomething();
        }
    }
}
```

But the client code does not change:

```
CalcSomething calc = CalcFactory.getCalculator();
double result = calc.calculate();
```

Given a choice, an experienced programmer uses a common interface for the implementation. It allows for a more flexible design, as a class in Java can implement multiple interfaces, but can extend (inherit from) one class.

Operator instanceof

Unfortunately, life is not always that easy and, once in a while, a programmer has to deal with a code that is assembled from unrelated classes, coming even from different frameworks. In such a case, using polymorphism may be not an option. Still, you can hide the complexity of an algorithm selection and even simulate polymorphic behavior using the instanceof operator, which returns true when the object is an instance of a certain class.

Let's assume we have two unrelated classes:

```
class CalcUsingAlg1 {
    public double calculate(CalcInput1 input){
        return 42. * input.getProp1();
    }
}

class CalcUsingAlg2{
    public double calculate(CalcInput2 input){
        return input.getProp2() * input.getProp1();
    }
}
```

Each of the classes expects as an input an object of a certain type:

```
class CalcInput1{
    private int prop1;
    public CalcInput1(int prop1) { this.prop1 = prop1; }
    public int getProp1() { return prop1; }
}

class CalcInput2{
    private int prop1;
    private double prop2;
    public CalcInput2(int prop1, double prop2) {
        this.prop1 = prop1;
```

```
            this.prop2 = prop2;
    }
    public int getProp1() { return prop1; }
    public double getProp2() { return prop2; }
}
```

And let's assume that the method we implement receives such an object:

```
void calculate(Object input) {
    double result = Calculator.calculate(input);
    //other code follows
}
```

We still use polymorphism here because we describe our input as the Object type. We can do it because the Object class is the base class for all Java classes.

Now let's look at how the Calculator class is implemented:

```
class Calculator{
    public static double calculate(Object input){
        if(input instanceof CalcInput1){
            return new CalcUsingAlg1().calculate((CalcInput1)input);
        } else if (input instanceof CalcInput2){
            return new CalcUsingAlg2().calculate((CalcInput2)input);
        } else {
            throw new RuntimeException("Unknown input type " +
                            input.getClass().getCanonicalName());
        }
    }
}
```

As you can see, it uses the instanceof operator for selecting the appropriate algorithm. By using the Object class as an input type, the Calculator class takes advantage of polymorphism too, but most of its implementation has nothing to do with it. Yet, from outside, it looks polymorphic and it is, but only to a degree.

Summary

This chapter introduced readers to the concepts of OOP and how they are implemented in Java. It provided an explanation of each concept and demonstrated how to use it in specific code examples. The Java language constructs of `class` and `interface` were discussed in detail. The reader also has learned what overloading, overriding, and hiding are and how to use the `final` keyword to protect methods from being overridden.

From the *Polymorphism in action* section, the reader learned about the powerful Java feature of polymorphism. This section brought all the presented material together and showed how polymorphism stays in the center of OOP.

In the next chapter, the reader will become familiar with Java language syntax, including packages, importing, access modifiers, reserved and restricted keywords, and some aspects of Java reference types. The reader will also learn how to use the `this` and `super` keywords, what widening and narrowing conversions of primitive types are, boxing and unboxing, primitive and reference types assignment, and how the `equals()` method of a reference type works.

Quiz

1. Select all the correct OOP concepts from the following list:

    a. Encapsulation

    b. Isolation

    c. Polynation

    d. Inheritance

2. Select all the correct statements from the following list:

    a. A Java object has status

    b. A Java object has behavior

    c. A Java object has state

    d. A Java object has methods

3. Select all the correct statements from the following list:

    a. A Java object behavior can be inherited
    b. A Java object behavior can be overridden
    c. A Java object behavior can be overloaded
    d. A Java object behavior can be overwhelmed

4. Select all the correct statements from the following list:

    a. Java objects of different classes can have the same behavior
    b. Java objects of different classes share a parent object state
    c. Java objects of different classes have as a parent an object of the same class
    d. Java objects of different classes can share behavior

5. Select all the correct statements from the following list:

    a. The method signature includes the return type
    b. The method signature is different if the return type is different
    c. The method signature changes if two parameters of the same type switch position
    d. The method signature changes if two parameters of different types switch position

6. Select all the correct statements from the following list:

    a. Encapsulation hides the class name
    b. Encapsulation hides behavior
    c. Encapsulation allows access to data only via methods
    d. Encapsulation does not allow direct access to the state

7. Select all the correct statements from the following list:

    a. The class is declared in the `.java` file
    b. The class bytecode is stored in the `.class` file
    c. The parent class is stored in the `.base` file
    d. The child class is stored in the `.sub` file

8. Select all the correct statements from the following list:

      a. A method defines an object state

      b. A method defines an object behavior

      c. A method without parameters is marked as `void`

      d. A method can have many `return` statements

9. Select all the correct statements from the following list:

      a. Varargs is declared as a `var` type

      b. Varargs stands for *various arguments*

      c. Varargs is a `String` array

      d. Varargs can act as an array of the specified type

10. Select all the correct statements from the following list:

      a. A constructor is a method that creates a state

      b. The primary responsibility of a constructor is to initialize a state

      c. JVM always provides a default constructor

      d. The parent class constructor can be called using the `parent` keyword

11. Select all the correct statements from the following list:

      a. The `new` operator allocates memory to an object

      b. The `new` operator assigns default values to the object properties

      c. The `new` operator creates a parent object first

      d. The `new` operator creates a child object first

12. Select all the correct statements from the following list:

      a. A class `Object` belongs to the `java.base` package

      b. A class `Object` belongs to the `java.lang` package

      c. A class `Object` belongs to a package of the Java Class Library

      d. A class `Object` is imported automatically

13. Select all the correct statements from the following list:

    a. An instance method is invoked using an object
    b. A static method is invoked using a class
    c. An instance method is invoked using a class
    d. A static method is invoked using an object

14. Select all the correct statements from the following list:

    a. Methods in an interface are implicitly `public`, `static`, and `final`
    b. An interface can have methods that can be invoked without being implemented in a class
    c. An interface can have fields that can be used without any class
    d. An interface can be instantiated

15. Select all the correct statements from the following list:

    a. The default method of an interface is always invoked by default
    b. The private method of an interface can be invoked only by the default method
    c. The interface static method can be invoked without being implemented in a class
    d. The default method can enhance a class that implements the interface

16. Select all the correct statements from the following list:

    a. An `Abstract` class can have a default method
    b. An `Abstract` class can be declared without an `abstract` method
    c. Any class can be declared abstract
    d. An interface is an `abstract` class without a constructor

17. Select all the correct statements from the following list:

    a. Overloading can be done only in an interface
    b. Overloading can be done only when one class extends another
    c. Overloading can be done in any class
    d. The overloaded method must have the same signature

18. Select all the correct statements from the following list:

       a. Overriding can be done only in a child class
       b. Overriding can be done in an interface
       c. The overridden method must have the same name
       d. No method of an `Object` class can be overridden

19. Select all the correct statements from the following list:

       a. Any method can be hidden
       b. A variable can hide a property
       c. A static method can be hidden
       d. A public instance property can be hidden

20. Select all the correct statements from the following list:

       a. Any variable can be declared final
       b. A public method cannot be declared final
       c. A protected method can be declared final
       d. A class can be declared protected

21. Select all the correct statements from the following list:

       a. Polymorphic behavior can be based on inheritance
       b. Polymorphic behavior can be based on overloading
       c. Polymorphic behavior can be based on overriding
       d. Polymorphic behavior can be based on an interface

3
Java Fundamentals

This chapter presents to the reader a more detailed view of Java as a language. It starts with the code organization in packages and the description of accessibility levels of classes (interfaces) and their methods and properties (fields). The reference types, as the main types of a Java object—oriented nature, are also presented in detail , followed by a list of reserved and restricted keywords and a discussion of their usage. The chapter ends with the methods of conversion between primitive types and from a primitive type to the corresponding reference type and back.

These are the Java language fundamental terms and features. The importance of their understanding cannot be overstated. Without them, you cannot write any Java program. So, try not to rush through this chapter and make sure you understand everything presented.

The following topics will be covered in this chapter:

- Packages, importing, and access
- Java reference types
- Reserved and restricted keywords
- The usage of `this` and `super` keywords
- Converting between primitive types
- Converting between primitive and reference types

Packages, importing, and access

As you already know, the package name reflects the directory structure, starting with the project directory that contains the `.java` files. The name of each `.java` file has to be the same as the name of the top—level class declared in it (this class can contain other classes). The first line of the `.java` file is the package statement that starts with the `package` keyword, followed by the actual package name – the directory path to this file in which slashes are replaced with dots.

A package name and the class name together compose a **fully qualified class name**. It uniquely identifies the class, but tends to be too long and inconvenient to use. That is when **importing** comes to the rescue, by allowing specification of the fully qualified name only once, and then referring to the class only by the class name.

Invoking a method of a class from the method of another class is possible only if the caller has access to that class and its methods. The access modifiers `public`, `protected`, and `private` define the level of accessibility and allow (or disallow) some methods, properties, or even the class itself to be visible to other classes.

All these aspects will be discussed in detail in the current section.

Packages

Let's look at the class we called `Packages`:

```
package com.packt.learnjava.ch03_fundamentals;
import com.packt.learnjava.ch02_oop.hiding.C;
import com.packt.learnjava.ch02_oop.hiding.D;
public class Packages {
    public void method(){
        C c = new C();
        D d = new D();
    }
}
```

The first line in the `Packages` class is a package declaration that identifies the class location on the source tree or, in other words, the `.java` file location in a filesystem. When the class is compiled and its `.class` file with bytecode is generated, the package name also reflects the `.class` file location in the filesystem.

Importing

After the package declaration, the `import` statements follow. As you can see from the previous example, they allow use of of the fully qualified class (or interface) name to be avoided anywhere else in the current class. When many classes (and interfaces) from the same package are imported, it is possible to import all classes and interfaces from the same package as a group, using the symbol `*`. In our example, it would look as follows:

```
import com.packt.learnjava.ch02_oop.hiding.*;
```

But that is not recommended practice, as it hides away the imported class (and interface) location when several packages are imported as a group. For example, look at this code snippet:

```
package com.packt.learnjava.ch03_fundamentals;
import com.packt.learnjava.ch02_oop.*;
import com.packt.learnjava.ch02_oop.hiding.*;
public class Packages {
    public void method(){
        C c = new C();
        D d = new D();
    }
}
```

In the preceding code, can you guess to which package a class C or class D belongs? Also, it is possible that two classes in different packages have the same name. If that is the case, group importing can create a degree confusion or even a problem difficult to nail down.

It is also possible to import individual static class (or interface) members. For example, if SomeInterface has a NAME property (to remind you, interface properties are public and static by default), you can typically refer to it as follows:

```
package com.packt.learnjava.ch03_fundamentals;
import com.packt.learnjava.ch02_oop.SomeInterface;
public class Packages {
    public void method(){
        System.out.println(SomeInterface.NAME);
    }
}
```

To avoid using even the interface name, you can use a static import:

```
package com.packt.learnjava.ch03_fundamentals;
import static com.packt.learnjava.ch02_oop.SomeInterface.NAME;
public class Packages {
    public void method(){
        System.out.println(NAME);
    }
}
```

Similarly, if SomeClass has a public static property, someProperty, and a public static method, someMethod(), it is possible to import them statically too:

```
package com.packt.learnjava.ch03_fundamentals;
import com.packt.learnjava.ch02_oop.StaticMembers.SomeClass;
import com.packt.learnjava.ch02_oop.hiding.C;
import com.packt.learnjava.ch02_oop.hiding.D;
```

```
import static com.packt.learnjava.ch02_oop.StaticMembers
                                    .SomeClass.someMethod;
import static com.packt.learnjava.ch02_oop.StaticMembers
                                    .SomeClass.SOME_PROPERTY;
public class Packages {
    public static void main(String... args){
        C c = new C();
        D d = new D();

        SomeClass obj = new SomeClass();
        someMethod(42);
        System.out.println(SOME_PROPERTY);     //prints: abc
    }
}
```

But this technique should be used wisely since it may create an impression that a statically imported method or property belongs to the current class.

Access modifiers

We have already used in our examples the three access modifiers– `public`, `protected`, and `private`–which regulate access to the classes, interfaces, and their members from outside–from other classes or interfaces. There is also a fourth implicit one (also called **default modifier package-private**) that is applied when none of the three explicit access modifiers is specified.

The effect of their usage is pretty straightforward:

- `public`: Accessible to other classes and interfaces of the current and other packages
- `protected`: Accessible only to other members of the same package and children of the class
- No access modifier means *accessible only to other members of the same package*
- `private`: Accessible only to members of the same class

From inside the class or an interface, all the class or interface members are always accessible. Besides, as we have stated several times already, all interface members are public by default, unless declared as `private`.

Also, please note that class accessibility supersedes the class members' accessibility because, if the class itself is not accessible from somewhere, no change in the accessibility of its methods or properties can make them accessible.

When people talk about access modifiers for classes and interfaces, they mean the classes and interfaces that are declared inside other classes or interfaces. The encompassing class or interface is called a *top-level class or interface,* while those inside them are called **inner classes or interfaces.** The static inner classes are also called **static nested classes.**

It does not make sense to declare a top-level class or interface `private` because it will not be accessible from anywhere. And the Java authors decided against allowing the top-level class or interface to be declared `protected` too. It is possible, though, to have a class without an explicit access modifier, thus making it accessible only to the members of the same package.

Here is an example:

```
public class AccessModifiers {
    String prop1;
    private String prop2;
    protected String prop3;
    public String prop4;
    void method1(){ }
    private void method2(){ }
    protected void method3(){ }
    public void method4(){ }

    class A1{ }
    private class A2{ }
    protected class A3{ }
    public class A4{ }

    interface I1 {}
    private interface I2 {}
    protected interface I3 {}
    public interface I4 {}
}
```

Please note that static nested classes do not have access to other members of the top-level class.

Another particular feature of an inner class is that it has access to all, even private members, of the top-level class, and vice versa. To demonstrate this feature, let's create the following private properties and methods in the top-level class and in a private inner class:

```
public class AccessModifiers {
    private String topLevelPrivateProperty = "Top-level private value";
    private void topLevelPrivateMethod(){
        var inner = new InnerClass();
        System.out.println(inner.innerPrivateProperty);
```

```
            inner.innerPrivateMethod();
    }

    private class InnerClass {
        //private static String PROP = "Inner static"; //error
        private String innerPrivateProperty = "Inner private value";
        private void innerPrivateMethod(){
            System.out.println(topLevelPrivateProperty);
        }
    }

    private static class InnerStaticClass {
        private static String PROP = "Inner private static";
        private String innerPrivateProperty = "Inner private value";
        private void innerPrivateMethod(){
            var top = new AccessModifiers();
            System.out.println(top.topLevelPrivateProperty);
        }
    }
}
```

As you can see, all the methods and properties in the previous classes are private, which means that normally, they are not accessible from outside the class. And that is true for the AccessModifiers class: its private methods and properties are not accessible for other classes that are declared outside of it. But the InnerClass class can access the private members of the top-level class, while the top-level class can access the private members of its inner classes. The only limitation is that a non-static inner class cannot have static members. By contrast, a static nested class can have both static and non-static members, which makes a static nested class much more usable.

To demonstrate all the possibilities described, we add the following main() method to the AccessModifiers class:

```
public static void main(String... args){
    var top = new AccessModifiers();
    top.topLevelPrivateMethod();
    //var inner = new InnerClass();   //error
    System.out.println(InnerStaticClass.PROP);
    var inner = new InnerStaticClass();
    System.out.println(inner.innerPrivateProperty);
    inner.innerPrivateMethod();
}
```

Naturally, a non-static inner class cannot be accessed from a static context of the top-level class, hence the comment in the preceding code. If we run it, the result will be as follows:

```
Inner private value
Top-level private value
Inner private static
Inner private value
Top-level private value
```

The first two lines of the output come from the `topLevelPrivateMethod()`, the rest from the `main()` method. As you can see, an inner and a top-level class can access each other's private state, inaccessible from outside.

Java reference types

A `new` operator creates an object of a class and returns the reference to the memory where the object resides. From a practical standpoint, the variable that holds this reference is treated in the code as if it is the object itself. Such a variable can be a class, an interface, an array, or a `null` literal that indicates that no memory reference is assigned to the variable. If the type of the reference is an interface, it can be assigned either `null` or a reference to the object of the class that implements this interface because the interface itself cannot be instantiated.

JVM watches for all the created objects and checks whether there are references to each of them in the currently executed code. If there is an object without any reference to it, JVM removes it from the memory in the process called **garbage collection**. We will describe this process in Chapter 9, *JVM Structure and Garbage Collection*. For example, an object was created during a method execution and was referred to by the local variable. This reference will disappear as soon as the method finishes its execution.

You have seen the examples of custom classes and interfaces, and we have talked about the `String` class already (see Chapter 1, *Getting Started with Java 12*). In this section, we will also describe two other Java reference types—array and enum—and demonstrate how to use them.

Class and interface

A variable of a class type is declared using the corresponding class name:

```
<Class name> identifier;
```

The value that can be assigned to such a variable can be one of the following:

- A reference type `null` literal (means the variable can be used but does not refer to any object)
- A reference to an object of the same class or any of its descendants (because a descendant inherits the types of all of its ancestors)

This last type of assignment is called a **widening assignment** because it forces a specialized reference to become less specialized. For example, since every Java class is a subclass of `java.lang.Object`, the following assignment can be done for any class:

```
Object obj = new AnyClassName();
```

Such an assignment is also called an **upcasting** because it moves the type of the variable up on the line of inheritance (which, like any family tree, is usually presented with the oldest ancestor at the top).

After such an upcasting, it is possible to make a narrowing assignment using a cast operator `(type)`:

```
AnyClassName anyClassName = (AnyClassName)obj;
```

Such an assignment is also called a **downcasting** and allows you to restore the descendant type. To apply this operation, you have to be sure that the identifier in fact refers to a descendant type. If in doubt, you can use the `instanceof` operator (see Chapter 2, *Java Object-Oriented Programming*) to check the reference type.

Similarly, if a class implements a certain interface, its object reference can be assigned to this interface or any ancestor of the interface:

```
interface C {}
interface B extends C {}
class A implements B { }
B b = new A();
C c = new A();
A a1 = (A)b;
A a2 = (A)c;
```

As you can see, as in the case with class reference upcasting and downcasting, it is possible to recover the original type of the object after its reference was assigned to a variable of one of the implemented interfaces types.

The material of this section can also be viewed as another demonstration of Java polymorphism in action.

Array

An **array** is a reference type and, as such, extends the `java.lang.Object` class too. The array elements have the same type as the declared array type. The number of elements may be zero, in which case the array is said to be an empty array. Each element can be accessed by an index, which is a positive integer or zero. The first element has an index of zero. The number of elements is called an array length. Once an array is created, its length never changes.

The following are examples of an array declaration:

```
int[] intArray;
float[][] floatArray;
String[] stringArray;
SomeClass[][][] arr;
```

Each bracket pair indicates another dimension. The number of bracket pairs is the nesting depth of the array:

```
int[] intArray = new int[10];
float[][] floatArray = new float[3][4];
String[] stringArray = new String[2];
SomeClass[][][] arr = new SomeClass[3][5][2];
```

The `new` operator allocates memory for each element that can be assigned (filled with) a value later. But the elements of an array are initialized to the default values at creation time in my case, as the following example demonstrates:

```
System.out.println(intArray[3]);        //prints: 0
System.out.println(floatArray[2][2]);   //prints: 0.0
System.out.println(stringArray[1]);     //prints: null
```

Another way to create an array is to use an array initializer–comma-separated list of values enclosed in braces for each dimension, for example:

```
int[] intArray = {1,2,3,4,5,6,7,8,9,10};
float[][] floatArray ={{1.1f,2.2f,3,2},{10,20.f,30.f,5},{1,2,3,4}};
String[] stringArray = {"abc", "a23"};

System.out.println(intArray[3]);        //prints: 4
System.out.println(floatArray[2][2]);   //prints: 3.0
System.out.println(stringArray[1]);     //prints: a23
```

A multidimensional array can be created without declaring the length of each dimension. Only the first dimension has to have the length specified:

```
float[][] floatArray = new float[3][];

System.out.println(floatArray.length);    //prints: 3
System.out.println(floatArray[0]);        //prints: null
System.out.println(floatArray[1]);        //prints: null
System.out.println(floatArray[2]);        //prints: null
//System.out.println(floatArray[3]);      //error
//System.out.println(floatArray[2][2]);   //error
```

The missing length of other dimensions can be specified later:

```
float[][] floatArray = new float[3][];
floatArray[0] = new float[4];
floatArray[1] = new float[3];
floatArray[2] = new float[7];
System.out.println(floatArray[2][5]);     //prints: 0.0
```

This way, it is possible to assign a different length to different dimensions. Using the array initializer, it is also possible to create dimensions of different lengths:

```
float[][] floatArray ={{1.1f},{10,5},{1,2,3,4}};
```

The only requirement is that a dimension has to be initialized before it can be used.

Enum

The **enum** reference type class extends the `java.lang.Enum` class, which, in turn, extends `java.lang.Object`. It allows the specification of a limited set of constants, each of them an instance of the same type. The declaration of such a set starts with the keyword, `enum`. Here is an example:

```
enum Season { SPRING, SUMMER, AUTUMN, WINTER }
```

Each of the listed items – SPRING, SUMMER, AUTUMN, and WINTER – is an instance of a `Season` type. They are the only four instances the `Season` class can have. They are created in advance and can be used everywhere as a value of a `Season` type. No other instance of the `Season` class can be created. And that is the reason for the creation of the `enum` type: it can be used for cases when the list of instances of a class has to be limited to the fixed set.

The `enum` declaration can also be written in a camel case:

```
enum Season { Spring, Summer, Autumn, Winter }
```

However, the all-capitals style is used more often, because, as we have mentioned earlier, there is a convention to express the static final constant's identifier in a capital case. It helps to distinguish constants from variables. The enum constants are static and final implicitly.

Because the enum values are constants, they exist uniquely in JVM and can be compared by reference:

```
Season season = Season.WINTER;
boolean b = season == Season.WINTER;
System.out.println(b);    //prints: true
```

The following are the most frequently used methods of the java.lang.Enum class:

- name(): returns the enum constant's identifier as it is spelled when declared (WINTER, for example).
- toString(): returns the same value as the name() method by default, but can be overridden to return any other String value.
- ordinal(): returns the position of the enum constant when declared (the first in the list has a 0 ordinal value).
- valueOf(Class enumType, String name): returns the enum constant object by its name expressed as a String literal.
- values(): a static method not described in the documentation of the java.lang.Enum class. In the *Java Language Specification, section 8.9.3* (https://docs.oracle.com/javase/specs/jls/se12/html/jls-8.html#jls-8.9.3), it is described as implicitly declared. *The Java™ Tutorials* (https://docs.oracle.com/javase/tutorial/java/javaOO/enum.html) states that the compiler automatically adds some special methods when it creates an enum, among them, a static values() method that returns an array containing all of the values of the enum in the order they are declared.

To demonstrate the preceding methods, we are going to use the already familiar enum, Season:

```
enum Season { SPRING, SUMMER, AUTUMN, WINTER }
```

And here is the demo code:

```
System.out.println(Season.SPRING.name());              //prints: SPRING
System.out.println(Season.WINTER.toString());          //prints: WINTER
System.out.println(Season.SUMMER.ordinal());           //prints: 1
Season season = Enum.valueOf(Season.class, "AUTUMN");
System.out.println(season == Season.AUTUMN);           //prints: true
```

```
for(Season s: Season.values()){
    System.out.print(s.name() + " ");
                            //prints: SPRING SUMMER AUTUMN WINTER
}
```

To override the `toString()` method, let's create `enum Season1`:

```
enum Season1 {
    SPRING, SUMMER, AUTUMN, WINTER;
    public String toString() {
        return this.name().charAt(0) +
                this.name().substring(1).toLowerCase();
    }
}
```

Here is how it works:

```
for(Season1 s: Season1.values()){
    System.out.print(s.toString() + " ");
                            //prints: Spring Summer Autumn Winter
}
```

It is possible to add any other property to each `enum` constant. For example, let's add an average temperature value to each `enum` instance:

```
SPRING(42), SUMMER(67), AUTUMN(32), WINTER(20);
private int temperature;
Season2(int temperature){
this.temperature = temperature;
}
public int getTemperature(){
return this.temperature;
}
public String toString() {
return this.name().charAt(0) +
this.name().substring(1).toLowerCase() +
"(" + this.temperature + ")";
}
}
```

If we iterate over values of `enum Season2`, the result will be as follows:

```
for(Season2 s: Season2.values()){
    System.out.print(s.toString() + " ");
            //prints: Spring(42) Summer(67) Autumn(32) Winter(20)
}
```

In the standard Java libraries, there are several `enum` classes, for example, `java.time.Month`, `java.time.DayOfWeek`, and `java.util.concurrent.TimeUnit`.

Default values and literals

As we have already seen, the default value of a reference type is `null`. Some sources call it **special type null**, but Java Language Specification qualifies it as a literal. When an instance property or an array of a reference type is initialized automatically (when a value is not assigned explicitly), the assigned value is `null`.

The only reference type that has a literal other than the `null` literal is the `String` class. We discussed strings in `Chapter 1`, *Getting Started with Java 12*.

Reference type as a method parameter

When a primitive type value is passed into a method, we use it. If we do not like the value passed into the method, we change it as we see fit and do not think twice about it:

```
void modifyParameter(int x){
    x = 2;
}
```

We have no concerns that the variable value outside the method may change:

```
int x = 1;
modifyParameter(x);
System.out.println(x);   //prints: 1
```

It is not possible to change the parameter value of a primitive type outside the method because a primitive type parameter is passed into the method *by value*. This means that the copy of the value is passed into the method, so even if the code inside the method assigns a different value to it, the original value is not affected.

Another issue with a reference type is that even though the reference itself is passed by value, it still points to the same original object in the memory, so the code inside the method can access the object and modify it. To demonstrate it, let's create a `DemoClass` and the method that uses it:

```
class DemoClass{
    private String prop;
    public DemoClass(String prop) { this.prop = prop; }
```

```
        public String getProp() { return prop; }
        public void setProp(String prop) { this.prop = prop; }
    }
    void modifyParameter(DemoClass obj){
        obj.setProp("Changed inside the method");
    }
```

If we use the preceding method, the result will be as follows:

```
    DemoClass obj = new DemoClass("Is not changed");
    modifyParameter(obj);
    System.out.println(obj.getProp()); //prints: Changed inside the method
```

That is a big difference, isn't it? So, you have to be careful not to modify the passed-in object in order to avoid an undesirable effect. However, this effect is occasionally used to return the result. But it does not belong to the list of best practices because it makes code less readable. Changing the passed-in object is like using a secret tunnel that is difficult to notice. So, use it only when you have to.

Even if the passed-in object is a class that wraps a primitive value, this effect still holds (we will talk about primitive values wrapping type in the *Conversion between primitive and reference types* section). Here is DemoClass1 and an overloaded version of the modifyParameter() method:

```
    class DemoClass1{
        private Integer prop;
        public DemoClass1(Integer prop) { this.prop = prop; }
        public Integer getProp() { return prop; }
        public void setProp(Integer prop) { this.prop = prop; }
    }
    void modifyParameter(DemoClass1 obj){
        obj.setProp(Integer.valueOf(2));
    }
```

If we use the preceding method, the result will be as follows:

```
    DemoClass1 obj = new DemoClass1(Integer.valueOf(1));
    modifyParameter(obj);
    System.out.println(obj.getProp());   //prints: 2
```

The only exception to this behavior of reference types is an object of the String class. Here is another overloaded version of the modifyParameter() method:

```
    void modifyParameter(String obj){
        obj = "Changed inside the method";
    }
```

If we use the preceding method, the result will be as follows:

```
String obj = "Is not changed";
modifyParameter(obj);
System.out.println(obj); //prints: Is not changed

obj = new String("Is not changed");
modifyParameter(obj);
System.out.println(obj); //prints: Is not changed
```

As you can see, whether we use a literal or a new `String` object, the result remains the same: the original `String` value is not changed after the method that assigns another value to it. That is exactly the purpose of the `String` value immutability feature we discussed in `Chapter 1`, *Getting Started with Java 12*.

equals() method

The equality operator (==), when applied to the variables of reference types, compares the references themselves, not the content (the state) of the objects. But two objects always have different memory references even if they have identical content. Even when used for `String` objects, the operator (==) returns `false` if at least one of them is created using a `new` operator (see the discussion about `String` value immutability in `Chapter 1`, *Getting Started with Java 12*).

To compare content, you can use the `equals()` method. Its implementation in the `String` class and numerical type wrapper classes (`Integer`, `Float`, and so on) do exactly that–compare the content of the objects.

However, the `equals()` method implementation in the `java.lang.Object` class compares only references , which is understandable because the variety of possible content the descendants can have is huge and the implementation of the generic content comparison is just not feasible. This means that every Java object that needs to have the `equals()` method comparing the objects' content–not just references–has to re-implement the `equals()` method and, thus, override its implementation in the `java.lang.Object` class, which appears as follows:

```
public boolean equals(Object obj) {
    return (this == obj);
}
```

By contrast, look how the same method is implemented in the `Integer` class:

```
private final int value;
public boolean equals(Object obj) {
    if (obj instanceof Integer) {
        return value == ((Integer)obj).intValue();
    }
    return false;
}
```

As you can see, it extracts the primitive `int` value from the input object and compares it to the primitive value of the current object. It does not compare object references at all.

The `String` class, on the other hand, compares the references first and, if the references are not the same value, compares the content of the objects:

```
private final byte[] value;
public boolean equals(Object anObject) {
        if (this == anObject) {
                return true;
        }
        if (anObject instanceof String) {
            String aString = (String)anObject;
            if (coder() == aString.coder()) {
                return isLatin1() ? StringLatin1.equals(value, aString.value)
                                  : StringUTF16.equals(value, aString.value);
            }
        }
        return false;
}
```

The `StringLatin1.equals()` and `StringUTF16.equals()` methods compare the values character by character, not just references.

Similarly, if the application code needs to compare two objects by their content, the `equals()` method in the corresponding class has to be overridden. For example, let's look at the familiar `DemoClass` class:

```
class DemoClass{
    private String prop;
    public DemoClass(String prop) { this.prop = prop; }
    public String getProp() { return prop; }
    public void setProp(String prop) { this.prop = prop; }
}
```

We could add to it the `equals()` method manually, but IDE can help us to do this as follows:

1. Right-click inside the class just before the closing brace (`}`)
2. Select **Generate**, and then follow the prompts

Eventually, two methods will be generated and added to the class:

```
@Override
public boolean equals(Object o) {
    if (this == o) return true;
    if (!(o instanceof DemoClass)) return false;
    DemoClass demoClass = (DemoClass) o;
    return Objects.equals(getProp(), demoClass.getProp());
}

@Override
public int hashCode() {
    return Objects.hash(getProp());
}
```

Looking at the generated code, we would like to attract your attention to the following points:

- Usage of the `@Override` annotation: this ensures that the method does override a method (with the same signature) in one of the ancestors. With this annotation in place, if you modify the method and change the signature (by mistake or intentionally), the compiler (and your IDE) will immediately raise an error telling you that there is no method with such a signature in any of the ancestor classes. So, it helps to detect an error early.
- Usage of the `java.util.Objects` class: this has quite a few very helpful methods, including the `equals()` static method that compares not only references, but also uses the `equals()` method:

```
public static boolean equals(Object a, Object b) {
    return (a == b) || (a != null && a.equals(b));
}
```

Since, as we have demonstrated earlier, the `equals()` method, implemented in the `String` class, compares strings by their content and serves our this purpose because the `getProp()` method of the `DemoClass` returns a string.

- The `hashCode()` method: the integer returned by this method uniquely identifies this particular object (but please do not expect it to be the same between different runs of the application). It is not necessary to have this method implemented if the only method needed is `equals()`. Nevertheless, it is recommended to have it just in case the object of this class is going to be collected in `Set` or another collection based on a hash code (we are going to talk about Java collections in `Chapter 6`, *Data Structures, Generics, and Popular Utilities*).

Both these methods are implemented in `Object` because many algorithms use the `equals()` and `hashCode()` methods, and your application may not work without these methods implemented. Meanwhile, your objects may not need them in your application. However, once you decide to implement the `equals()` method, it is a good idea to implement the `hasCode()` method too. Besides, as you have seen, an IDE can do this without any overhead.

Reserved and restricted keywords

The **keywords** are the words that have particular meaning for a compiler and cannot be used as identificators. There are 51 reserved keywords and 10 restricted keywords. The reserved keywords cannot be used as identificators anywhere in the Java code, while the restricted keywords cannot be used as identificators only in the context of a module declaration.

Reserved keywords

The following is the list of all Java-reserved keywords:

abstract	assert	boolean	break	byte
case	catch	char	class	const
continue	default	do	double	else
enum	extends	final	finally	float
for	if	goto	implements	import
instanceof	int	interface	long	native
new	package	private	protected	public
return	short	static	strictfp	super
switch	synchronized	this	throw	throws
transient	try	void	volatile	while

An underscore (_) is a reserved word too.

By now, you should feel at home with most of the preceding keywords. By way of an exercise, you can go through the list and check how many of them you remember. We did not talk only about the following eight keywords:

- `const` and `goto` are reserved but not used, so far
- The `assert` keyword is used in an `assert` statement (we will talk about this in `Chapter 4`, *Exception Handling*).
- The `synchronized` keyword is used in concurrent programming (we will talk about this in `Chapter 8`, *Multithreading and Concurrent Processing*).
- The `volatile` keyword makes the value of a variable uncacheable.
- The `transient` keyword makes the value of a variable not serializable
- The `strictfp` keyword restricts floating-point calculation, making it the same result on every platform while performing operations in the floating-point variable
- The `native` keyword declares a method implemented in platform-dependent code, such as C or C++

Restricted keywords

The 10 restricted keywords in Java are as follows:

- `open`
- `module`
- `requires`
- `transitive`
- `exports`
- `opens`
- `to`
- `uses`
- `provides`
- `with`

They are called *restricted* because they cannot be identifiers in the context of a module declaration, which we will not discuss in this book. In all other places, it is possible to use them as identifiers, for example:

```
String to = "To";
String with = "abc";
```

Although you can, this is a good practice not to use them as identifiers even outside module declaration.

Usage of the this and super keywords

The this keyword provides a reference to the current object. The super keyword refers to the parent class object. These keywords allow us to refer to a variable or method that has the same name in the current context and in the parent object.

Usage of the this keyword

Here is the most popular example:

```
class A {
    private int count;
    public void setCount(int count) {
        count = count;          // 1
    }
    public int getCount(){
        return count;           // 2
    }
}
```

The first line looks ambiguous, but, in fact, it is not: the local variable, int count , hides the private property instance int count. We can demonstrate this by running the following code:

```
A a = new A();
a.setCount(2);
System.out.println(a.getCount());      //prints: 0
```

Using the this keyword fixes the problem:

```
class A {
    private int count;
    public void setCount(int count) {
        this.count = count;          // 1
    }
    public int getCount(){
        return this.count;           // 2
    }
}
```

Adding this to line 1 allows the value to be assigned the instance property. Adding this to line 2 does not make a difference, but it is good practice to use the this keyword every time with the instance property. It makes the code more readable and helps avoid difficult-to-trace errors, such as the one we have just demonstrated.

We have also seen this keyword usage in the equals() method:

```
@Override
public boolean equals(Object o) {
    if (this == o) return true;
    if (!(o instanceof DemoClass)) return false;
    DemoClass demoClass = (DemoClass) o;
    return Objects.equals(getProp(), demoClass.getProp());
}
```

And, just to remind you, here are the examples of a constructor we presented in Chapter 2, *Java Object-Oriented Programming (OOP)*:

```
class TheChildClass extends TheParentClass{
    private int x;
    private String prop;
    private String anotherProp = "abc";
    public TheChildClass(String prop){
        super(42);
        this.prop = prop;
    }
    public TheChildClass(int arg1, String arg2){
        super(arg1);
        this.prop = arg2;
    }
    // methods follow
}
```

In the preceding code, you can see not only the this keyword, but also the usage of the super keyword, which we are going to discuss next.

Usage of the super keyword

The super keyword refers to the parent object. We saw its usage in the *Usage of the this keyword* section in a constructor already, where it has to be used only in the first line because the parent class object has to be created before the current object can be created. If the first line of the constructor is not super(), this means the parent class has a constructor without parameters.

The super keyword is especially helpful when a method is overridden and the method of the parent class has to be called:

```
class B  {
    public void someMethod() {
        System.out.println("Method of B class");
    }
}
class C extends B {
    public void someMethod() {
        System.out.println("Method of C class");
    }
    public void anotherMethod() {
        this.someMethod();      //prints: Method of C class
        super.someMethod();     //prints: Method of B class
    }
}
```

As we progress through this book, we will see more examples of using the this and super keywords.

Converting between primitive types

The maximum numeric value a numeric type can hold depends on the number of bits allocated to it. The following are the number of bits for each numeric type of representation:

- byte: 8 bit
- char: 16 bit
- short: 16 bit
- int: 32 bit

- `long`: 64 bit
- `float`: 32 bit
- `double`: 64 bit

When a value of one numeric type is assigned to a variable of another numeric type and the new type can hold a bigger number, such a conversion is called a **widening conversion**. Otherwise, it is a **narrowing conversion**, which usually requires a typecasting, using a `cast` operator.

Widening conversion

According to Java Language Specification, there are 19 widening primitive type conversions:

- `byte` to `short, int, long, float,` or `double`
- `short` to `int, long, float,` or `double`
- `char` to `int, long, float,` or `double`
- `int` to `long, float,` or `double`
- `long` to `float` or `double`
- `float` to `double`

During widening conversion between integral types, and from some integral types to a floating-point type, the resulting value matches the original one exactly. However, conversion from `int` to `float`, or from `long` to `float`, or from `long` to `double`, may result in a loss of precision. The resulting floating-point value may be correctly rounded using `IEEE 754 round-to-nearest mode`, according to Java Language Specification. Here are a few examples that demonstrate the loss of precision:

```
int i = 123456789;
double d = (double)i;
System.out.println(i - (int)d);     //prints: 0

long l1 = 12345678L;
float f1 = (float)l1;
System.out.println(l1 - (long)f1);     //prints: 0

long l2 = 123456789L;
float f2 = (float)l2;
System.out.println(l2 - (long)f2);     //prints: -3

long l3 = 1234567891111111L;
```

```
double d3 = (double)13;
System.out.println(13 - (long)d3);      //prints: 0

long 14 = 12345678999999999L;
double d4 = (double)14;
System.out.println(14 - (long)d4);      //prints: -1
```

As you can see, conversion from int to double preserves the value, but long to float , or long to double , may lose precision. It depends on how big the value is. So, be aware and allow for some loss of precision if it is important for your calculations.

Narrowing conversion

Java Language Specification identifies 22 narrowing primitive conversions:

- short to byte or char
- char to byte or short
- int to byte, short, or char
- long to byte, short, char, or int
- float to byte, short, char, int, or long
- double to byte, short, char, int, long, or float

Similar to the widening conversion, a narrowing conversion may result in a loss of precision, or even in a loss of the value magnitude. The narrowing conversion is more complicated than a widening one and we are not going to discuss it in this book. It is important to remember that before performing a narrowing, you must make sure that the original value is smaller than the maximum value of the target type. Otherwise, you can get a completely different value (with lost magnitude). Look at the following example:

```
System.out.println(Integer.MAX_VALUE); //prints: 2147483647
double d1 = 1234567890.0;
System.out.println((int)d1);            //prints: 1234567890

double d2 = 12345678909999999999999.0;
System.out.println((int)d2);            //prints: 2147483647
```

As you can see from the examples, without checking first whether the target type can accommodate the value, you can get the result to just equal to the maximum value of the target type. The rest will be just lost, no matter how big the difference is.

Before performing a narrowing conversion, check whether the maximum value of the target type can hold the original value.

Please note that the conversion between the char type and the byte or short types is an even more complicated procedure because the char type is an unsigned numeric type, while the byte and short types are signed numeric types, so some loss of information is possible even when a value may look as though it fits in the target type.

Methods of conversion

In addition to the casting, each primitive type has a corresponding reference type (called a **wrapper class**) that has methods that convert the value of this type to any other primitive type, except boolean and char. All the wrapper classes belong to the java.lang package:

- java.lang.Boolean
- java.lang.Byte
- java.lang.Character
- java.lang.Short
- java.lang.Integer
- java.lang.Long
- java.lang.Float
- java.lang.Double

Each of them—except the Boolean and Character classes—extends the abstract class java.lang.Number, which has the following abstract methods:

- byteValue()
- shortValue()
- intValue()
- longValue()
- floatValue()
- doubleValue()

Such design forces the descendants of the Number class to implement all of them. The results they produce are the same as the cast operator in the previous examples:

```java
int i = 123456789;
double d = Integer.valueOf(i).doubleValue();
System.out.println(i - (int)d);              //prints: 0


long l1 = 12345678L;
float f1 = Long.valueOf(l1).floatValue();
System.out.println(l1 - (long)f1);           //prints: 0


long l2 = 123456789L;
float f2 = Long.valueOf(l2).floatValue();
System.out.println(l2 - (long)f2);           //prints: -3


long l3 = 1234567891111111L;
double d3 = Long.valueOf(l3).doubleValue();
System.out.println(l3 - (long)d3);           //prints: 0


long l4 = 12345678999999999L;
double d4 = Long.valueOf(l4).doubleValue();
System.out.println(l4 - (long)d4);           //prints: -1


double d1 = 1234567890.0;
System.out.println(Double.valueOf(d1)
                        .intValue());    //prints: 1234567890


double d2 = 12345678909999999999999.0;
System.out.println(Double.valueOf(d2)
                        .intValue());    //prints: 2147483647
```

In addition, each of the wrapper classes has methods that allow the conversion of String representation of a numeric value to the corresponding primitive numeric type or reference type, for example:

```java
byte b1 = Byte.parseByte("42");
System.out.println(b1);                   //prints: 42
Byte b2 = Byte.decode("42");
System.out.println(b2);                   //prints: 42

boolean b3 = Boolean.getBoolean("property");
System.out.println(b3);                   //prints: false
Boolean b4 = Boolean.valueOf("false");
System.out.println(b4);                   //prints: false

int i1 = Integer.parseInt("42");
System.out.println(i1);                   //prints: 42
Integer i2 = Integer.getInteger("property");
```

```
System.out.println(i2);                 //prints: null

double d1 = Double.parseDouble("3.14");
System.out.println(d1);                 //prints: 3.14
Double d2 = Double.valueOf("3.14");
System.out.println(d2);                 //prints: 3.14
```

In the examples, please note the two methods that accept the `property` parameter. These two and similar methods of other wrapper classes convert system property (if such exists) to the corresponding primitive type.

Each of the wrapper classes has the `toString(primitive value)` static method to convert the primitive type value to its `String` representation, for example:

```
String s1 = Integer.toString(42);
System.out.println(s1);                 //prints: 42
String s2 = Double.toString(3.14);
System.out.println(s2);                 //prints: 3.14
```

The wrapper classes have many other useful methods of conversion from one primitive type to another and to different formats. So, if you need to do something such as that, look into the corresponding wrapper class first.

Converting between primitive and reference types

The conversion of a primitive type value to an object of the corresponding wrapper class is called **boxing**. Also, the conversion from an object of a wrapper class to the corresponding primitive type value is called **unboxing**.

Boxing

The boxing of a primitive type can be done either automatically (called **autoboxing**) or explicitly using the `valueOf()` method available in each wrapper type:

```
int i1 = 42;
Integer i2 = i1;                 //autoboxing
//Long 12 = i1;                  //error
System.out.println(i2);          //prints: 42

i2 = Integer.valueOf(i1);
System.out.println(i2);          //prints: 42
```

```
Byte b = Byte.valueOf((byte)i1);
System.out.println(b);         //prints: 42

Short s = Short.valueOf((short)i1);
System.out.println(s);         //prints: 42

Long l = Long.valueOf(i1);
System.out.println(l);         //prints: 42

Float f = Float.valueOf(i1);
System.out.println(f);         //prints: 42.0

Double d = Double.valueOf(i1);
System.out.println(d);         //prints: 42.0
```

Notice that autoboxing is only possible in relation to a corresponding wrapper type. Otherwise, the
compiler generates an error.

The input value of the `valueOf()` method of `Byte` and `Short` wrappers required casting because it was a narrowing of a primitive type we have discussed in the previous section.

Unboxing

Unboxing can be accomplished using methods of the `Number` class implemented in each wrapper class:

```
Integer i1 = Integer.valueOf(42);
int i2 = i1.intValue();
System.out.println(i2);        //prints: 42

byte b = i1.byteValue();
System.out.println(b);         //prints: 42

short s = i1.shortValue();
System.out.println(s);         //prints: 42

long l = i1.longValue();
System.out.println(l);         //prints: 42

float f = i1.floatValue();
System.out.println(f);         //prints: 42.0

double d = i1.doubleValue();
System.out.println(d);         //prints: 42.0
```

```
Long l1 = Long.valueOf(42L);
long l2 = l1;                    //implicit unboxing
System.out.println(l2);         //prints: 42

double d2 = l1;                  //implicit unboxing
System.out.println(d2);         //prints: 42

long l3 = i1;                    //implicit unboxing
System.out.println(l3);         //prints: 42

double d3 = i1;                  //implicit unboxing
System.out.println(d3);         //prints: 42
```

As you can see from the comment in the example, the conversion from a wrapper type to the corresponding primitive type is not called **auto-unboxing**; it is called **implicit unboxing** instead. In contrast to autoboxing, it is possible to use implicit unboxing even between wrapping and primitive types that do not match.

Summary

In this chapter, you have learned what Java packages are and the role they play in organizing code and class accessibility, including the `import` statement and access modifiers. You also became familiar with reference types: classes, interfaces, arrays, and enums. The default value of any reference type is `null`, including the `String` type.

You should now understand that the reference type is passed into a method by reference and how the `equals()` method is used and can be overridden. You also had an opportunity to study the full list of reserved and restricted keywords and learned the meaning and usage of the `this` and `super` keywords.

The chapter concluded by describing the process and methods of conversion between primitive types, wrapping types, and `String` literals.

In the next chapter, we will talk about the Java exceptions framework, checked and unchecked (runtime) exceptions, `try-catch-finally` blocks, `throws` and `throw` statements, and the best practices of exception handling.

Quiz

1. Select all the statements that are correct:
   a. The `Package` statement describes the class or interface location
   b. The `Package` statement describes the class or interface name
   c. `Package` is a fully qualified name
   d. The `Package` name and class name compose a fully qualified name of the class

2. Select all the statements that are correct:
   a. The `Import` statement allows use of the fully qualified name
   b. The `Import` statement has to be the first in the `.java` file
   c. The `Group import` statement brings in the classes (and interfaces) of one package only
   d. The `Import statement` allows use of of the fully qualified name to be avoided

3. Select all the statements that are correct:
   a. Without an access modifier, the class is accessible only by other classes and interfaces of the same package
   b. The private method of a private class is accessible to other classes declared in the same `.java` file
   c. The public method of a private class is accessible to other classes not declared in the same `.java` file but from the same package
   d. The protected method is accessible only to the descendants of the class

4. Select all the statements that are correct:
   a. Private methods can be overloaded but not overridden
   b. Protected methods can be overridden but not overloaded
   c. Methods without an access modifier can be both overridden and overloaded
   d. Private methods can access private properties of the same class

5. Select all the statements that are correct:
   a. Narrowing and downcasting are synonyms
   b. Widening and downcasting are synonyms
   c. Widening and upcasting are synonyms
   d. Widening and narrowing have nothing in common with upcasting and downcasting

6. Select all the statements that are correct:
   a. `Array` is an object
   b. `Array` has a length that is a number of the elements it can hold
   c. The first element of an array has the index 1
   d. The second element of an array has the index 1

7. Select all the statements that are correct:
   a. `Enum` contains constants.
   b. `Enum` always has a constructor, be it default or explicit
   c. An `enum` constant can have properties
   d. `Enum` can have constants of any reference type

8. Select all the statements that are correct:
   a. Any reference type passed in as a parameter can be modified
   b. A `new String()` object passed in as a parameter can be modified
   c. An object reference value passed in as a parameter cannot be modified
   d. An array passed in as a parameter can have elements assigned to different values

9. Select all the statements that are correct:
   a. Reserved keywords cannot be used
   b. Restricted keywords cannot be used as identifiers
   c. A reserved keyword `identifier` cannot be used as an identifier
   d. A reserved keyword cannot be used as an identifier

10. Select all the statements that are correct:
    a. The `this` keyword refers to the `current` class
    b. The `super` keyword refers to the `super` class
    c. The `this` and `super` keywords refer to objects
    d. The `this` and `super` keywords refer to methods

11. Select all the statements that are correct:
    a. The widening of a primitive type makes the value bigger
    b. The narrowing of a primitive type always changes the type of the value
    c. The widening of a primitive type can be done only after narrowing a conversion
    d. Narrowing makes the value smaller

12. Select all the statements that are correct:
    a. Boxing puts a limit on the value
    b. Unboxing creates a new value
    c. Boxing creates a reference type object
    d. Unboxing deletes a reference type object

Section 2: Building Blocks of Java

2

The second part of the book constitutes the bulk of the Java presentation. It discusses the main Java components and constructs, as well as the algorithms and data structures. Java's system of exceptions is reviewed in detail, the `String` class and I/O streams are presented as well, along with classes that allow manage files.

Java collections and the three main interfaces – `List`, `Set`, and `Map` – are discussed and demonstrated, along with an explanation of generics, followed by utility classes for managing arrays, objects, and time/date values. These classes belong to the **Java Class Library (JCL)**, the most popular packages of which are discussed as well. They are complemented by the third-party libraries that are popular among programming professionals.

The provided material instigates discussions on aspects of programming, such as performance, concurrent processing, and garbage collection, that are at the core of Java design. Together with the dedicated chapters on the graphical user interface and database management, it covers all three tiers – front, middle, and backend – of any powerful Java application. The chapter about the network protocols and the ways applications can talk to each other completes the picture of all the main interactions an application can have.

This section contains the following chapters:

Chapter 4, *Exception Handling*

Chapter 5, *Strings, Input/Output, and Files*

Chapter 6, *Data Structures, Generics, and Popular Utilities*

Chapter 7, *Java Standard and External Libraries*

Chapter 8, *Multithreading and Concurrent Processing*

Chapter 9, *JVM Structure and Garbage Collection*

Chapter 10, *Managing Data in a Database*

Chapter 11, *Network Programming*

Chapter 12, *Java GUI Programming*

4
Exception Handling

We have introduced exceptions briefly in `Chapter 1`, *Getting Started with Java 12*. In this chapter, we will treat this topic more systematically. There are two kinds of exceptions in Java: checked and unchecked exceptions. Both of them will be demonstrated and the difference between the two will be explained. The reader will also learn about the syntax of the Java constructs related to exceptions handling and the best practices to address (handle) the exceptions. The chapter will end with the related topic of an assertion statement that can be used to debug the code in production.

The following topics will be covered in this chapter:

- Java exceptions framework
- Checked and unchecked (runtime) exceptions
- The `try`, `catch`, and `finally` blocks
- The `throws` statement
- The `throw` statement
- The `assert` statement
- Best practices of exceptions handling

Java exceptions framework

As we have described in Chapter 1, *Getting Started with Java 12*, an unexpected condition can cause the **Java Virtual Machine (JVM)** to create and throw an exception object, or the application code can do it. As soon as it happens, the control flow is transferred to the catch clause, if the exception was thrown inside a try block. Let's see an example. Consider the following method:

```
void method(String s){
    if(s.equals("abc")){
        System.out.println("Equals abc");
    } else {
        System.out.println("Not equal");
    }
}
```

If the input parameter value is null, one could expect to see the output as Not equal. Unfortunately, that is not the case. The s.equals("abc") expression calls the equals() method on an object referred by the s variable, but, in case the s variable is null, it does not refer to any object. Let's see what happens then.

Let's run the following code:

```
try {
    method(null);
} catch (Exception ex){
    System.out.println(ex.getClass().getCanonicalName());
                            //prints: java.lang.NullPointerException
    ex.printStackTrace();       //prints: see the screenshot
    if(ex instanceof NullPointerException){
        //do something
    } else {
        //do something else
    }
}
```

The output of this code is as follows:

```
java.lang.NullPointerException
java.lang.NullPointerException
    at com.packt.learnjava.ch04_exceptions.Framework.method(Framework.java:12)
    at com.packt.learnjava.ch04_exceptions.Framework.catchException1(Framework.java:22)
    at com.packt.learnjava.ch04_exceptions.Framework.main(Framework.java:6)
```

What you see in red on the screenshot is called a **stack trace**. The name comes from the way the method calls are stored (as a stack) in JVM memory: one method calls another, which in turns calls another, and so on. After the most inner method returns, the stack is walked back, and the returned method (**stack frame**) is removed from the stack. We will talk more about JVM memory structure in `Chapter 9`, *JVM Structure and Garbage Collection*. When an exception happens, all the stack content (stack frames) are returned as the stack trace. It allows us to track down the line of code that caused the problem.

In our preceding code sample, different blocks of code were executed depending on the type of the exception. In our case, it was `java.lang.NullPointerException`. If the application code did not catch it, this exception would propagate all the way through the stack of the called methods into the JVM, which then stops executing the application. To avoid this happening, the exception can be caught and some code is executed to recover from the exceptional condition.

The purpose of the exception handling framework in Java is to protect the application code from an unexpected condition and recover from it, if possible. In the following sections, we will dissect it in more detail and re-write the given example using the framework capability.

Checked and unchecked exceptions

If you look up the documentation of the `java.lang` package API, you will discover that the package contains almost three dozen exception classes and a couple of dozen error classes. Both groups extend the `java.lang.Throwable` class, inherit all the methods from it, and do not add other methods. The most often used methods of the `java.lang.Throwable` class are the following:

- `void printStackTrace()`: Outputs the stack trace (stack frames) of the method calls
- `StackTraceElement[] getStackTrace()`: Returns the same information as `printStackTrace()`, but allows programmatic access of any frame of the stack trace
- `String getMessage()`: Retrieves the message that often contains a user-friendly explanation of the reason for the exception or error
- `Throwable getCause()`: Retrieves an optional object of `java.lang.Throwable` that was the original reason for the exception (but the author of the code decided to wrap it in another exception or error)

All errors extend the `java.lang.Error` class that, in turn,
extends the `java.lang.Throwable` class. An error is typically thrown by JVM and,
according to the official documentation, *indicates serious problems that a reasonable application
should not try to catch.* Here are a few examples:

- `OutOfMemoryError`: Thrown when JVM runs out memory and cannot clean it using garbage collection
- `StackOverflowError`: Thrown when the memory allocated for the stack of the method calls is not enough to store another stack frame
- `NoClassDefFoundError`: Thrown when JVM cannot find the definition of the class requested by the currently loaded class

The authors of the framework assumed that an application cannot recover from these errors automatically, which proved to be a largely correct assumption. That is why programmers typically do not catch errors and we are not going to talk about them anymore.

The exceptions, on the other hand, are typically related to the application-specific problems and often do not require us to shut down the application and allow the recovery. That is why programmers typically catch them and implement an alternative (to the main flow) path of the application logic, or at least report the problem without shutting down the application. Here are a few examples:

- `ArrayIndexOutOfBoundsException`: Thrown when the code tries to access the element by the index that is equal to, or bigger than, the array length (remember that the first element of an array has index 0, so the index equals the array length points outside of the array)
- `ClassCastException`: Thrown when the code casts a reference to a class or an interface not associated with the object referred to by the variable
- `NumberFormatException`: Thrown when the code tries to convert a string to a numeric type, but the string does not contain the necessary number format

All exceptions extend the `java.lang.Exception` class that, in turn,
extends the `java.lang.Throwable` class. That is why by catching an object of
the `java.lang.Exception` class, the code catches an object of any exception type. We
have demonstrated it in the *Java exceptions framework* section by
catching `java.lang.NullPointerException` this way.

One of the exceptions is `java.lang.RuntimeException`. The exceptions that extend it are called **runtime exceptions** or **unchecked exceptions**. We have already mentioned some of them: `NullPointerException`, `ArrayIndexOutOfBoundsException`, `ClassCastException`, and `NumberFormatException`. Why they are called runtime exceptions is clear; while why they are called unchecked exceptions will become clear in the next paragraph.

Those that do not have `java.lang.RuntimeException` among their ancestors are called **checked exceptions**. The reason for such a name is that a compiler makes sure (checks) that these exceptions are either caught or listed in the `throws` clause of the method (see *Throws statement* section). This design forces the programmer to make a conscious decision, either to catch the checked exception, or inform the client of the method that this exception may be thrown by the method and has to be processed (handled) by the client. Here are a few examples of checked exceptions:

- `ClassNotFoundException`: Thrown when an attempt to load a class using its string name with the `forName()` method of the `Class` class failed
- `CloneNotSupportedException`: Thrown when the code tried to clone an object that does not implement the `Cloneable` interface
- `NoSuchMethodException`: Thrown when there is no method called by the code

Not all exceptions reside in the `java.lang` package. Many other packages contain exceptions related to the functionality that is supported by the package. For example, there is a `java.util.MissingResourceException` runtime exception and `java.io.IOException` checked exception.

Despite not being forced to, programmers often catch runtime (unchecked) exceptions too, in order to have better control of the program flow, making the behavior of an application more stable and predictable. By the way, all errors are also runtime (unchecked) exceptions, but, as we have said already, typically, it is not possible to handle them programmatically, so there is no point in catching descendants of the `java.lang.Error` class.

The try, catch, and finally blocks

When an exception is thrown inside a `try` block, it redirects control flow to the first `catch` clause. If there is no `catch` block that can capture the exception (but then `finally` block has to be in place), the exception propagates all the way up and out of the method. If there is more than one `catch` clause, the compiler forces you to arrange them so that the child exception is listed before the parent exception. Let's look at the following example:

```
void someMethod(String s){
    try {
        method(s);
    } catch (NullPointerException ex){
        //do something
    } catch (Exception ex){
        //do something else
    }
}
```

In the preceding example, a `catch` block with `NullPointerException` is placed before the block with `Exception` because `NullPointerException` extends `RuntimeException`, which, in turn, extends `Exception`. We could even implement this example as follows:

```
void someMethod(String s){
    try {
        method(s);
    } catch (NullPointerException ex){
        //do something
    } catch (RuntimeException ex){
        //do something else
    } catch (Exception ex){
        //do something different
    }
}
```

The first `catch` clause catches `NullPointerException` only. Other exceptions that extend `RuntimeException` will be caught by the second `catch` clause. The rest of the exception types (all checked exceptions) will be caught by the last `catch` block. Note that errors will not be caught by any of these `catch` clauses. To catch them, one should add `catch` clause for `Error` (in any position) or `Throwable` (after the last `catch` clause in the previous example), but programmers usually do not do it and allow errors to propagate all the way into the JVM.

Having a `catch` block for each exception type allows us to provide an exception type-specific processing. However, if there is no difference in the exception processing, one can have just one `catch` block with the `Exception` base class to catch all types of exceptions:

```
void someMethod(String s){
    try {
        method(s);
    } catch (Exception ex){
        //do something
    }
}
```

If none of the clauses catch the exception, it is thrown further up until it is either handled by a `try...catch` statement in one of the methods-callers or propagates all the way out of the application code. In such a case, JVM terminates the application and exits.

Adding a `finally` block does not change the described behavior. If present, it is always executed, whether an exception was generated or not. A `finally` block is usually used to release the resources, to close a database connection, a file, and similar. However, if the resource implements the `Closeable` interface, it is better to use the try-with-resources statement that allows releasing the resources automatically. Here is how it can be done with Java 7:

```
try (Connection conn = DriverManager.getConnection("dburl",
                                    "username", "password");
     ResultSet rs = conn.createStatement()
                        .executeQuery("select * from some_table")) {
    while (rs.next()) {
        //process the retrieved data
    }
} catch (SQLException ex) {
    //Do something
    //The exception was probably caused by incorrect SQL statement
}
```

This example creates the database connection, retrieves data and processes it, then closes (calls the `close()` method) the `conn` and `rs` objects.

Java 9 enhanced try-with-resources statement capabilities by allowing the creation of objects that represent resources outside the `try` block and then the use of them in a try-with-resources statement, as follows:

```
void method(Connection conn, ResultSet rs) {
    try (conn; rs) {
        while (rs.next()) {
            //process the retrieved data
```

```
        }
    } catch (SQLException ex) {
        //Do something
        //The exception was probably caused by incorrect SQL statement
    }
}
```

The preceding code looks much cleaner, although, in practice, programmers prefer to create and release (close) resources in the same context. If that is your preference too, consider using the `throws` statement in conjunction with the try-with-resources statement.

The throws statement

The previous example of using a try-with-resources statement can be re-written with resource objects created in the same context as follows:

```
Connection conn;
ResultSet rs;
try {
    conn = DriverManager.getConnection("dburl", "username", "password");
    rs = conn.createStatement().executeQuery("select * from some_table");
} catch (SQLException e) {
    e.printStackTrace();
    return;
}

try (conn; rs) {
    while (rs.next()) {
        //process the retrieved data
    }
} catch (SQLException ex) {
    //Do something
    //The exception was probably caused by incorrect SQL statement
}
```

We have to deal with `SQLException` because it is a checked exception, and `getConnection()`, `createStatement()`, `executeQuery()`, and `next()` methods declare it in their `throws` clause. Here is an example:

```
Statement createStatement() throws SQLException;
```

It means that the method's author warns the method's users that it may throw such an exception and forces them either to catch the exception or to declare it in the throws clause of their methods. In our preceding example, we have chosen to catch it and had to use two try...catch statements. Alternatively, we can list the exception in the throws clause too and, thus, remove the clutter by effectively pushing the burden of the exception handling onto the users of our method:

```
void throwsDemo() throws SQLException {
    Connection conn = DriverManager.getConnection("url","user","pass");
    ResultSet rs = conn.createStatement().executeQuery("select * ...");
    try (conn; rs) {
        while (rs.next()) {
            //process the retrieved data
        }
    } finally { }
}
```

We got rid of the catch clause but Java syntax requires that either a catch or finally block has to follow the try block, so we added an empty finally block.

The throws clause allows but does not require us to list unchecked exceptions. Adding unchecked exceptions does not force the method's users to handle them.

And, finally, if the method throws several different exceptions, it is possible to list the base Exception exception class instead of listing all of them. That will make the compiler happy, but is not considered a good practice because it hides details of particular exceptions a method's user may expect.

Please notice that the compiler does not check what kind of exception the code in the method's body can throw. So, it is possible to list any exception in the throws clause, which may lead to unnecessary overhead. If, by mistake, a programmer includes a checked exception in the throws clause that is never actually thrown by the method, the method's user may write a catch block for it that is never executed.

The throw statement

The throw statement allows throwing any exception that a programmer deems necessary. One can even create their own exception. To create a checked exception, extend the java.lang.Exception class:

```
class MyCheckedException extends Exception{
    public MyCheckedException(String message){
        super(message);
```

```
        }
        //add code you need to have here
    }
```

Also, to create an unchecked exception, extend the `java.lang.RunitmeException` class, as follows:

```
class MyUncheckedException extends RuntimeException{
    public MyUncheckedException(String message){
        super(message);
    }
    //add code you need to have here
}
```

Notice the comment *add code you need to have here*. You can add methods and properties to the custom exception as to any other regular class, but programmers rarely do it. The best practices even explicitly recommend avoiding using exceptions for driving the business logic. Exceptions should be what the name implies, covering only exceptional, very rare situations.

But if you need to announce an exceptional condition use the `throw` keyword and `new` operator to create and trigger propagation of an exception object. Here are a few examples:

```
throw new Exception("Something happend");
throw new RunitmeException("Something happened");
throw new MyCheckedException("Something happened");
throw new MyUncheckedException("Something happened");
```

It is even possible to throw `null` as follows:

```
throw null;
```

The result of the preceding statement is the same as the result of this one:

```
throw new NullPointerException;
```

In both cases, an object of an unchecked `NullPointerException` begins to propagate through the system, until it is caught either by the application or the JVM.

The assert statement

Once in a while, a programmer needs to know if a particular condition happens in the code, even after the application has been already deployed to production. At the same time, there is no need to run the check all the time. That is where the branching `assert` statement comes in handy. Here is an example:

```
public someMethod(String s){
    //any code goes here
    assert(assertSomething(x, y, z));
    //any code goes here
}

boolean assertSomething(int x, String y, double z){
  //do something and return boolean
}
```

In the preceding code, the `assert()` method takes input from the `assertSomething()` method. If the `assertSomething()` method returns `false`, the program stops executing.

The `assert()` method is executed only when the JVM is run with the `-ea` option. The `-ea` flag should not be used in production, except maybe temporarily for testing purposes, because it creates the overhead that affects the application performance.

Best practices of exceptions handling

The checked exceptions were designed to be used for the recoverable conditions when an application can do something automatically to amend or work around the problem. In practice, it doesn't happen very often. Typically, when an exception is caught, the application logs the stack trace and aborts the current action. Based on the logged information, the application support team modifies the code to address the unaccounted-for condition or to prevent it from occurring in the future.

Each application is different, so best practices depend on the particular application requirements, design, and context. In general, it seems that there is an agreement in the development community to avoid using checked exceptions and to minimize their propagation in the application code. And here are a few other recommendations that have proved to be useful:

- Always catch all checked exceptions close to the source
- If in doubt, catch unchecked exceptions close to the source too

- Handle the exception as close to the source as possible, because it is where the context is the most specific and where the root cause resides
- Do not throw checked exceptions unless you have to, because you force building extra code for a case that may never happen
- Convert third-party checked exceptions into unchecked ones by re-throwing them as `RuntimeException` with the corresponding message if you have to
- Do not create custom exceptions unless you have to
- Do not drive business logic by using the exception handling mechanism unless you have to
- Customize generic `RuntimeException` by using the system of messages and, optionally, enum type, instead of using exception type for communicating the cause of the error

Summary

In this chapter, readers were introduced to the Java exception handling framework, learned about two kinds of exceptions – checked and unchecked (runtime) – and how to handle them using `try-catch-finally` and `throws` statements. Readers have also learned how to generate (throw) exceptions and how to create their own (custom) exceptions. The chapter concluded with the best practices of exception handling.

In the next chapter, we will talk about strings and their processing in detail, as well as input/output streams and file reading and writing techniques.

Quiz

1. What is a stack trace? Select all that apply:
   a. A list of classes currently loaded
   b. A list of methods currently executing
   c. A list of code lines currently executing
   d. A list of variables currently used

2. What kinds of exceptions are there? Select all that apply:
   a. Compilation exceptions
   b. Runtime exceptions
   c. Read exceptions
   d. Write exceptions

3. What is the output of the following code?

```
try {
    throw null;
} catch (RuntimeException ex) {
    System.out.print("RuntimeException ");
} catch (Exception ex) {
    System.out.print("Exception ");
} catch (Error ex) {
    System.out.print("Error ");
} catch (Throwable ex) {
    System.out.print("Throwable ");
} finally {
    System.out.println("Finally ");
}
```

   a. A RuntimeException Error
   b. Exception Error Finally
   c. RuntimeException Finally
   d. Throwable Finally

4. Which of the following methods will compile without an error?

```
void method1() throws Exception { throw null; }
void method2() throws RuntimeException { throw null; }
void method3() throws Throwable { throw null; }
void method4() throws Error { throw null; }
```

   a. method1()
   b. method2()
   c. method3()
   d. method4()

5. Which of the following statements will compile without an error?

```
throw new NullPointerException("Hi there!"); //1
throws new Exception("Hi there!");           //2
throw RuntimeException("Hi there!");         //3
throws RuntimeException("Hi there!");        //4
```

    a. 1

    b. 2

    c. 3

    d. 4

6. Assuming that int x = 4, which of the following statements will compile without an error?

```
assert (x > 3); //1
assert (x = 3); //2
assert (x < 4); //3
assert (x = 4); //4
```

    a. 1

    b. 2

    c. 3

    d. 4

7. What are the best practices from the following list?

    1. Always catch all exceptions and errors

    2. Always catch all exceptions

    3. Never throw unchecked exceptions

    4. Try not to throw checked exceptions unless you have to

5

Strings, Input/Output, and Files

In this chapter, a reader will be presented with `String` class methods in more detail. We will also discuss popular string utilities from standard libraries and Apache Commons project. An overview of Java input/output streams and related classes of the `java.io` packages will follow along with some classes of the `org.apache.commons.io` package. The file managing classes and their methods are described in the dedicated section.

The following topics will be covered in this chapter:

- Strings processing
- I/O streams
- Files management
- Apache Commons utilities `FileUtils` and `IOUtils`

Strings processing

In a mainstream programming, `String` probably is the most popular class. In Chapter 1, *Getting Started with Java 12*, we have learned about this class, its literals and its specific feature called **string immutability**. In this section, we will explain how a string can be processed using `String` class methods and utility classes from the standard library and the `StringUtils` class from the `org.apache.commons.lang3` package in particular.

Methods of the String class

The `String` class has more than 70 methods that enable analyzing, modifying, comparing strings, and converting numeric literals into the corresponding string literals. To see all the methods of the `String` class, please refer the Java API online at https://docs.oracle.com/en/java/javase.

Strings analysis

The `length()` method returns the count of characters in the string as shown in the following code:

```
String s7 = "42";
System.out.println(s7.length());      //prints: 2
System.out.println("0 0".length()); //prints: 3
```

The following `isEmpty()` method returns true when the length of the string (count of characters) is 0:

```
System.out.println("".isEmpty());    //prints: true
System.out.println(" ".isEmpty());   //prints: false
```

The `indexOf()` and `lastIndexOf()` methods return the position of the specified substring in the string shown in this code snippet:

```
String s6 = "abc42t%";
System.out.println(s6.indexOf(s7));          //prints: 3
System.out.println(s6.indexOf("a"));         //prints: 0
System.out.println(s6.indexOf("xyz"));       //prints: -1
System.out.println("ababa".lastIndexOf("ba")); //prints: 3
```

As you can see, the first character in the string has a position (index) 0, and the absence of the specified substring results in the index −1.

The `matches()` method applies the regular expression (passed as an argument) to the string as follows:

```
System.out.println("abc".matches("[a-z]+"));   //prints: true
System.out.println("ab1".matches("[a-z]+"));   //prints: false
```

Regular expressions are outside the scope of this book. You can learn about them at https://www.regular-expressions.info. In the preceding example, the expression [a-z]+ matches one or more letters only.

Strings comparison

In Chapter 3, *Java Fundamentals*, we have already talked about the `equals()` method that returns `true` only when two `String` objects or literals are spelled exactly the same way. The following code snippet demonstrates how it works:

```
String s1 = "abc";
String s2 = "abc";
```

```
String s3 = "acb";
System.out.println(s1.equals(s2));      //prints: true
System.out.println(s1.equals(s3));      //prints: false
System.out.println("abc".equals(s2));   //prints: true
System.out.println("abc".equals(s3));   //prints: false
```

Another `String` class `equalsIgnoreCase()` method does a similar job, but ignores the difference in the characters' case as follows:

```
String s4 = "aBc";
String s5 = "Abc";
System.out.println(s4.equals(s5));              //prints: false
System.out.println(s4.equalsIgnoreCase(s5));    //prints: true
```

The `contentEquals()` method acts similar to the `equals()` method shown here:

```
String s1 = "abc";
String s2 = "abc";
System.out.println(s1.contentEquals(s2));       //prints: true
System.out.println("abc".contentEquals(s2));    //prints: true
```

The difference is that `equals()` method checks if both values are represented by the `String` class,
while `contentEquals()` compares only characters (content) of the character sequence. The character sequence can be represented
by `String`, `StringBuilder`, `StringBuffer`, `CharBuffer`, or any other class that implements a `CharSequence` interface. Nevertheless, the `contentEquals()` method will return `true` if both sequences contain the same characters, while the `equals()` method will return `false` if one of the sequences is not created by the `String` class.

The `contains()` method returns `true` if the `string` contains a certain substring as follows:

```
String s6 = "abc42t%";
String s7 = "42";
String s8 = "xyz";
System.out.println(s6.contains(s7));    //prints: true
System.out.println(s6.contains(s8));    //prints: false
```

The `startsWith()` and `endsWith()` methods perform a similar check but only at the start of the string or at the end of the string value as shown in the following code:

```
String s6 = "abc42t%";
String s7 = "42";

System.out.println(s6.startsWith(s7));          //prints: false
```

```
System.out.println(s6.startsWith("ab"));      //prints: true
System.out.println(s6.startsWith("42", 3));  //prints: true

System.out.println(s6.endsWith(s7));          //prints: false
System.out.println(s6.endsWith("t%"));        //prints: true
```

The `compareTo()` and `compareToIgnoreCase()` methods compare strings lexicographically—based on the Unicode value of each character in the strings. They return the value 0 if the strings are equal, a negative integer value if the first string is lexicographically less (has a smaller Unicode value) than the second string, and a positive integer value if the first string is lexicographically greater than the second string (has a bigger Unicode value). For example:

```
String s4 = "aBc";
String s5 = "Abc";
System.out.println(s4.compareTo(s5));             //prints: 32
System.out.println(s4.compareToIgnoreCase(s5));   //prints: 0
System.out.println(s4.codePointAt(0));            //prints: 97
System.out.println(s5.codePointAt(0));            //prints: 65
```

From this code snippet, you can see that the `compareTo()` and `compareToIgnoreCase()` methods are based on the code points of the characters that compose the strings. The reason the string s4 is bigger than the string s5 by 32 because the code point of the character a (97) is bigger than the code point of the character A (65) by 32.

The given example also shows that the `codePointAt()` method returns the code point of the character located in the string at the specified position. The code points were described in the *Integral types* section of `Chapter 1`, *Getting Started with Java 12*.

Strings transformation

The `substring()` method returns the substring starting with the specified position (index) as follows:

```
System.out.println("42".substring(0));    //prints: 42
System.out.println("42".substring(1));    //prints: 2
System.out.println("42".substring(2));    //prints:
System.out.println("42".substring(3));    //error: index out of range: -1
String s6 = "abc42t%";
System.out.println(s6.substring(3));       //prints: 42t%
System.out.println(s6.substring(3, 5));    //prints: 42
```

The `format()` method uses the passed-in first argument as a template and inserts the other arguments in the corresponding position of the template sequentially. The following code example prints the sentence, *"Hey, Nick! Give me 2 apples, please!"* three times:

```
String t = "Hey, %s! Give me %d apples, please!";
System.out.println(String.format(t, "Nick", 2));

String t1 = String.format(t, "Nick", 2);
System.out.println(t1);

System.out.println(String
        .format("Hey, %s! Give me %d apples, please!", "Nick", 2));
```

The `%s` and `%d` symbols are called **format specifiers**. There are many specifiers and various flags, that allow the programmer to fine-control the result. You can read about them in the API of the `java.util.Formatter` class.

The `concat()` method works the same way as the arithmetic operator (+) as shown:

```
String s7 = "42";
String s8 = "xyz";
String newStr1 = s7.concat(s8);
System.out.println(newStr1);     //prints: 42xyz

String newStr2 = s7 + s8;
System.out.println(newStr2);     //prints: 42xyz
```

The following `join()` method acts similarly but allows the addition of a delimiter:

```
String newStr1 = String.join(",", "abc", "xyz");
System.out.println(newStr1);            //prints: abc,xyz

List<String> list = List.of("abc","xyz");
String newStr2 = String.join(",", list);
System.out.println(newStr2);            //prints: abc,xyz
```

The following group of `replace()`, `replaceFirst()`, and `replaceAll()` methods replace certain characters in the string with the provided ones:

```
System.out.println("abcbc".replace("bc", "42"));        //prints: a4242
System.out.println("abcbc".replaceFirst("bc", "42"));   //prints: a42bc
System.out.println("ab11bcd".replaceAll("[a-z]+", "42"));//prints: 421142
```

The first line of the preceding code replaces all the instances of `"bc"` with `"42"`. The second replaces only the first instance of `"bc"` with `"42"`. And the last one replaces all the substrings that match the provided regular expression with `"42"`.

The `toLowerCase()` and `toUpperCase()` methods change the case of the whole string as shown here:

```
System.out.println("aBc".toLowerCase());    //prints: abc
System.out.println("aBc".toUpperCase());    //prints: ABC
```

The `split()` method breaks the string into substrings, using the provided character as the delimiter, as follows:

```
String[] arr = "abcbc".split("b");
System.out.println(arr[0]);    //prints: a
System.out.println(arr[1]);    //prints: c
System.out.println(arr[2]);    //prints: c
```

There are several `valueOf()` methods that transform values of a primitive type to a `String` type. For example:

```
float f = 23.42f;
String sf = String.valueOf(f);
System.out.println(sf);            //prints: 23.42
```

There are also `()` and `getChars()` methods that transform a string to an array of a corresponding type, while the `chars()` method creates an `IntStream` of characters (their code points). We will talk about streams in `Chapter 14`, *Java Standard Streams*.

Methods added with Java 11

Java 11 introduced several new methods in the `String` class.

The `repeat()` method allows you to create a new String value based on multiple concatenations of the same string as shown in the following code:

```
System.out.println("ab".repeat(3)); //prints: ababab
System.out.println("ab".repeat(1)); //prints: ab
System.out.println("ab".repeat(0)); //prints:
```

The `isBlank()` method returns true if the string has length 0 or consists of white spaces only. For example:

```
System.out.println("".isBlank());       //prints: true
System.out.println("   ".isBlank());    //prints: true
System.out.println(" a ".isBlank());    //prints: false
```

The `stripLeading()` method removes leading white spaces from the string, the `stripTrailing()` method removes trailing white spaces, and `strip()` method removes both, as shown here:

```
String sp = "   abc   ";
System.out.println("'" + sp + "'");                    //prints: '   abc   '
System.out.println("'" + sp.stripLeading() + "'");     //prints: 'abc   '
System.out.println("'" + sp.stripTrailing() + "'");    //prints: '   abc'
System.out.println("'" + sp.strip() + "'");            //prints: 'abc'
```

And, finally, the `lines()` method breaks the string by line terminators and returns a `Stream<String>` of resulting lines. A line terminator is an escape sequence line feed \n (\u000a), or a carriage return \r (\u000d), or a carriage return followed immediately by a line feed \r\n (\u000d\u000a). For example:

```
String line = "Line 1\nLine 2\rLine 3\r\nLine 4";
line.lines().forEach(System.out::println);
```

The output of the preceding code is as follows:

We will talk about streams in Chapter 14, *Java Standard Streams*.

String utilities

In addition to the `String` class, there are many other classes that have methods that process `String` values. Among the most useful is the `StringUtils` class of the `org.apache.commons.lang3` package from a project called an **Apache Commons**, maintained by an open source community of programmers called **Apache Software Foundation**. We will talk more about this project and its libraries in Chapter 7, *Java Standard and External Libraries*. To use it in your project, add the following dependency in the `pom.xml` file:

```
<dependency>
    <groupId>org.apache.commons</groupId>
    <artifactId>commons-lang3</artifactId>
    <version>3.8.1</version>
</dependency>
```

The `StringUtils` class is the favorite of many programmers. It complements methods of the `String` class by providing the following null-safe operations:

- `isBlank(CharSequence cs)`: Returns `true` if the input value is white space, empty (`""`), or `null`

- `isNotBlank(CharSequence cs)`: Returns `false` when the preceding method returns `true`

- `isEmpty(CharSequence cs)`: Returns `true` if the input value is empty (`""`) or `null`

- `isNotEmpty(CharSequence cs)`: Returns `false` when the preceding method returns `true`

- `trim(String str)`: Removes leading and trailing white space from the input value and processes `null`, empty (`""`), and white space, as follows:

```
System.out.println("'" + StringUtils.trim(" x ") + "'"); //prints:
'x'
System.out.println(StringUtils.trim(null));                 //prints:
null
System.out.println("'" + StringUtils.trim("") + "'");     //prints:
''
System.out.println("'" + StringUtils.trim("   ") + "'"); //prints:
''
```

- `trimToNull(String str)`: Removes leading and trailing white space from the input value and processes `null`, empty (`""`), and white space, as follows:

```
System.out.println("'" + StringUtils.trimToNull(" x ") + "'");   //
'x'
System.out.println(StringUtils.trimToNull(null));           //prints:
null
System.out.println(StringUtils.trimToNull(""));             //prints:
null
System.out.println(StringUtils.trimToNull("   "));          //prints:
null
```

- `trimToEmpty(String str)`: Removes leading and trailing white space from the input value and processes `null`, empty (`""`), and white space, as follows:

```
System.out.println("'" + StringUtils.trimToEmpty(" x ") + "'");
// 'x'
System.out.println("'" + StringUtils.trimToEmpty(null) + "'");
// ''
System.out.println("'" + StringUtils.trimToEmpty("") + "'");
// ''
System.out.println("'" + StringUtils.trimToEmpty("   ") + "'");
```

```
// ''
```

- `strip(String str)`, `stripToNull(String str)`, `stripToEmpty(String str)`: Produce the same result as the preceding `trim*(String str)` methods but use a more extensive definition of white space (based on `Character.isWhitespace(int codepoint)`) and thus remove the same characters as `trim*(String str)` does, and more

- `strip(String str, String stripChars)`, `stripAccents(String input)`, `stripAll(String... strs)`, `stripAll(String[] strs, String stripChars)`, `stripEnd(String str, String stripChars)`, `stripStart(String str, String stripChars)`: Remove particular characters from particular parts of a `String` or a `String[]` array elements

- `startsWith(CharSequence str, CharSequence prefix)`, `startsWithAny(CharSequence string, CharSequence... searchStrings)`, `startsWithIgnoreCase(CharSequence str, CharSequence prefix)`, and similar `endsWith*()` methods: Check if a `String` value starts (or ends) with a certain prefix (or suffix)

- `indexOf`, `lastIndexOf`, `contains`: Check index in a null-safe manner

- `indexOfAny`, `lastIndexOfAny`, `indexOfAnyBut`, `lastIndexOfAnyBut`: Return index

- `containsOnly`, `containsNone`, `containsAny`: Check if the value contains or not certain characters

- `substring`, `left`, `right`, `mid`: Return substring in a null-safe manner

- `substringBefore`, `substringAfter`, `substringBetween`: Return substring from relative position

- `split`, `join`: Split or join a value (correspondingly)

- `remove`, `delete`: Eliminate substring

- `replace`, `overlay`: Replace a value

- `chomp`, `chop`: Remove the end

- `appendIfMissing`: Adds a value if not present

- `prependIfMissing`: Prepends a prefix to the start of the `String` value if not present

- `leftPad`, `rightPad`, `center`, `repeat`: Add padding

- `upperCase`, `lowerCase`, `swapCase`, `capitalize`, `uncapitalize`: Change the case

- countMatches: Returns the number of the substring occurrences
- isWhitespace, isAsciiPrintable, isNumeric, isNumericSpace, isAlpha, isAlphaNumeric, isAlphaSpace, isAlphaNumericSpace: Check the presence of certain type of characters
- isAllLowerCase, isAllUpperCase: Check the case
- defaultString, defaultIfBlank, defaultIfEmpty: Returns default value if null
- rotate: Rotates characters using a circular shift
- reverse, reverseDelimited: Reverse characters or delimited groups of characters
- abbreviate, abbreviateMiddle: Abbreviate value using an ellipsis or another value
- difference: Returns the differences in values
- getLevenshteinDistance: Returns the number of changes needed to transform one value to another

As you can see, the StringUtils class has a very rich (we have not listed everything) set of methods for strings analysis, comparison, and transformation that compliment the methods of the String class.

I/O streams

Any software system has to receive and produce some kind of data that can be organized as a set of isolated inputs/outputs or as a stream of data. A stream can be limited or endless. A program can read from a stream (then it is called an **input stream**), or write to a stream (then it is called an **output stream**). The Java I/O stream is either byte-based or character-based, meaning that its data is interpreted either as raw bytes or as characters.

The java.io package contains classes that support many, but not all, possible data sources. It is built for the most part around the input from and to files, network streams, and internal memory buffers. It does not contain many classes necessary for network communication. They belong to the java.net, javax.net, and other packages of Java Networking API. Only after the networking source or destination is established (a network socket, for example), can a program read and write data using InputStream and OutputStream classes of the java.io package.

Classes of the `java.nio` package have pretty much the same functionality as the classes of `java.io` packages. But, in addition, they can work in *a non-blocking* mode, which can substantially increase the performance in certain situations. We will talk about non-blocking processing in Chapter 15, *Reactive Programming*.

Stream data

The data a program understands has to be binary—expressed in 0s and 1s—at the very basis. The data can be read or written one byte at a time or an array of several bytes at a time. These bytes can remain binary or can be interpreted as characters.

In the first case, they can be read as bytes or byte arrays by the descendants of `InputStream` and `OutputStream` classes. For example (we omit the package name if the class belongs to the `java.io` package): `ByteArrayInputStream`, `ByteArrayOutputStream`, `FileInputStream`, `FileOutputStream`, `ObjectInputStream`, `ObjectOutputStream`, `javax.sound.sampled.AudioInputStream`, and `org.omg.CORBA.portable.OutputStream`; which one to use depends on the source or destination of the data. The `InputStream` and `OutputStream` classes themselves are abstract and cannot be instantiated.

In the second case, the data that can be interpreted as characters are called **text data**, and there are character-oriented reading and writing classes based on the `Reader` and `Writer`, which are abstract classes too. Examples of their sub-classes are: `CharArrayReader`, `CharArrayWriter`, `InputStreamReader`, `OutputStreamWriter`, `PipedReader`, `PipedWriter`, `StringReader`, and `StringWriter`.

You may have noticed that we listed the classes in pairs. But not every input class has a matching output specialization. For example, there are `PrintStream` and `PrintWriter` classes that support output to a printing device, but there is no corresponding input partner, not by name at least. However, there is a `java.util.Scanner` class that parses input text in a known format.

There is also a set of buffer-equipped classes that help to improve performance by reading or writing a bigger chunk of data at a time, especially in the cases when access to the source or destination takes a long time.

In the rest of this section, we will review classes of the `java.io` package and some popular related classes from other packages.

Class InputStream and its subclasses

In Java Class Library, the InputStream abstract class has the following direct implementations: ByteArrayInputStream, FileInputStream, ObjectInputStream, PipedInputStream, SequenceInputStream, FilterInputStream, and javax.sound.sampled.AudioInputStream.

All of them either use as-is or override the following methods of the InputStream class:

- int available(): Returns the number of bytes available for reading
- void close(): Closes the stream and releases the resources
- void mark(int readlimit): Marks a position in the stream and defines how many bytes can be read
- boolean markSupported(): Returns true if the marking is supported
- static InputStream nullInputStream(): Creates an empty stream
- abstract int read(): Reads the next byte in the stream
- int read(byte[] b): Reads data from the stream into the b buffer
- int read(byte[] b, int off, int len): Reads len or fewer bytes from the stream into the b buffer
- byte[] readAllBytes(): Reads all the remaining bytes from the stream
- int readNBytes(byte[] b, int off, int len): Reads len or fewer bytes into the b buffer at the off offset
- byte[] readNBytes(int len): Reads len or fewer bytes into the b buffer
- void reset(): Resets the reading location to the position where the mark() method was last called
- long skip(long n): Skips n or fewer bytes of the stream; returns the actual number of bytes skipped
- long transferTo(OutputStream out): Reads from the input stream and writes to the provided output stream byte by byte; returns the actual number of bytes transferred

The abstract int read() is the only method that has to be implemented, but most of the descendants of this class override many of the other methods too.

ByteArrayInputStream

The ByteArrayInputStream class allows reading a byte array as an input stream. It has the following two constructors that create an object of the class and defines the buffer used to read the input stream of bytes:

- ByteArrayInputStream(byte[] buffer)
- ByteArrayInputStream(byte[] buffer, int offset, int length)

The second of the constructors allows setting, in addition to the buffer, the offset and the length of the buffer too. Let's look at the example and see how this class can be used. We assume there is a source of byte[] array with data:

```
byte[] bytesSource(){
    return new byte[]{42, 43, 44};
}
```

Then we can write the following:

```
byte[] buffer = bytesSource();
try(ByteArrayInputStream bais = new ByteArrayInputStream(buffer)){
    int data = bais.read();
    while(data != -1) {
        System.out.print(data + " ");    //prints: 42 43 44
        data = bais.read();
    }
} catch (Exception ex){
    ex.printStackTrace();
}
```

The bytesSource() method produces the array of bytes that fills the buffer that is passed into the constructor of the ByteArrayInputStream class as a parameter. The resulting stream is then read byte by byte using the read() method until the end of the stream is reached (and read() method returns −1). Each new byte is printed out (without line feed and with white space after it, so all the read bytes are displayed in one line separated by the white space).

The preceding code is usually expressed in a more compact form as follows:

```
byte[] buffer = bytesSource();
try(ByteArrayInputStream bais = new ByteArrayInputStream(buffer)){
    int data;
    while ((data = bais.read()) != -1) {
        System.out.print(data + " ");    //prints: 42 43 44
    }
} catch (Exception ex){
```

```
        ex.printStackTrace();
    }
```

Instead of just printing the bytes, they can be processed in any other manner necessary, including interpreting them as characters. For example:

```
byte[] buffer = bytesSource();
try(ByteArrayInputStream bais = new ByteArrayInputStream(buffer)){
    int data;
    while ((data = bais.read()) != -1) {
        System.out.print(((char)data) + " ");    //prints: * + ,
    }
} catch (Exception ex){
    ex.printStackTrace();
}
```

But, in such a case, it is better to use one of the `Reader` classes that are specialized for characters processing. We will talk about them in the *Reader and Writer classes and their subclasses* section.

FileInputStream

The `FileInputStream` class gets data from a file in a filesystem, the raw bytes of an image, for example. It has the following three constructors:

- `FileInputStream(File file)`
- `FileInputStream(String name)`
- `FileInputStream(FileDescriptor fdObj)`

Each constructor opens the file specified as the parameter. The first constructor accepts `File` object, the second, the path to the file in the filesystem, and the third, the file descriptor object that represents an existing connection to an actual file in the filesystem. Let's look at the following example:

```
String filePath = "src/main/resources/hello.txt";
try(FileInputStream fis=new FileInputStream(filePath)){
    int data;
    while ((data = fis.read()) != -1) {
        System.out.print(((char)data) + " ");    //prints: H e l l o !
    }
} catch (Exception ex){
    ex.printStackTrace();
}
```

In the `src/main/resources` folder, we have created the `hello.txt` file that has only one line in it—`Hello!`. The output of the preceding example looks as follows:

Since we are running this example inside the IDE, it is executed in the project root directory. In order to find where your code is executed, you can always print it out like this:

```
File f = new File(".");                    //points to the current directory
System.out.println(f.getAbsolutePath()); //prints the directory path
```

After reading bytes from the `hello.txt` file, we decided, for demo purposes, to cast each `byte` to `char`, so you can see that our code does read from the specified file, but the `FileReader` class is a better choice for text file processing (we will discuss it shortly). Without the cast, the result would be the following:

```
System.out.print((data) + " ");    //prints: 72 101 108 108 111 33
```

By the way, because the `src/main/resources` folder is placed by the IDE (using Maven) on the classpath, a file placed in it can also be accessed via a classloader that creates a stream using its own `InputStream` implementation:

```
try(InputStream is =
InputOutputStream.class.getResourceAsStream("/hello.txt")){
    int data;
    while ((data = is.read()) != -1) {
        System.out.print((data) + " ");    //prints: 72 101 108 108 111 33
    }
} catch (Exception ex){
    ex.printStackTrace();
}
```

The `InputOutputStream` class in the preceding example is not a class from some library. It is just the main class we used to run the example.
The `InputOutputStream.class.getResourceAsStream()` construct allows using the same classloader that has loaded the `InputOutputStream` class for the purpose of finding a file on the classpath and creating a stream that contains its content. In the *Files Management* section, we will present other ways of reading a file too.

ObjectInputStream

The set of methods of the `ObjectInputStream` class is much bigger than the set of methods of any other `InputStream` implementation. The reason for that is that it is built around reading the values of the object fields that can be of various types. In order for the `ObjectInputStream` to be able to construct an object from the input stream of data, the object has to be *deserializable*, which means it has to be *serializable* in the first place—that is to be convertible into a byte stream. Usually, it is done for the purpose of transporting objects over a network. At the destination, the serialized objects are deserialized and the values of the original objects are restored.

Primitive types and most of Java classes, including `String` class and primitive type wrappers, are serializable. If a class has fields of custom types, they have to be made serializable by implementing `java.io.Serizalizable`. How to do it is outside the scope of this book. For now, we are going to use only the serializable types. Let's look at this class:

```
class SomeClass implements Serializable {
    private int field1 = 42;
    private String field2 = "abc";
}
```

We have to tell the compiler that it is serializable. Otherwise, the compilation will fail. It is done in order to make sure that, before stating that the class is serializable, the programmer either reviewed all the fields and made sure they are serializable or has implemented the methods necessary for the serialization.

Before we can create an input stream and use `ObjectInputStream` for deserialization, we need to serialize the object first. That is why we first use `ObjectOutputStream` and `FileOutputStream` to serialize an object and write it into the `someClass.bin` file. We will talk more about them in the *Class OutputStream and its subclasses* section. Then we read from the file using `FileInputStream` and deserialize the file content using `ObjectInputStream`:

```
String fileName = "someClass.bin";
try (ObjectOutputStream objectOutputStream =
            new ObjectOutputStream(new FileOutputStream(fileName));
    ObjectInputStream objectInputStream =
            new ObjectInputStream(new FileInputStream(fileName))){
    SomeClass obj = new SomeClass();
    objectOutputStream.writeObject(obj);
    SomeClass objRead = (SomeClass) objectInputStream.readObject();
    System.out.println(objRead.field1);  //prints: 42
    System.out.println(objRead.field2);  //prints: abc
} catch (Exception ex){
```

```
        ex.printStackTrace();
    }
```

Note that, the file has to be created first before the preceding code is run. We will show how it can be done in the *Creating files and directories* section. And, to remind you, we have used the try-with-resources statement because `InputStream` and `OutputStream` both implement the `Closeable` interface.

PipedInputStream

A piped input stream has very particular specialization; it is used as one of the mechanisms of communication between threads. One thread reads from a `PipedInputStream` object and passes data to another thread that writes data to a `PipedOutputStream` object. Here is an example:

```
PipedInputStream pis = new PipedInputStream();
PipedOutputStream pos = new PipedOutputStream(pis);
```

Alternatively, data can be moved in the opposite direction when one thread reads from a `PipedOutputStream` object and another thread writes to a `PipedInputStream` object as follows:

```
PipedOutputStream pos = new PipedOutputStream();
PipedInputStream pis = new PipedInputStream(pos);
```

Those who work in this area are familiar with the message, "*Broken pipe*", which means that the supplying data pipe stream has stopped working.

The piped streams can also be created without any connection and connected later as shown here:

```
PipedInputStream pis = new PipedInputStream();
PipedOutputStream pos = new PipedOutputStream();
pos.connect(pis);
```

For example, here are two classes that are going to be executed by different threads. First, the `PipedOutputWorker` class as follows:

```
class PipedOutputWorker implements Runnable{
    private PipedOutputStream pos;
    public PipedOutputWorker(PipedOutputStream pos) {
        this.pos = pos;
    }
    @Override
    public void run() {
```

```
        try {
            for(int i = 1; i < 4; i++){
                pos.write(i);
            }
            pos.close();
        } catch (Exception ex) {
            ex.printStackTrace();
        }
    }
}
```

The `PipedOutputWorker` class has `run()` method (because it implements a `Runnable` interface) that writes into the stream the three numbers 1, 2, and 3, and then closes. Now let's look at `PipedInputWorker` class as shown here:

```
class PipedInputWorker implements Runnable{
    private PipedInputStream pis;
    public PipedInputWorker(PipedInputStream pis) {
        this.pis = pis;
    }
    @Override
    public void run() {
        try {
            int i;
            while((i = pis.read()) > -1){
                System.out.print(i + " ");
            }
            pis.close();
        } catch (Exception ex) {
            ex.printStackTrace();
        }
    }
}
```

It also has a `run()` method (because it implements a `Runnable` interface) that reads from the stream and prints out each byte until the stream ends (indicated by –1). Now let's connect these pipes and execute a `run()` methods of these classes:

```
PipedOutputStream pos = new PipedOutputStream();
PipedInputStream pis = new PipedInputStream();
try {
    pos.connect(pis);
    new Thread(new PipedOutputWorker(pos)).start();
    new Thread(new PipedInputWorker(pis)).start(); //prints: 1 2 3
} catch (Exception ex) {
    ex.printStackTrace();
}
```

As you can see, the objects of the workers were passed into the constructor of
the Thread class. The start() method of the Thread object executes the run() method of
the passed in Runnable. And we see the results we have expected; the PipedInputWorker
prints all the bytes written to the piped stream by the PipedOutputWorker. We will get
into more details about threads in Chapter 8, *Multithreading and Concurrent Processing*.

SequenceInputStream

The SequenceInputStream class concatenates input streams passed into one of the
following constructors as parameters:

- SequenceInputStream(InputStream s1, InputStream s2)
- SequenceInputStream(Enumeration<InputStream> e)

Enumeration is a collection of objects of the type indicated in the angle brackets, called
generics, meaning *of type T*. The SequenceInputStream class reads from the first input
string until it ends, whereupon it reads from the second one, and so on, until the end of the
last of the streams. For example, let's create a howAreYou.txt file (with the text How are
you?) in the resources folder next to
the hello.txt file. The SequenceInputStream class can then be used as follows:

```
try(FileInputStream fis1 =
                new FileInputStream("src/main/resources/hello.txt");
    FileInputStream fis2 =
                new FileInputStream("src/main/resources/howAreYou.txt");
    SequenceInputStream sis=new SequenceInputStream(fis1, fis2)){
    int i;
    while((i = sis.read()) > -1){
        System.out.print((char)i);        //prints: Hello!How are you?
    }
} catch (Exception ex) {
    ex.printStackTrace();
}
```

Similarly, when an enumeration of input streams is passed in, each of the streams is read
(and printed in our case) until the end.

FilterInputStream

The `FilterInputStream` class is a wrapper around the `InputStream` object passed as a parameter in the constructor. Here is the constructor and two `read()` methods of the `FilterInputStream` class:

```
protected volatile InputStream in;
protected FilterInputStream(InputStream in) { this.in = in; }
public int read() throws IOException { return in.read(); }
public int read(byte b[]) throws IOException {
    return read(b, 0, b.length);
}
```

All the other methods of the `InputStream` class are overridden similarly; the function is delegated to the object assigned to the `in` property.

As you can see, the constructor is protected, which means that only the child has access to it. Such a design hides from the client the actual source of the stream and forces the programmer to use one of the `FilterInputStream` class extensions: `BufferedInputStream`, `CheckedInputStream`, `DataInputStream`, `PushbackInputStream`, `javax.crypto.CipherInputStream`, `java.util.zip.DeflaterInputStream`, `java.util.zip.InflaterInputStream`, `java.security.DigestInputStream`, or `javax.swing.ProgressMonitorInputStream`. Alternatively, one can create a custom extension. But, before creating your own extension, look at the listed classes and see if one of them fits your needs. Here is an example of using a `BufferedInputStream` class:

```
try(FileInputStream  fis =
        new FileInputStream("src/main/resources/hello.txt");
    FilterInputStream filter = new BufferedInputStream(fis)){
    int i;
    while((i = filter.read()) > -1){
        System.out.print((char)i);      //prints: Hello!
    }
} catch (Exception ex) {
    ex.printStackTrace();
}
```

The `BufferedInputStream` class uses the buffer to improve the performance. When the bytes from the stream are skipped or read, the internal buffer is automatically refilled with as many bytes as necessary at the time, from the contained input stream.

The `CheckedInputStream` class adds a checksum of the data being read that allows the verification of the integrity of the input data using `getChecksum()` method.

The `DataInputStream` class reads and interprets input data as primitive Java data types in a machine-independent way.

The `PushbackInputStream` class adds the ability to push back the read data using the `unread()` method. It is useful in situations when the code has the logic of analyzing the just read data and deciding to unread it, so it can be re-read at the next step.

The `javax.crypto.CipherInputStream` class adds a `Cipher` to the `read()` methods. If the `Cipher` is initialized for decryption, the `javax.crypto.CipherInputStream` will attempt to decrypt the data before returning.

The `java.util.zip.DeflaterInputStream` class compresses data in the deflate compression format.

The `java.util.zip.InflaterInputStream` class uncompresses data in the deflate compression format.

The `java.security.DigestInputStream` class updates the associated message digest using the bits going through the stream. The `on (boolean on)` method turns the digest function on or off. The calculated digest can be retrieved using the `getMessageDigest()` method.

The `javax.swing.ProgressMonitorInputStream` class provides a monitor of the progress of reading from the `InputStream`. The monitor object can be accessed using the `getProgressMonitor()` method.

javax.sound.sampled.AudioInputStream

The `AudioInputStream` class represents an input stream with a specified audio format and length. It has the following two constructors:

- `AudioInputStream (InputStream stream, AudioFormat format, long length)`: Accepts the stream of audio data, the requested format, and the length in sample frames
- `AudioInputStream (TargetDataLine line)`: Accepts the target data line indicated

The `javax.sound.sampled.AudioFormat` class describes audio-format properties such as channels, encoding, frame rate, and similar.

The `javax.sound.sampled.TargetDataLine` class has `open()` method that opens the line with the specified format and `read()` method that reads audio data from the data line's input buffer.

There is also the `javax.sound.sampled.AudioSystem` class and its methods handle `AudioInputStream` objects. They can be used for reading from an audio file, a stream, or a URL, and write to an audio file. They also can be used to convert an audio stream to another audio format.

Class OutputStream and its subclasses

The `OutputStream` class is a peer of the `InputStream` class that writes data instead of reading. It is an abstract class that has the following direct implementations in the **Java Class Library (JCL)**: `ByteArrayOutputStream`, `FilterOutputStream`, `ObjectOutputStream`, `PipedOutputStream`, and `FileOutputStream`.

The `FileOutputStream` class has the following direct extensions: `BufferedOutputStream`, `CheckedOutputStream`, `DataOutputStream`, `PrintStream`, `javax.crypto.CipherOutputStream`, `java.util.zip.DeflaterOutputStream`, `java.security.DigestOutputStream`, and `java.util.zip.InflaterOutputStream`.

All of them either use as-is or override the following methods of the `OutputStream` class:

- `void close()`: Closes the stream and releases the resources
- `void flush()`: Forces the remaining bytes to be written out
- `static OutputStream nullOutputStream()`: Creates a new `OutputStream` that writes nothing
- `void write(byte[] b)`: Writes the provided byte array to the output stream
- `void write(byte[] b, int off, int len)`: Writes `len` bytes of the provided byte array starting at `off` offset to the output stream
- `abstract void write(int b)`: Writes the provided byte to the output stream

The only method that has to be implemented is `abstract void write(int b)`, but most of the descendants of `OutputStream` class override many of the other methods too.

After learning about the input streams in the *Class InputStream and its subclasses* section, all of the `OutputStream` implementations, except the `PrintStream` class, should be intuitively familiar to you. So, we will discuss here only the `PrintStream` class.

PrintStream

The `PrintStream` class adds to another output stream the ability to print data as characters. We have actually used it already many times. The `System` class has an object of `PrintStream` class set as a `System.out` public static property. This means that a every time we print something using `System.out`, we are using the `PrintStream` class:

```
System.out.println("Printing a line");
```

Let's look at another example of the `PrintStream` class usage:

```
String fileName = "output.txt";
try(FileOutputStream  fos = new FileOutputStream(fileName);
    PrintStream ps = new PrintStream(fos)){
    ps.println("Hi there!");
} catch (Exception ex) {
    ex.printStackTrace();
}
```

As you can see, the `PrintStream` class takes `FileOutputStream` object and prints the characters generated by it. In this case, it prints out all the bytes the `FileOutputStream` writes to the file. By the way, there is no need to create the destination file explicitly. If absent, it will be created automatically inside the `FileOutputStream` constructor. If we open the file after the preceding code is run, we will see one line in it: `"Hi there!"`.

Alternatively, the same result can be achieved using another `PrintStream` constructor that takes `File` object as follows:

```
String fileName = "output.txt";
File file = new File(fileName);
try(PrintStream ps = new PrintStream(file)){
    ps.println("Hi there!");
} catch (Exception ex) {
    ex.printStackTrace();
}
```

An even simpler solution can be created using the third variation of the `PrintStream` constructor that takes the file name as a parameter:

```
String fileName = "output.txt";
try(PrintStream ps = new PrintStream(fileName)){
    ps.println("Hi there!");
} catch (Exception ex) {
    ex.printStackTrace();
}
```

The last two preceding examples are possible because the `PrintStream` constructor uses the `FileOutputStream` class behind the scenes, exactly as we did it in the first example of the `PrintStream` class usage. So, the `PrintStream` class has several constructors just for the convenience, but all of them essentially have the same functionality:

- `PrintStream(File file)`
- `PrintStream(File file, String csn)`
- `PrintStream(File file, Charset charset)`
- `PrintStream(String fileName)`
- `PrintStream(String fileName, String csn)`
- `PrintStream(String fileName, Charset charset)`
- `PrintStream(OutputStream out)`
- `PrintStream(OutputStream out, boolean autoFlush)`
- `PrintStream(OutputStream out, boolean autoFlush, String encoding)`
- `PrintStream(OutputStream out, boolean autoFlush, Charset charset)`

Some of the constructors also take a `Charset` instance or just its name (`String csn`), which allows applying a different mapping between sequences of sixteen-bit Unicode code units and sequences of bytes. You can see all available charsets by just printing them out as shown here:

```
for (String chs : Charset.availableCharsets().keySet()) {
    System.out.println(chs);
}
```

Other constructors take a `boolean autoFlush` as a parameter. This parameter indicates (when `true`) that the output buffer should be flushed automatically when an array is written or the symbol end-of-line is encountered.

Once an object of `PrintStream` is created, it provides a variety of methods as listed here:

- `void print(T value)`: Prints the value of any `T` primitive type passed in without moving to another line
- `void print(Object obj)`: Calls the `toString()` method on the passed in object and prints the result without moving to another line; does not generate a `NullPointerException` in case the passed-in object is `null` and prints `null` instead

- `void println(T value)`: Prints the value of any `T` primitive type passed in and moves to another line
- `void println(Object obj)`: Calls the `toString()` method on the passed-in object, prints the result, and moves to another line; does not generate a `NullPointerException` in case the passed-in object is `null` and prints `null` instead
- `void println()`: Just moves to another line
- `PrintStream printf(String format, Object... values)`: Substitutes the placeholders in the provided `format` string with the provided `values` and writes the result into the stream
- `PrintStream printf(Locale l, String format, Object... args)`: Same as the preceding method but applies localization using the provided `Local` object; if the provided `Local` object is `null`, no localization is applied and this method behaves exactly like the preceding one
- `PrintStream format(String format, Object... args)` and `PrintStream format(Locale l, String format, Object... args)`: Behave the same way as `PrintStream printf(String format, Object... values)` and `PrintStream printf(Locale l, String format, Object... args)` (already described in the list) for example:

    ```
    System.out.printf("Hi, %s!%n", "dear reader"); //prints: Hi, dear
    reader!
    System.out.format("Hi, %s!%n", "dear reader"); //prints: Hi, dear
    reader!
    ```

In the preceding example, (`%`) indicates a formatting rule. The following symbol (`s`) indicates a `String` value. Other possible symbols in this position can be (`d`) (decimal), (`f`) (floating-point), and so on. The symbol (`n`) indicates a new line (same as (`\n`) escape character). There are many formatting rules. All of them are described in the documentation for `java.util.Formatter` class.

- `PrintStream append(char c)`, `PrintStream append(CharSequence c)`, `PrintStream append(CharSequence c, int start, int end)`: Append the provided character to the stream. For example:

    ```
    System.out.printf("Hi %s", "there").append("!\n"); //prints: Hi
    there!
    System.out.printf("Hi ")
                .append("one there!\n two", 4, 11); //prints: Hi
    there!
    ```

With this, we conclude the discussion of the `OutputStream` subclass and now turn our attention to another class hierarchy—the `Reader` and `Writer` classes and their subclasses from the JCL.

Reader and Writer classes and their subclasses

As we mentioned several times already, `Reader` and `Writer` classes are very similar in their function to `InputStream` and `OutputStream` classes but specialize in processing texts. They interpret stream bytes as characters and have their own independent `InputStream` and `OutputStream` class hierarchy. It is possible to process stream bytes as characters without `Reader` and `Writer` or any of their subclasses. We have seen such examples in the preceding sections describing `InputStream` and `OutputStream` classes. However, using `Reader` and `Writer` classes makes text processing simpler and the code easier to read.

Reader and its subclasses

The class `Reader` is an abstract class that reads streams as characters. It is an analog to `InputStream` and has the following methods:

- `abstract void close()`: Closes the stream and other used resources
- `void mark(int readAheadLimit)`: Marks the current position in the stream
- `boolean markSupported()`: Returns `true` if the stream supports the `mark()` operation
- `static Reader nullReader()`: Creates an empty Reader that reads no characters
- `int read()`: Reads one character
- `int read(char[] buf)`: Reads characters into the provided `buf` array and returns the count of the read characters
- `abstract int read(char[] buf, int off, int len)`: Reads `len` characters into an array starting from the `off` index
- `int read(CharBuffer target)`: Attempts to read characters into the provided `target` buffer
- `boolean ready()`: Returns `true` when the stream is ready to be read
- `void reset()`: Resets the mark; however, not all streams support this operation, while some support it, but do not support setting a mark

- `long skip(long n)`: Attempts to skip n characters; returns the count of skipped characters
- `long transferTo(Writer out)`: Reads all characters from this reader and writes the characters to the provided `Writer` object

As you can see, the only methods that need to be implemented are two abstract `read()` and `close()` methods. Nevertheless, many children of this class override other methods too, sometimes for better performance or different functionality. The `Reader` subclasses in the JCL are: `CharArrayReader`, `InputStreamReader`, `PipedReader`, `StringReader`, `BufferedReader`, and `FilterReader`. The `BufferedReader` class has a `LineNumberReader` subclass, and the `FilterReader` class has a `PushbackReader` subclass.

Writer and its subclasses

The abstract `Writer` class writes to character streams. It is an analog to `OutputStream` and has the following methods:

- `Writer append(char c)`: Appends the provided character to the stream
- `Writer append(CharSequence c)`: Appends the provided character sequence to the stream
- `Writer append(CharSequence c, int start, int end)`: Appends a subsequence of the provided character sequence to the stream
- `abstract void close()`: Flushes and closes the stream and related system resources
- `abstract void flush()`: Flushes the stream
- `static Writer nullWriter()`: Creates a new `Writer` object that discards all characters
- `void write(char[] c)`: Writes an array of c characters
- `abstract void write(char[] c, int off, int len)`: Writes len elements of an array of c characters starting from the off index
- `void write(int c)`: Writes one character
- `void write(String str)`: Writes the provided string
- `void write(String str, int off, int len)`: Writes a substring of len length from the provided str string starting from the off index

As you can see, the three abstract methods: `write(char[], int, int)`, `flush()`, and `close()` must be implemented by the children of this class. They also typically override other methods too.

The `Writer` subclasses in the JCL are: `CharArrayWriter`, `OutputStreamWriter`, `PipedWriter`, `StringWriter`, `BufferedWriter`, `FilterWriter`, and `PrintWriter`. The `OutputStreamWriter` class has a `FileWriter` subclass.

Other classes of java.io package

Other classes of the `java.io` package include the following:

- `Console`: Allows interacting with the character-based console device, associated with the current JVM instance
- `StreamTokenizer`: Takes an input stream and parses it into `tokens`
- `ObjectStreamClass`: The serialization's descriptor for classes
- `ObjectStreamField`: A description of a serializable field from a serializable class
- `RandomAccessFile`: Allows random access for reading from and writing to a file, but its discussion is outside the scope of this book
- `File`: Allows creating and managing files and directories; described in the *Files management* section

Console

There are several ways to create and run a **Java Virtual Machine (JVM)** instance that executes an application. If the JVM is started from a command line, a console window is automatically opened. It allows typing on the display from the keyboard; however, the JVM can be started by a background process too. In such a case, a console is not created.

To check programmatically if a console exists, one can invoke the `System.console()` static method. If no console device is available then an invocation of that method will return `null`. Otherwise, it will return an object of the `Console` class that allows interacting with the console device and the application user.

Let's create the following `ConsoleDemo` class:

```
package com.packt.learnjava.ch05_stringsIoStreams;
import java.io.Console;
public class ConsoleDemo {
```

```
public static void main(String... args)  {
    Console console = System.console();
    System.out.println(console);
}
}
```

If we run it from the IDE, as we usually do, the result will be as follows:

That is because the JVM was launched not from the command line. In order to do it, let's compile our application and create a `.jar` file by executing the `mvn clean package` Maven command in the root directory of the project. It will delete the `target` folder, then recreate it, and compile all the `.java` files to the corresponding `.class` files in the `target` folder, and then will archive them in a `.jar` file `learnjava-1.0-SNAPSHOT.jar`.

Now we can launch the `ConsoleDemo` application from the same project root directory using the following command:

java -cp ./target/learnjava-1.0-SNAPSHOT.jar
com.packt.learnjava.ch05_stringsIoStreams.ConsoleDemo

The preceding command is shown in two lines because the page width cannot accommodate it. But if you would like to run it, make sure you do it as one line. The result will be as follows:

java.io.Console@33c7353a

It tells us that we have the `Console` class object now. Let's see what we can do with it. The class has the following methods:

- `String readLine()`: Waits until the user hits the *Enter* key and reads the line of text from the console
- `String readLine(String format, Object... args)`: Displays prompt (the message produced after the provided format got the placeholders substituted with the provided arguments), waits until the user hits key *Enter*, and reads the line of text from the console; if no arguments `args` are provided, displays the format as the prompt

- `char[] readPassword()`: Performs the same function as `readLine()` function but without echoing the typed characters
- `char[] readPassword(String format, Object... args)`: Performs the same function as `readLine(String format, Object... args)` but without echoing the typed characters

Let's demonstrate the preceding methods with the following example:

```
Console console = System.console();

String line = console.readLine();
System.out.println("Entered 1: " + line);
line = console.readLine("Enter something 2: ");
System.out.println("Entered 2: " + line);
line = console.readLine("Enter some%s", "thing 3: ");
System.out.println("Entered 3: " + line);

char[] password = console.readPassword();
System.out.println("Entered 4: " + new String(password));
password = console.readPassword("Enter password 5: ");
System.out.println("Entered 5: " + new String(password));
password = console.readPassword("Enter pass%s", "word 6: ");
System.out.println("Entered 6: " + new String(password));
```

The result of the preceding example is as follows:

```
abc
Entered 1: abc
Enter something 2: xyz
Entered 2: xyz
Enter something 3: 123
Entered 3: 123

Entered 4: abc
Enter password 5:
Entered 5: xyz
Enter password 6:
Entered 6: 123
```

Another group of `Console` class methods can be used in conjunction with the just demonstrated methods:

- `Console format(String format, Object... args)`: Substitutes the placeholders in the provided `format` string with the provided `args` values and displays the result
- `Console printf(String format, Object... args)`: Behaves the same way as the `format()` method

For example, look at the following line:

```
String line = console.format("Enter some%s", "thing:").readLine();
```

It produces the same result as this line:

```
String line = console.readLine("Enter some%s", "thing:");
```

And, finally, the last three methods of the Console class are as follows:

- `PrintWriter writer()`: Creates a `PrintWriter` object associated with this console that can be used for producing an output stream of characters
- `Reader reader()`: Creates a `Reader` object associated with this console that can be used for reading the input as a stream of characters
- `void flush()`: Flushes the console and forces any buffered output to be written immediately

Here is an example of their usage:

```
try (Reader reader = console.reader()){
    char[] chars = new char[10];
    System.out.print("Enter something: ");
    reader.read(chars);
    System.out.print("Entered: " + new String(chars));
} catch (IOException e) {
    e.printStackTrace();
}

PrintWriter out = console.writer();
out.println("Hello!");

console.flush();
```

The result of the preceding code looks as follows:

```
Enter something: Hi!
Entered: Hi!
Hello!
```

`Reader` and `PrintWriter` can also be used to create other `Input` and `Output` streams we have been talking about in this section.

StreamTokenizer

The `StreamTokenizer` class parses the input stream and produces tokens. Its `StreamTokenizer(Reader r)` constructor accepts a `Reader` object that is the source of the tokens. Every time the `int nextToken()` method is called on the `StreamTokenizer` object, the following happens:

1. The next token is parsed.
2. The `StreamTokenizer` instance field `ttype` is populated by the value that indicates the token type:
   - The `ttype` value can be one of the following integer constants: `TT_WORD`, `TT_NUMBER`, `TT_EOL` (end of line), or `TT_EOF` (end of stream).
   - If the `ttype` value is `TT_WORD`, the `StreamTokenizer` instance `sval` field is populated by the `String` value of the token.
   - If the `ttype` value is `TT_NUMBER`, the `StreamTokenizer` instance field `nval` is populated by the `double` value of the token.
3. The `lineno()` method of the `StreamTokenizer` instance returns the current line number.

Let's look at an example before talking about other methods of the `StreamTokenizer` class. Let's assume that, in the project `resources` folder, there is a `tokens.txt` file that contains the following four lines of text:

```
There
happened
42
events.
```

The following code will read the file and tokenize its content:

```
String filePath = "src/main/resources/tokens.txt";
try(FileReader fr = new FileReader(filePath);
 BufferedReader br = new BufferedReader(fr)){
 StreamTokenizer st = new StreamTokenizer(br);
    st.eolIsSignificant(true);
    st.commentChar('e');
    System.out.println("Line " + st.lineno() + ":");
    int i;
    while ((i = st.nextToken()) != StreamTokenizer.TT_EOF) {
        switch (i) {
            case StreamTokenizer.TT_EOL:
                System.out.println("\nLine " + st.lineno() + ":");
                break;
            case StreamTokenizer.TT_WORD:
                System.out.println("TT_WORD => " + st.sval);
                break;
            case StreamTokenizer.TT_NUMBER:
                System.out.println("TT_NUMBER => " + st.nval);
                break;
            default:
                System.out.println("Unexpected => " + st.ttype);
        }
    }
} catch (Exception ex){
    ex.printStackTrace();
}
```

If we run this code, the result will be the following:

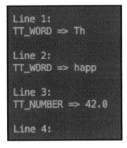

We have used the `BufferedReader` class, which is a good practice for higher efficiency, but in our case, we could easily avoid it like this:

```
FileReader fr = new FileReader(filePath);
StreamTokenizer st = new StreamTokenizer(fr);
```

The result would not change. We also used the following three methods we have not described yet:

- `void eolIsSignificant (boolean flag)`: Indicates whether the end-of-line should be treated as a token
- `void commentChar (int ch)`: Indicates which character starts a comment, so the rest of the line is ignored
- `int lineno ()`: Returns the current line number

The following methods can be invoked using the `StreamTokenizer` object:

- `void lowerCaseMode (boolean fl)`: Indicates whether a word token should be lowercased
- `void ordinaryChar (int ch)`, `void ordinaryChars (int low, int hi)`: Indicate a specific character or the range of characters that have to be treated as *ordinary* (not as a comment character, word component, string delimiter, white space, or number character)
- `void parseNumbers ()`: Indicates that a word token that has the format of a double precision floating-point number has to be interpreted as a number, rather than a word
- `void pushBack ()`: Forces the `nextToken ()` method to return the current value of the `ttype` field
- `void quoteChar (int ch)`: Indicates that the provided character has to be interpreted as the beginning and the end of the string value that has to be taken as-is (as a quote)
- `void resetSyntax ()`: Resets this tokenizer's syntax table so that all characters are *ordinary*
- `void slashSlashComments (boolean flag)`: Indicates that C++ style comments have to be recognized
- `void slashStarComments (boolean flag)`: Indicates that C style comments have to be recognized
- `String toString ()`: Returns the string representation of the token and the line number
  `void whitespaceChars (int low, int hi)`: Indicates the range of characters that have to be interpreted as white space
- `void wordChars (int low, int hi)`: Indicates the range of characters that have to be interpreted as a word

As you can see, using the wealth of the preceding methods allows fine-tuning of the text interpretation.

ObjectStreamClass and ObjectStreamField

The `ObjectStreamClass` and `ObjectStreamField` class provide access to the serialized data of a class loaded in the JVM. The `ObjectStreamClass` object can be found/created using one of the following lookup methods:

- `static ObjectStreamClass lookup(Class cl)`: Finds the descriptor of a serializable class
- `static ObjectStreamClass lookupAny(Class cl)`: Finds the descriptor for any class, whether serializable or not

After the `ObjectStreamClass` is found and the class is serializable (implements `Serializable` interface), it can be used to access `ObjectStreamField` objects, each containing information about one serialized field. If the class is not serializable, there is no `ObjectStreamField` object associated with any of the fields.

Let's look at an example. Here is the method that displays information obtained from the `ObjectStreamClass` and `ObjectStreamField` objects :

```
void printInfo(ObjectStreamClass osc) {
    System.out.println(osc.forClass());
    System.out.println("Class name: " + osc.getName());
    System.out.println("SerialVersionUID: " + osc.getSerialVersionUID());
    ObjectStreamField[] fields = osc.getFields();
    System.out.println("Serialized fields:");
    for (ObjectStreamField osf : fields) {
        System.out.println(osf.getName() + ": ");
        System.out.println("\t" + osf.getType());
        System.out.println("\t" + osf.getTypeCode());
        System.out.println("\t" + osf.getTypeString());
    }
}
```

To demonstrate how it works, we created a serializable `Person1` class:

```
package com.packt.learnjava.ch05_stringsIoStreams;
import java.io.Serializable;
public class Person1 implements Serializable {
    private int age;
    private String name;
```

```
    public Person1(int age, String name) {
        this.age = age;
        this.name = name;
    }
}
```

We did not add methods because only the object state is serializable, not the methods. Now let's run the following code:

```
ObjectStreamClass osc1 = ObjectStreamClass.lookup(Person1.class);
printInfo(osc1);
```

The result will be as follows:

```
class com.packt.learnjava.ch05_stringsIoStreams.Person1
Class name: com.packt.learnjava.ch05_stringsIoStreams.Person1
SerialVersionUID: -2546904836625458265
Serialized fields:
age:
    int
    I
    null
name:
    class java.lang.String
    L
    Ljava/lang/String;
```

As you can see, there is information about the class name and all field names and types. There are also two other methods that can be called using the `ObjectStreamField` object:

- `boolean isPrimitive()`: Returns `true` if this field has a primitive type
- `boolean isUnshared()`: Returns `true` if this field is unshared (private or accessible only from the same package)

Now let's create a non-serializable `Person2` class:

```
package com.packt.learnjava.ch05_stringsIoStreams;
public class Person2 {
    private int age;
    private String name;
    public Person2(int age, String name) {
        this.age = age;
        this.name = name;
    }
}
```

This time, we will run the code that only looks up the class as follows:

```
ObjectStreamClass osc2 = ObjectStreamClass.lookup(Person2.class);
System.out.println("osc2: " + osc2);      //prints: null
```

As has been expected, the non-serializable object was not found using the `lookup()` method. In order to find a non-serializable object, we need to use the `lookupAny()` method:

```
ObjectStreamClass osc3 = ObjectStreamClass.lookupAny(Person2.class);
printInfo(osc3);
```

If we run the preceding example, the result will be as follows:

```
class com.packt.learnjava.ch05_stringsIoStreams.Person2
Class name: com.packt.learnjava.ch05_stringsIoStreams.Person2
SerialVersionUID: 0
Serialized fields:
```

From a non-serializable object, we were able to extract information about the class, but not about the fields.

Class java.util.Scanner

The `java.util.Scanner` class is typically used to read an input from a keyboard but can read text from any object that implements the `Readable` interface (this interface only has `int read(CharBuffer buffer)` method). It breaks the input value by a delimiter (white space is a default delimiter) into tokens that are processed using different methods.

For example, we can read an input from `System.in`—a standard input stream, which typically represents the keyboard input:

```
Scanner sc = new Scanner(System.in);
System.out.print("Enter something: ");
while(sc.hasNext()){
    String line = sc.nextLine();
    if("end".equals(line)){
        System.exit(0);
    }
    System.out.println(line);
}
```

It accepts many lines (each line ends after the key *Enter* is pressed) until the line *end* is entered as follows:

```
Enter something: type
type
one line
one line
at a time
at a time
end

Process finished with exit code 0
```

Alternatively, Scanner can read lines from a file:

```
String filePath = "src/main/resources/tokens.txt";
try(Scanner sc = new Scanner(new File(filePath))){
    while(sc.hasNextLine()){
        System.out.println(sc.nextLine());
    }
} catch (Exception ex){
    ex.printStackTrace();
}
```

As you can see, we have used the tokens.txt file again. The results are as follows:

```
There
happened
42
events.
```

To demonstrate Scanner breaking the input by a delimiter, let's run the following code:

```
String input = "One two three";
Scanner sc = new Scanner(input);
while(sc.hasNext()){
    System.out.println(sc.next());
}
```

The result is as follows:

```
One
two
three
```

To use another delimiter, it can be set as follows:

```
String input = "One,two,three";
Scanner sc = new Scanner(input).useDelimiter(",");
while(sc.hasNext()){
    System.out.println(sc.next());
}
```

The result remains the same:

It is also possible to use a regular expression for extracting the tokens, but this topic is outside the scope of this book.

The `Scanner` class has many other methods that make its usage applicable to a variety of sources and required results. The `findInLine()`, `findWithinHorizon()`, `skip()`, and `findAll()` methods do not use the delimiter; they just try to match the provided pattern. For more information, refer to Scanner documentation (`https://docs.oracle.com/en/java/javase/12/docs/api/java.base/java/util/Scanner.html`).

File management

We have already used some methods for finding, creating, reading, and writing files using JCL classes. We had to do it in order to support a demo code of input/output streams. In this section, we are going to talk about file management using JCL in more detail.

The `File` class from the `java.io` package represents the underlying filesystem. An object of the `File` class can be created with one of the following constructors:

- `File(String pathname)`: Creates a new `File` instance based on the provided pathname
- `File(String parent, String child)`: Creates a new `File` instance based on the provided parent pathname and a child pathname
- `File(File parent, String child)`: Creates a new `File` instance based on the provided parent `File` object and a child pathname
- `File(URI uri)`: Creates a new `File` instance based on the provided `URI` object that represents the pathname

We will now see the examples of the constructors' usage while talking about creating and deleting files.

Creating and deleting files and directories

To create a file or directory in the filesystem, one needs first to construct a new `File` object using one of the constructors listed in the *Files management* section. For example, assuming that the file name is `FileName.txt`, the `File` object can be created as `new File("FileName.txt")`. If the file has to be created inside a directory, then either a path has to be added in front of the file name (when it is passed into the constructor) or one of the other three constructors has to be used. For example:

```
String path = "demo1" + File.separator + "demo2" + File.separator;
String fileName = "FileName.txt";
File f = new File(path + fileName);
```

Note the usage of `File.separator` instead of the slash symbol (/) or (\). That is because the `File.separator` returns the platform-specific slash symbol. And here is an example of another `File` constructor usage:

```
String path = "demo1" + File.separator + "demo2" + File.separator;
String fileName = "FileName.txt";
File f = new File(path, fileName);
```

Yet another constructor can be used as follows:

```
String path = "demo1" + File.separator + "demo2" + File.separator;
String fileName = "FileName.txt";
File f = new File(new File(path), fileName);
```

However, if you prefer or have to use a **Universal Resource Identifier (URI)**, you can construct a `File` object like this:

```
String path = "demo1" + File.separator + "demo2" + File.separator;
String fileName = "FileName.txt";
URI uri = new File(path + fileName).toURI();
File f = new File(uri);
```

Then, one of the following methods has to be invoked on the newly created `File` object:

- `boolean createNewFile()`: If a file with this name does not yet exist, creates a new file and returns `true`; otherwise, returns `false`

- static File createTempFile(String prefix, String suffix): Creates a file in the temporary-file directory
- static File createTempFile(String prefix, String suffix, File directory): Creates the directory; the provided prefix and suffix are used to generate the directory name

If the file you would like to create has to be placed inside a directory that does not exist yet, one of the following methods has to be used first, invoked on the File object that represents the filesystem path to the file:

- boolean mkdir(): creates the directory with the provided name
- boolean mkdirs(): Creates the directory with the provided name, including any necessary but nonexistent parent directories

And, before we look at a code example, we need to explain how the delete() method works:

- boolean delete(): Deletes the file or empty directory, which means you can delete the file but not all of the directories as follows:

```
String path = "demo1" + File.separator + "demo2" + File.separator;
String fileName = "FileName.txt";
File f = new File(path + fileName);
f.delete();
```

Let's look at how to overcome this limitation in the following example:

```
String path = "demo1" + File.separator + "demo2" + File.separator;
String fileName = "FileName.txt";
File f = new File(path + fileName);
try {
    new File(path).mkdirs();
    f.createNewFile();
    f.delete();
    path = StringUtils.substringBeforeLast(path, File.separator);
    while (new File(path).delete()) {
        path = StringUtils.substringBeforeLast(path,
File.separator);
    }
} catch (Exception e) {
    e.printStackTrace();
}
```

This example creates and deletes a file and all related directories. Notice our usage of the `org.apache.commons.lang3.StringUtils` class, which we have discussed in the *String utilities* section. It allowed us to remove from the path the just deleted directory and to continue doing it until all the nested directories are deleted, and the top level directory is deleted last.

Listing files and directories

The following methods can be used for listing directories and the files in them:

- `String[] list()`: Returns names of the files and directories in the directory
- `File[] listFiles()`: Returns `File` objects that represent the files and directories in the directory
- `static File[] listRoots()`: Lists the available filesystem roots

In order to demonstrate the preceding methods, let's assume we have created the directories and two files in them, as follows:

```
String path1 = "demo1" + File.separator;
String path2 = "demo2" + File.separator;
String path = path1 + path2;
File f1 = new File(path + "file1.txt");
File f2 = new File(path + "file2.txt");
File dir1 = new File(path1);
File dir = new File(path);
dir.mkdirs();
f1.createNewFile();
f2.createNewFile();
```

After that, we should be able to run the following code:

```
System.out.print("\ndir1.list(): ");
for(String d: dir1.list()){
    System.out.print(d + " ");
}
System.out.print("\ndir1.listFiles(): ");
for(File f: dir1.listFiles()){
    System.out.print(f + " ");
}
System.out.print("\ndir.list(): ");
for(String d: dir.list()){
    System.out.print(d + " ");
}
System.out.print("\ndir.listFiles(): ");
```

```
for(File f: dir.listFiles()){
    System.out.print(f + " ");
}
System.out.print("\nFile.listRoots(): ");
for(File f: File.listRoots()){
    System.out.print(f + " ");
}
```

The result should be as follows:

```
dir1.list(): demo2
dir1.listFiles(): demo1/demo2
dir.list(): file1.txt file2.txt
dir.listFiles(): demo1/demo2/file1.txt demo1/demo2/file2.txt
File.listRoots(): /
```

The demonstrated methods can be enhanced by adding the following filters to them, so they will list only the files and directories that match the filter:

- `String[] list(FilenameFilter filter)`
- `File[] listFiles(FileFilter filter)`
- `File[] listFiles(FilenameFilter filter)`

However, discussion of the file filters is outside the scope of this book.

Apache Commons utilities FileUtils and IOUtils

The popular companion of JCL is the Apache Commons project (`https://commons.apache.org`) that provides many libraries that compliment the JCL functionality. The classes of the `org.apache.commons.io` package are contained in the following root package and sub-packages:

- The `org.apache.commons.io` root package contains utility classes with static methods for common tasks, like the popular `FileUtils` and `IOUtils` classes described in the sections *Class FileUtils* and *Class IOUtils*, respectively.
- The `org.apache.commons.io.input` package contains classes that support input based on `InputStream` and `Reader` implementations, like `XmlStreamReader` or `ReversedLinesFileReader`.

- The `org.apache.commons.io.output` package contains classes that support output based on `OutputStream` and `Writer` implementations, like `XmlStreamWriter` or `StringBuilderWriter`.
- The `org.apache.commons.io.filefilter` package contains classes that serve as file filters, like `DirectoryFileFilter` or `RegexFileFilter`.
- The `org.apache.commons.io.comparator` package contains various implementations of `java.util.Comparator` for files, like `NameFileComparator`.
- The `org.apache.commons.io.serialization` package provides a framework for controlling the deserialization of classes.
- The `org.apache.commons.io.monitor` package allows monitoring filesystems and checking for a directory or file creating, updating, or deleting; one can launch the `FileAlterationMonitor` object as a thread and create an object of `FileAlterationObserver` that performs a check of the changes in the filesystem at the specified interval.

Refer to Apache Commons project documentation (`https://commons.apache.org`) for more details.

Class FileUtils

A popular `org.apache.commons.io.FileUtils` class allows doing all possible operations with files, as follows:

- Writing to a file
- Reading from a file
- Making a directory including parent directories
- Copying files and directories
- Deleting files and directories
- Converting to and from a URL
- Listing files and directories by filter and extension
- Comparing files content
- Getting a file last-changed date
- Calculating a checksum

If you plan to manage files and directories programmatically, it is imperative that you study the documentation of this class on the Apache Commons project website (`https://commons.apache.org/proper/commons-io/javadocs/api-2.5/org/apache/commons/io/FileUtils.html`).

Class IOUtils

The `org.apache.commons.io.IOUtils` is another very useful utility class that provides the following general IO streams manipulation methods:

- The `closeQuietly` methods that close a stream ignoring nulls and exceptions
- `toXxx/read` methods that read data from a stream
- `write` methods that write data to a stream
- `copy` methods that copy all the data from one stream to another
- `contentEquals` methods that compare the content of two streams

All the methods in this class that read a stream are buffered internally, so there is no need to use the `BufferedInputStream` or `BufferedReader` class. The `copy` methods all use `copyLarge` methods behind the scene that substantially increase their performance and efficiency.

This class is indispensable for managing the IO streams. See more details about this class and its methods on the Apache Commons project website (`https://commons.apache.org/proper/commons-io/javadocs/api-2.5/org/apache/commons/io/IOUtils.html`).

Summary

In this chapter, we have discussed the `String` class methods that allow analyzing strings, comparing, and transforming them. We have also discussed popular string utilities from JCL and the Apache Commons project. Two big sections of this chapter were dedicated to the input/output streams and the supporting classes in JCL and the Apache Commons project. The file managing classes and their methods were also discussed and demonstrated in specific code examples.

In the next chapter, we will present the Java collections framework and its three main interfaces, `List`, `Set`, and `Map`, including discussion and demonstration of generics. We will also discuss utility classes for managing arrays, objects, and time/date values.

Quiz

1. What does the following code print?

```
String str = "&8a!L";
System.out.println(str.indexOf("a!L"));
```

   a. 3

   b. 2

   c. 1

   d. 0

2. What does the following code print?

```
String s1 = "x12";
String s2 = new String("x12");
System.out.println(s1.equals(s2));
```

   a. Error

   b. Exception

   c. true

   d. false

3. What does the following code print?

```
System.out.println("%wx6".substring(2));
```

   a. wx

   b. x6

   c. %w

   d. Exception

4. What does the following code print?

```
System.out.println("ab"+"42".repeat(2));
```

   a. ab4242

   b. ab42ab42

   c. ab422

   d. Error

5. What does the following code print?

```
String s = "   ";
System.out.println(s.isBlank()+" "+s.isEmpty());
```

   a. false false
   b. false true
   c. true true
   d. true false

6. Select all correct statements:
   a. A stream can represent a data source
   b. An input stream can write to a file
   c. A stream can represent a data destination
   d. An output stream can display data on a screen

7. Select all correct statements about classes of java.io package:
   a. Reader extends InputStream
   b. Reader extends OutputStream
   c. Reader extends java.lang.Object
   d. Reader extends java.lang.Input

8. Select all correct statements about classes of java.io package:
   a. Writer extends FilterOutputStream
   b. Writer extends OutputStream
   c. Writer extends java.lang.Output
   d. Writer extends java.lang.Object

9. Select all correct statements about classes of java.io package:
   a. PrintStream extends FilterOutputStream
   b. PrintStream extends OutputStream
   c. PrintStream extends java.lang.Object
   d. PrintStream extends java.lang.Output

10. What does the following code do?

```
String path = "demo1" + File.separator + "demo2" + File.separator;
String fileName = "FileName.txt";
File f = new File(path, fileName);
try {
    new File(path).mkdir();
    f.createNewFile();
} catch (Exception e) {
    e.printStackTrace();
}
```

    a. Creates two directories and a file in `demo2` directory

    b. Creates one directory and a file in it

    c. Does not create any directory

    d. Exception

6
Data Structures, Generics, and Popular Utilities

This chapter presents the Java collections framework and its three main interfaces: `List`, `Set`, and `Map`, including a discussion and demonstration of generics. The `equals()` and `hashCode()` methods are also discussed in the context of Java collections. Utility classes for managing arrays, objects, and time/date values have corresponding dedicated sections too.

The following topics will be covered in this chapter:

- `List`, `Set`, and `Map` interfaces
- Collections utilities
- Arrays utilities
- Object utilities
- The `java.time` package

List, Set, and Map interfaces

The **Java collections framework** consists of the classes and interfaces that implement a collection data structure. Collections are similar to arrays in that respect as they can hold references to objects and can be managed as a group. The difference is that arrays require their capacity being defined before they can be used, while collections can increase and decrease their size automatically as needed. You just add or remove an object reference to a collection, and the collection changes its size accordingly. Another difference is that the collections cannot have their elements to be primitive types, such as `short`, `int`, or `double`. If you need to store such type values, the elements must be of a corresponding wrapper type, such as `Short`, `Integer`, or `Double`, for example.

Java collections support various algorithms of storing and accessing the elements of a collection: an ordered list, a unique set, a dictionary called a **map** in Java, a **stack**, a **queue**, and some others. All the classes and interfaces of the Java collections framework belong to the `java.util` package of the Java Class Library. The `java.util` package contains the following:

- The interfaces that extend the `Collection` interface: `List`, `Set`, and `Queue`, to name the most popular ones
- The classes that implement the previously listed interfaces: `ArrayList`, `HashSet`, `Stack`, `LinkedList`, and some others
- The `Map` interface and its sub-interfaces, `ConcurrentMap`, and `SortedMap`, to name a couple
- The classes that implement the `Map`-related interfaces: `HashMap`, `HashTable`, and `TreeMap`, to name the three most frequently used

To review all the classes and interfaces of the `java.util` package would require a dedicated book. So, in this section, we will just have a brief overview of the three main interfaces: `List`, `Set`, and `Map`—and one implementation class for each of them—`ArrayList`, `HashSet`, and `HashMap`. We start with methods that are shared by the `List` and `Set` interfaces. The principal difference between `List` and `Set` is that `Set` does not allow duplication of the elements. Another difference is that `List` preserves the order of the elements and also allows them to be sorted.

To identify an element inside a collection, the `equals()` method is used. To improve performance, the classes that implement the `Set` interface often use the `hashCode()` method too. This facilitates rapid calculation of an integer (called a **hash value** or **hash code**) that is, most of the time (but not always), unique to each element. The elements with the same hash value are placed in the same *bucket*. While establishing whether there is already a certain value in the set, it is enough to check the internal hash table and see whether such a value has already been used. If not, the new element is unique. If yes, then the new element can be compared (using the `equals()` method) with each of the elements with the same hash value. Such a procedure is faster than comparing a new element with each element of the set one by one.

That is why we often see that the name of the classes has a "hash" prefix, indicating that the class uses the hash value, so the element must implement the `hashCode()` method. While doing this, you must make sure that it is implemented so that every time the `equals()` method returns `true` for two objects, the hash values of these two objects returned by the `hashCode()` method are equal too. Otherwise, the just described algorithm of using the hash value will not work.

And finally, before talking about the `java.util` interfaces, a few words about generics.

Generics

You can see these most often in such declarations:

```
List<String> list = new ArrayList<String>();
Set<Integer> set = new HashSet<Integer>();
```

In the preceding examples, **generics** is the element nature declaration surrounded by the angle brackets. As you can see, they are redundant, as they are repeated in the left- and right-hand sides of the assignment statement. That is why Java allows replacement of the generics on the right side with empty brackets (<>) called a **diamond**:

```
List<String> list = new ArrayList<>();
Set<Integer> set = new HashSet<>();
```

Generics inform the compiler about the expected type of the collection elements. This way, the compiler can check whether the element a programmer tries to add to the declared collection is of a compatible type. Observe the following, for example:

```
List<String> list = new ArrayList<>();
list.add("abc");
list.add(42);    //compilation error
```

This helps to avoid runtime errors. It also tips off the programmer (because an IDE compiles the code when a programmer writes it) about possible manipulations of the collection elements.

We will also see these other types of generics:

- `<? extends T>` means *a type that is either T or a child of T*, where T is the type used as the generics of the collection.
- `<? super T>` means *a type T or any of its base (parent) class*, where T is the type used as the generics of the collection.

With that, let's start with the way an object of the class that implements the `List` or `Set` interface can be created, or, in other words, the `List` or `Set` type of variable can be initialized. To demonstrate the methods of these two interfaces, we will use two classes: an `ArrayList` (implements `List`) and `HashSet` (implements `Set`).

How to initialize List and Set

Since Java 9, the `List` or `Set` interfaces have static `of()` factory methods that can be used to initialize a collection:

- `of()`: Returns an empty collection.
- `of(E... e)`: Returns a collection with as many elements as are passed in during the call. They can be passed in a comma-separated list or as an array.

Here are a few examples:

```
//Collection<String> coll = List.of("s1", null); //does not allow null
Collection<String> coll = List.of("s1", "s1", "s2");
//coll.add("s3");                      //does not allow add element
//coll.remove("s1");                   //does not allow remove element
((List<String>) coll).set(1, "s3");   //does not allow modify element
System.out.println(coll);             //prints: [s1, s1, s2]

//coll = Set.of("s3", "s3", "s4");    //does not allow duplicate
//coll = Set.of("s2", "s3", null);    //does not allow null
coll = Set.of("s3", "s4");
System.out.println(coll);             //prints: [s3, s4]

//coll.add("s5");                      //does not allow add element
//coll.remove("s2");                   //does not allow remove
```

As you might expect, the factory method for `Set` does not allow duplicates, so we have commented the line out (otherwise, the preceding example would stop running at that line). What was less expected is that you cannot have a `null` element, and you cannot add/remove/modify elements of a collection after it was initialized using one of the `of()` methods. That's why we have commented out some lines of the preceding example. If you need to add elements after the collection was initialized, you have to initialize it using a constructor or some other utilities that create a modifiable collection (we will see an example of `Arrays.asList()` shortly).

The interface `Collection` provides two methods for adding elements to an object that implements `Collection` (the parent interface of `List` and `Set`) that looks as follows:

- `boolean add(E e)`: This attempts to add the provided element `e` to the collection; returns `true` in case of success, and `false` in case of not being able to accomplish it (for example, when such an element already exists in the `Set`).

- `boolean addAll(Collection<? extends E> c)`: This attempts to add all of the elements in the provided collection to the collection; it returns `true` if at least one element was added, and `false` in case of not able to add an element to the collection (for example, when all the elements of the provided collection c already exist in the Set).

Here is an example of using the `add()` method:

```
List<String> list1 = new ArrayList<>();
list1.add("s1");
list1.add("s1");
System.out.println(list1);        //prints: [s1, s1]

Set<String> set1 = new HashSet<>();
set1.add("s1");
set1.add("s1");
System.out.println(set1);        //prints: [s1]
```

And here is an example of using the `addAll()` method:

```
List<String> list1 = new ArrayList<>();
list1.add("s1");
list1.add("s1");
System.out.println(list1);        //prints: [s1, s1]

List<String> list2 = new ArrayList<>();
list2.addAll(list1);
System.out.println(list2);        //prints: [s1, s1]

Set<String> set = new HashSet<>();
set.addAll(list1);
System.out.println(set);         //prints: [s1]
```

The following is an example of the `add()` and `addAll()` methods' functionality:

```
List<String> list1 = new ArrayList<>();
list1.add("s1");
list1.add("s1");
System.out.println(list1);        //prints: [s1, s1]

List<String> list2 = new ArrayList<>();
list2.addAll(list1);
System.out.println(list2);        //prints: [s1, s1]

Set<String> set = new HashSet<>();
set.addAll(list1);
System.out.println(set);         //prints: [s1]
```

```
Set<String> set1 = new HashSet<>();
set1.add("s1");

Set<String> set2 = new HashSet<>();
set2.add("s1");
set2.add("s2");

System.out.println(set1.addAll(set2));  //prints: true
System.out.println(set1);                //prints: [s1, s2]
```

Notice how, in the last example in the preceding code snippet, the `set1.addAll(set2)` method returns `true`, although not all elements were added. To see the case of the `add()` and `addAll()` methods returning `false`, look at the following examples:

```
Set<String> set = new HashSet<>();
System.out.println(set.add("s1"));  //prints: true
System.out.println(set.add("s1"));  //prints: false
System.out.println(set);             //prints: [s1]

Set<String> set1 = new HashSet<>();
set1.add("s1");
set1.add("s2");

Set<String> set2 = new HashSet<>();
set2.add("s1");
set2.add("s2");

System.out.println(set1.addAll(set2));  //prints: false
System.out.println(set1);                //prints: [s1, s2]
```

The `ArrayList` and `HashSet` classes also have constructors that accept a collection:

```
Collection<String> list1 = List.of("s1", "s1", "s2");
System.out.println(list1);         //prints: [s1, s1, s2]

List<String> list2 = new ArrayList<>(list1);
System.out.println(list2);         //prints: [s1, s1, s2]

Set<String> set = new HashSet<>(list1);
System.out.println(set);           //prints: [s1, s2]

List<String> list3 = new ArrayList<>(set);
System.out.println(list3);         //prints: [s1, s2]
```

Now, after we have learned how a collection can be initialized, we can turn to other methods in the `List` and `Set` interfaces.

java.lang.Iterable interface

The `Collection` interface extends the `java.lang.Iterable` interface, which means that those classes that implement the `Collection` interface, directly or not, also implement the `java.lang.Iterable` interface. There are only three methods in the `Iterable` interface:

- `Iterator<T> iterator()`: This returns an object of a class that implements the `java.util.Iterator` interface; it allows the collection to be used in FOR statements, for example:

```
Iterable<String> list = List.of("s1", "s2", "s3");
System.out.println(list);        //prints: [s1, s2, s3]

for(String e: list){
    System.out.print(e + " ");   //prints: s1 s2 s3
}
```

- `default void forEach (Consumer<? super T> function)`: This applies the provided function of the `Consumer` type to each element of the collection until all elements have been processed or the function throws an exception. What function is, we will discuss this in Chapter 13, *Functional Programming*; for now, we will just provide an example:

```
Iterable<String> list = List.of("s1", "s2", "s3");
System.out.println(list);                        //prints: [s1, s2,
s3]
list.forEach(e -> System.out.print(e + " ")); //prints: s1 s2 s3
```

- `default Spliterator<T> splititerator()`: This returns an object of a class that implements the `java.util.Spliterator` interface; it is used primarily for implementing methods that allow parallel processing and is outside the scope of this book.

Collection interface

As we have mentioned already, the `List` and `Set` interfaces extend the `Collection` interface, which means that all the methods of the `Collection` interface are inherited by `List` and `Set`. These methods are as follows:

- `boolean add(E e)`: This attempts to add an element to the collection.

- `boolean addAll(Collection<? extends E> c)`: This attempts to add all of the elements in the collection provided.
- `boolean equals(Object o)`: This compares the collection with the o object provided; if the object provided is not a collection, this object returns `false`; otherwise, it compares the composition of the collection with the composition of the collection provided (as o object); in the case of `List`, it also compares the order of the elements. Let's illustrate this with a few examples:

```
Collection<String> list1 = List.of("s1", "s2", "s3");
System.out.println(list1);          //prints: [s1, s2, s3]

Collection<String> list2 = List.of("s1", "s2", "s3");
System.out.println(list2);          //prints: [s1, s2, s3]

System.out.println(list1.equals(list2));   //prints: true

Collection<String> list3 = List.of("s2", "s1", "s3");
System.out.println(list3);          //prints: [s2, s1, s3]

System.out.println(list1.equals(list3));   //prints: false

Collection<String> set1 = Set.of("s1", "s2", "s3");
System.out.println(set1);    //prints: [s2, s3, s1] or different
order

Collection<String> set2 = Set.of("s2", "s1", "s3");
System.out.println(set2);    //prints: [s2, s1, s3] or different
order

System.out.println(set1.equals(set2));   //prints: true

Collection<String> set3 = Set.of("s4", "s1", "s3");
System.out.println(set3);    //prints: [s4, s1, s3] or different
order

System.out.println(set1.equals(set3));   //prints: false
```

- `int hashCode()`: This returns the hash value for the collection; it is used in the case where the collection is an element of a collection that requires the `hashCode()` method implementation.

- `boolean isEmpty()`: This returns `true` if the collection does not have any elements.
- `int size()`: This returns the count of elements of the collection; when the `isEmpty()` method returns `true`, this method returns 0.
- `void clear()`: This removes all of the elements from the collection; after this method is called, the `isEmpty()` method returns `true`, and the `size()` method returns 0.
- `boolean contains(Object o)`: This returns `true` if the collection contains the provided object `o`; for this method to work correctly, each element of the collection and the provided object must implement `equals()` method and, in the case of `Set`, `hashCode()` method should be implemented.
- `boolean containsAll(Collection<?> c)`: This returns `true` if the collection contains all of the elements in the collection provided; for this method to work correctly, each element of the collection and each element of the collection provided must implement the `equals()` method and, in the case of `Set`, the `hashCode()` method should be implemented.
- `boolean remove(Object o)`: This attempts to remove the specified element from this collection and returns `true` if it was present; for this method to work correctly, each element of the collection and the object provided must implement the `equals()` method and, in the case of `Set`, the `hashCode()` method should be implemented
- `boolean removeAll(Collection<?> c)`: This attempts to remove from the collection all the elements of the collection provided; similar to the `addAll()` method, this method returns `true` if at least one of the elements was removed; otherwise, it returns `false`; for this method to work correctly, each element of the collection and each element of the collection provided must implement the `equals()` method and, in the case of `Set`, the `hashCode()` method should be implemented
- `default boolean removeIf(Predicate<? super E> filter)`: This attempts to remove from the collection all the elements that satisfy the given predicate; it is a function we are going to describe in Chapter 13, *Functional Programming*; it returns `true` if at least one element was removed

- `boolean retainAll(Collection<?> c)`: This attempts to retain in the collection just the elements contained in the collection provided; similar to the `addAll()` method, this method returns `true` if at least one of the elements is retained; otherwise, it returns `false`; for this method to work correctly, each element of the collection and each element of the collection provided must implement the `equals()` method and, in the case of `Set`, the `hashCode()` method should be implemented.

- `Object[] toArray()`, `T[] toArray(T[] a)`: This converts the collection to an array.

- `default T[] toArray(IntFunction<T[]> generator)`: This converts the collection to an array, using the function provided; we are going to explain functions in Chapter 13, *Functional Programming*.

- `default Stream<E> stream()`: This returns a `Stream` object (we talk about streams in Chapter 14, *Java Standard Streams*).

- `default Stream<E> parallelStream()`: This returns a possibly parallel `Stream` object (we talk about streams in Chapter 14, *Java Standard Streams*).

List interface

The `List` interface has several other methods that do not belong to any of its parent interfaces:

- Static factory `of()` methods described in the *How to initialize List and Set* subsection

- `void add(int index, E element)`: This inserts the element provided at the provided position in the list.

- `static List<E> copyOf(Collection<E> coll)`: This returns an unmodifiable `List` containing the elements of the given `Collection` and preserves their order; here is the code that demonstrates the functionality of this method:

```
Collection<String> list = List.of("s1", "s2", "s3");
System.out.println(list);          //prints: [s1, s2, s3]

List<String> list1 = List.copyOf(list);
//list1.add("s4");                  //run-time error
//list1.set(1, "s5");               //run-time error
//list1.remove("s1");               //run-time error
```

```
Set<String> set = new HashSet<>();
System.out.println(set.add("s1"));
System.out.println(set);            //prints: [s1]

Set<String> set1 = Set.copyOf(set);
//set1.add("s2");                    //run-time error
//set1.remove("s1");                 //run-time error

Set<String> set2 = Set.copyOf(list);
System.out.println(set2);           //prints: [s1, s2, s3]
```

- E get(int index): This returns the element located at the position specified in the list
- List<E> subList(int fromIndex, int toIndex): Extracts a sublist between the fromIndex, inclusive, and the toIndex, exclusive
- int indexOf(Object o): This returns the first index (position) of the specified element in the list; the first element in the list has an index (position), 0.
- int lastIndexOf(Object o): This returns the last index (position) of the specified element in the list; the final element in the list has list.size() - 1 index position.
- E remove(int index): This removes the element located at the specified position in the list; it returns the element removed.
- E set(int index, E element): This replaces the element located at the position specified in the list; it returns the element replaced.
- default void replaceAll(UnaryOperator<E> operator): This transforms the list by applying the function provided to each element; the UnaryOperator function will be described in Chapter 13, *Functional Programming.*
- ListIterator<E> listIterator(): Returns a ListIterator object that allows the list to be traversed backward.
- ListIterator<E> listIterator(int index): Returns a ListIterator object that allows the sublist (starting from the provided position) to be traversed backward. Observe the following, for example:

```
List<String> list = List.of("s1", "s2", "s3");
ListIterator<String> li = list.listIterator();
while(li.hasNext()){
    System.out.print(li.next() + " ");          //prints: s1 s2 s3
}
while(li.hasPrevious()){
    System.out.print(li.previous() + " ");      //prints: s3 s2 s1
}
```

```
ListIterator<String> li1 = list.listIterator(1);
while(li1.hasNext()){
    System.out.print(li1.next() + " ");          //prints: s2 s3
}
ListIterator<String> li2 = list.listIterator(1);
while(li2.hasPrevious()){
    System.out.print(li2.previous() + " ");      //prints: s1
}
```

- default void sort(Comparator<? super E> c): This sorts the list according to the order generated by the Comparator provided. Observe the following, for example:

```
List<String> list = new ArrayList<>();
list.add("S2");
list.add("s3");
list.add("s1");
System.out.println(list);                    //prints: [S2, s3, s1]

list.sort(String.CASE_INSENSITIVE_ORDER);
System.out.println(list);                    //prints: [s1, S2, s3]

//list.add(null);               //causes NullPointerException
list.sort(Comparator.naturalOrder());
System.out.println(list);                    //prints: [S2, s1, s3]

list.sort(Comparator.reverseOrder());
System.out.println(list);                    //prints: [s3, s1, S2]

list.add(null);
list.sort(Comparator.nullsFirst(Comparator.naturalOrder()));
System.out.println(list);             //prints: [null, S2, s1, s3]

list.sort(Comparator.nullsLast(Comparator.naturalOrder()));
System.out.println(list);             //prints: [S2, s1, s3, null]

Comparator<String> comparator = (s1, s2) ->
 s1 == null ? -1 : s1.compareTo(s2);
list.sort(comparator);
System.out.println(list);             //prints: [null, S2, s1, s3]
```

There are principally two ways to sort a list:

- Using Comparable interface implementation (called **natural order**)
- Using Comparator interface implementation

The `Comparable` interface only has a `compareTo()` method. In the preceding example, we have implemented the `Comparator` interface basing it on the `Comparable` interface implementation in the `String` class. As you can see, this implementation provided the same sort order as `Comparator.nullsFirst(Comparator.naturalOrder())`. This style of implementing is called **functional programming**, which we will discuss in more detail in `Chapter 13`, *Functional Programming*.

Set interface

The `Set` interface has the following methods that do not belong to any of its parent interfaces:

- Static `of()` factory methods described in the *How to initialize List and Set* subsection.
- The `static Set<E> copyOf(Collection<E> coll)` method: This returns an unmodifiable `Set` containing the elements of the given `Collection`; it works the same way as the `static <E> List<E> copyOf(Collection<E> coll)` method described in the *Interface List* section.

Map interface

The `Map` interface has many methods similar to the `List` and `Set` methods:

- `int size()`
- `void clear()`
- `int hashCode()`
- `boolean isEmpty()`
- `boolean equals(Object o)`
- `default void forEach(BiConsumer<K,V> action)`
- Static factory methods: `of()`, `of(K k, V v)`, `of(K k1, V v1, K k2, V v2)`, and many other methods besides

`Map` interface, however, does not extend `Iterable`, `Collection`, or any other interface for that matter. It is designed to be able to store **values** by their **keys**. Each key is unique. While several equal values can be stored with different keys on the same map. The combination of key and value constitutes an `Entry`, which is an internal interface of `Map`. Both value and key objects must implement the `equals()` method. A key object must also implement the `hashCode()` method.

Many methods of `Map` interface have exactly the same signature and functionality as in the `List` and `Set` interfaces, so we are not going to repeat them here. We will only walk through the `Map`-specific methods:

- `V get(Object key)`: This retrieves the value according to the key provided; it returns `null` if there is no such key.
- `Set<K> keySet()`: This retrieves all the keys from the map.
- `Collection<V> values()`: This retrieves all the values from the map.
- `boolean containsKey(Object key)`: This returns `true` if the key provided exists in the map.
- `boolean containsValue(Object value)`: This returns `true` if the value provided exists in the map.
- `V put(K key, V value)`: This adds the value and its key to the map; it returns the previous value stored with the same key.
- `void putAll(Map<K,V> m)`: This copies from the map provided all the key-value pairs.
- `default V putIfAbsent(K key, V value)`: This stores the value provided and maps to the key provided if such a key is not already used by the map; returns the value mapped to the key provided—either the existing or the new one.
- `V remove(Object key)`: This removes both the key and value from the map; it returns a value or `null` if there is no such key, or if the value is `null`.
- `default boolean remove(Object key, Object value)`: This removes the key-value pair from the map if such a pair exists in the map.
- `default V replace(K key, V value)`: This replaces the value if the key providedis currently mapped to the value provided; it returns the old value if it was replaced; otherwise, it returns `null`.
- `default boolean replace(K key, V oldValue, V newValue)`: This replaces the value `oldValue` with the `newValue` provided if the key provided is currently mapped to the `oldValue`; it returns `true` if the `oldValue` was replaced; otherwise, it returns `false`.
- `default void replaceAll(BiFunction<K,V,V> function)`: This applies the function provided to each key-value pair in the map and replaces it with the result, or throws an exception if this is not possible.

- Set<Map.Entry<K,V>> entrySet(): This returns a set of all key-value pairs as the objects of Map.Entry.
- default V getOrDefault(Object key, V defaultValue): This returns the value mapped to the key provided or the defaultValue if the map does not have the key provided.
- static Map.Entry<K,V> entry(K key, V value): This returns an unmodifiable Map.Entry object with the key and value provided in it.
- static Map<K,V> copy(Map<K,V> map): This converts the Map provided to an unmodifiable one.

The following Map methods are much too complicated for the scope of this book, so we are just mentioning them for the sake of completeness. They allow multiple values to be combined or calculated and aggregated in a single existing value in the Map, or a new one to be created:

- default V merge(K key, V value, BiFunction<V,V,V> remappingFunction): If the provided key-value pair exists and the value is not null, the provided function is used to calculate a new value; it removes the key-value pair if the newly calculated value is null; if the key-value pair provided does not exist or the value is null, the non-null value provided replaces the current one; this method can be used for aggregating several values; for example, it can be used for concatenating the following string values: map.merge(key, value, String::concat); we will explain what String::concat means in Chapter 13, *Functional Programming.*
- default V compute(K key, BiFunction<K,V,V> remappingFunction): It computes a new value using the function provided.
- default V computeIfAbsent(K key, Function<K,V> mappingFunction): It computes a new value using the function provided only if the provided key is not already associated with a value, or the value is null.
- default V computeIfPresent(K key, BiFunction<K,V,V> remappingFunction): This computes a new value using the function provided only if the provided key is already associated with a value and the value is not null.

This last group of *computing* and *merging* methods is rarely used. The most popular, by far, are the `V put(K key, V value)` and `V get(Object key)` methods, which allow the use of the main `Map` function of storing key-value pairs and retrieving the value using the key. The `Set<K> keySet()` method is often used for iterating over the map's key-value pairs, although the `entrySet()` method seems a more natural way of doing that. Here is an example:

```
Map<Integer, String> map = Map.of(1, "s1", 2, "s2", 3, "s3");

for(Integer key: map.keySet()){
    System.out.print(key + ", " + map.get(key) + ", ");
                                //prints: 3, s3, 2, s2, 1, s1,
}
for(Map.Entry e: map.entrySet()){
    System.out.print(e.getKey() + ", " + e.getValue() + ", ");
                                //prints: 2, s2, 3, s3, 1, s1,
}
```

The first of the `for` loops in the preceding code example uses the more widespread way to access the key-pair values of a map by iterating over the keys. The second `for` loop iterates over the set of entries, which, in our opinion, is a more natural way to do it. Notice that the printed-out values are not in the same order we have put them in the map. That is because, since Java 9, the unmodifiable collections (that is, what `of()` factory methods produce) have added randomization to the order of `Set` elements. It changes the order of the elements between different code executions. Such a design was done to make sure a programmer does not rely on a certain order of `Set` elements, which is not guaranteed for a set.

Unmodifiable collections

Please note that collections produced by the `of()` factory methods used to be called **immutable** in Java 9, and **unmodifiable** since Java 10. That is because an immutable implies that you cannot change anything in them, while, in fact, the collection elements can be changed if they are modifiable objects. For example, let's build a collection of objects of the `Person1` class that appears as follows:

```
class Person1 {
    private int age;
    private String name;
    public Person1(int age, String name) {
        this.age = age;
        this.name = name == null ? "" : name;
    }
```

```
    public void setName(String name){ this.name = name; }
    @Override
    public String toString() {
        return "Person{age=" + age +
                ", name=" + name + "}";
    }
}
```

For simplicity, we will create a list with one element only and will then try to modify the element:

```
Person1 p1 = new Person1(45, "Bill");
List<Person1> list = List.of(p1);
//list.add(new Person1(22, "Bob")); //UnsupportedOperationException
System.out.println(list);          //prints: [Person{age=45, name=Bill}]
p1.setName("Kelly");
System.out.println(list);          //prints: [Person{age=45, name=Kelly}]
```

As you can see, although it is not possible to add an element to the list created by the of() factory method, its element can still be modified if the reference to the element exists outside the list.

Collections utilities

There are two classes with static methods handling collections that are very popular and helpful:

- java.util.Collections
- org.apache.commons.collections4.CollectionUtils

The fact that the methods are static means they do not depend on the object state, so they are also called **stateless methods**, or **utilities methods**.

java.util.Collections class

There are many methods in the `Collections` class that manage collections, and analyze, sort, and compare them. There are more than 70 of them, so we do not have a chance to talk about all of them. Instead, we are going to look at the ones most often used by mainstream application developers:

- `static copy(List<T> dest, List<T> src)`: This copies elements of the `src` list to the `dest` list and preserves the order of the elements and their position in the list; the destination `dest` list size has to be equal to, or bigger than, the `src` list size, otherwise a runtime exception is raised; here is an example of this method usage:

  ```
  List<String> list1 = Arrays.asList("s1","s2");
  List<String> list2 = Arrays.asList("s3", "s4", "s5");
  Collections.copy(list2, list1);
  System.out.println(list2);     //prints: [s1, s2, s5]
  ```

- `static void sort(List<T> list)`: This sorts the list in the order according to the `compareTo(T)` method implemented by each element (called **natural ordering**); only accepts lists with elements that implement the `Comparable` interface (which requires implementation of the `compareTo(T)` method); in the example that follows, we use `List<String>` because the `String` class implements `Comparable`:

  ```
  //List<String> list = List.of("a", "X", "10", "20", "1", "2");
  List<String> list = Arrays.asList("a", "X", "10", "20", "1", "2");
  Collections.sort(list);
  System.out.println(list);            //prints: [1, 10, 2, 20, X, a]
  ```

Note that we could not use the `List.of()` method to create a list because the list would be unmodifiable and its order could not be changed. Also, look at the resulting order: numbers come first, then capital letters, followed by lowercase letters. That is because the `compareTo()` method in the `String` class uses code points of the characters to establish the order. Here is the code that demonstrates it:

```
List<String> list = Arrays.asList("a", "X", "10", "20", "1", "2");
Collections.sort(list);
System.out.println(list);      //prints: [1, 10, 2, 20, X, a]
list.forEach(s -> {
    for(int i = 0; i < s.length(); i++){
        System.out.print(" " + Character.codePointAt(s, i));
    }
```

```
        if(!s.equals("a")) {
            System.out.print(",");    //prints: 49, 49 48, 50, 50 48,
    88, 97
        }
});
```

As you can see, the order is defined by the value of the code points of the characters that compose the string.

- `static void sort(List<T> list, Comparator<T> comparator)`: This sorts the order of the list according to the `Comparator` object provided, irrespective of whether the list elements implement the `Comparable` interface or not; as an example, let's sort a list that consists of objects in the `Person` class:

```
class Person  {
    private int age;
    private String name;
    public Person(int age, String name) {
        this.age = age;
        this.name = name == null ? "" : name;
    }
    public int getAge() { return this.age; }
    public String getName() { return this.name; }
    @Override
    public String toString() {
        return "Person{name=" + name + ", age=" + age + "}";
    }
}
```

And here is a `Comparator` class to sort the list of `Person` objects:

```
class ComparePersons implements Comparator<Person> {
    public int compare(Person p1, Person p2){
        int result = p1.getName().compareTo(p2.getName());
        if (result != 0) { return result; }
        return p1.age - p2.getAge();
    }
}
```

Now, we can use the `Person` and `ComparePersons` classes as follows:

```
List<Person> persons = Arrays.asList(new Person(23, "Jack"),
        new Person(30, "Bob"), new Person(15, "Bob"));
Collections.sort(persons, new ComparePersons());
System.out.println(persons);    //prints: [Person{name=Bob,
age=15},
                                    Person{name=Bob,
```

```
age=30},
                                                   Person{name=Jack,
age=23}]
```

As we have mentioned already, there are many more utilities in the `Collections` class, so we recommend you look through the related documentation at least once and understand all its capabilities.

CollectionUtils class

The `org.apache.commons.collections4.CollectionUtils` class in the Apache commons project contains static stateless methods that complement the methods of the `java.util.Collections` class. They help to search, process, and compare Java collections.

To use this class, you would need to add the following dependency to the Maven `pom.xml` configuration file:

```
<dependency>
    <groupId>org.apache.commons</groupId>
    <artifactId>commons-collections4</artifactId>
    <version>4.1</version>
</dependency>
```

There are many methods in this class, and more methods will probably be added over time. These utilities are created in addition to the `Collections` methods, so they are more complex and nuanced and do not fit the scope of this book. To give you an idea of the methods available in the `CollectionUtils` class, here are some brief descriptions of the methods grouped according to their functionality:

- Methods that retrieve an element from a collection
- Methods that add an element or a group of elements to a collection
- Methods that merge `Iterable` elements into a collection
- Methods that remove or retain elements with or without criteria
- Methods that compare two collections
- Methods that transform a collection
- Methods that select from, and filter, a collection
- Methods that generate the union, intersection, or difference of two collections
- Methods that create an immutable empty collection
- Methods that check collection size and emptiness
- A method that reverses an array

This last method should probably belong to the utility class that handles arrays. And that is what we are going to discuss now.

Arrays utilities

There are two classes with static methods handling collections that are very popular and helpful:

- `java.util.Arrays`
- `org.apache.commons.lang3.ArrayUtils`

We will briefly review each of them.

java.util.Arrays class

We have already used the `java.util.Arrays` class several times. It is the primary utility class for arrays management. This utility class used to be very popular because of the `asList(T...a)` method. It was the most compact way of creating and initializing a collection:

```
List<String> list = Arrays.asList("s0", "s1");
Set<String> set = new HashSet<>(Arrays.asList("s0", "s1");
```

It is still a popular way of creating a modifiable list. We used it, too. However, after a `List.of()` factory method was introduced and the `Arrays` class declined substantially.

Nevertheless, if you need to manage arrays, then the `Arrays` class may be a big help. It contains more than 160 methods. Most of them are overloaded with different parameters and array types. If we group them by method name, there will be 21 groups. And if we further group them by functionality, only the following 10 groups will cover all the `Arrays` class functionality:

- `asList()`: This creates an `ArrayList` object based on the provided array or comma-separated list of parameters.
- `binarySearch()`: The searches an array or only the specified (according to the range of indices) part of it
- `compare()`, `mismatch()`, `equals()`, and `deepEquals()`: This compares two arrays or their elements (according to the range of indices).

- `copyOf()` and `copyOfRange()`: This copies all arrays or only the specified (according to the range of indices) part of it.
- `hashcode()` and `deepHashCode()`: This generates the hash code value based on the array provided.
- `toString()` and `deepToString()`: This creates a `String` representation of an array.
- `fill()`, `setAll()`, `parallelPrefix()`, and `parallelSetAll()`: This sets a value (fixed or generated by the function provided) for every element of an array or those specified according to the range of indices.
- `sort()` and `parallelSort()`: This sorts elements of an array or only part of it (specified according to the range of indices).
- `splititerator()`: This returns a `Splititerator` object for parallel processing of an array or part of it (specified according to the range of indices).
- `stream()`: This generates a stream of array elements or some of them (specified according to the range of indices); see Chapter 14, *Java Standard Streams*.

All of these methods are helpful, but we would like to draw your attention to `equals(a1, a2)` methods and `deepEquals(a1, a2)`. They are particularly helpful for the array comparison because an array object cannot implement an `equals()` custom method and uses the implementation of the `Object` class instead (that compares only references). The `equals(a1, a2)` and `deepEquals(a1, a2)` methods allow a comparison of not just a1 and a2 references, but use the `equals()` method to compare elements as well. The following are the code examples that demonstrate how these methods work:

```
String[] arr1 = {"s1", "s2"};
String[] arr2 = {"s1", "s2"};
System.out.println(arr1.equals(arr2));              //prints: false
System.out.println(Arrays.equals(arr1, arr2));      //prints: true
System.out.println(Arrays.deepEquals(arr1, arr2));  //prints: true

String[][] arr3 = {{"s1", "s2"}};
String[][] arr4 = {{"s1", "s2"}};
System.out.println(arr3.equals(arr4));              //prints: false
System.out.println(Arrays.equals(arr3, arr4));      //prints: false
System.out.println(Arrays.deepEquals(arr3, arr4));  //prints: true
```

As you can see, `Arrays.deepEquals()` returns `true` every time two equal arrays are compared when every element of one array equals the element of another array in the same position, while the `Arrays.equals()` method does the same, but for one-dimensional arrays only.

ArrayUtils class

The `org.apache.commons.lang3.ArrayUtils` class complements the `java.util.Arrays` class by adding new methods to the array managing the toolkit and the ability to handle `null` in cases when, otherwise, `NullPointerException` could be thrown. To use this class, you would need to add the following dependency to the Maven `pom.xml` configuration file:

```
<dependency>
    <groupId>org.apache.commons</groupId>
    <artifactId>commons-lang3</artifactId>
    <version>3.8.1</version>
</dependency>
```

The `ArrayUtils` class has around 300 overloaded methods that can be collected in the following 12 groups:

- `add()`, `addAll()`, and `insert()`: This adds elements to an array.
- `clone()`: This clones an array, similar to the `copyOf()` method of the `Arrays` class and the `arraycopy()` method of `java.lang.System`.
- `getLength()`: This returns an array length or 0, when the array itself is `null`.
- `hashCode()`: This calculates the hash value of an array, including nested arrays.
- `contains()`, `indexOf()`, and `lastIndexOf()`: The searches an array.
- `isSorted()`, `isEmpty`, and `isNotEmpty()`: The checks an array and handles `null`.
- `isSameLength()` and `isSameType()`: this compares arrays.
- `nullToEmpty()`: This converts a `null` array to an empty one.
- `remove()`, `removeAll()`, `removeElement()`, `removeElements()`, and `removeAllOccurances()`: This removes certain or all elements.
- `reverse()`, `shift()`, `shuffle()`, `swap()`: This changes the order of array elements.
- `subarray()`: This extracts part of an array according to the range of indices.
- `toMap()`, `toObject()`, `toPrimitive()`, `toString()`, `toStringArray()`: This converts an array to another type and handles `null` values.

Object utilities

The two utilities described in this section are as follows:

- `java.util.Objects`
- `org.apache.commons.lang3.ObjectUtils`

They are especially useful during class creation, so we will concentrate largely on the methods related to this task.

java.util.Objects class

The `Objects` class has only 17 methods that are all static. Let's look at some of them while applying them to the `Person` class. Let's assume this class will be an element of a collection, which means it has to implement the `equals()` and `hashCode()` methods:

```
class Person {
    private int age;
    private String name;
    public Person(int age, String name) {
        this.age = age;
        this.name = name;
    }
    public int getAge(){ return this.age; }
    public String getName(){ return this.name; }
    @Override
    public boolean equals(Object o) {
        if (this == o) return true;
        if (o == null) return false;
        if(!(o instanceof Person)) return false;
        Person person = (Person)o;
        return age == person.getAge() &&
                Objects.equals(name, person.getName());
    }
    @Override
    public int hashCode(){
        return Objects.hash(age, name);
    }
}
```

Notice that we do not check `name` property for `null` because `Object.equals()` does not break when any of the parameters is `null`. It just does the job of comparing the objects. If only one of them is `null`, it returns `false`. If both are null, it returns `true`.

Using `Object.equals()` is a safe way to implement the `equals()` method. But if you need to compare to objects that may be arrays, it is better to use the `Objects.deepEquals()` method because it not only handles `null`, as the `Object.equals()` method does, but also compares values of all array elements, even if the array is multidimensional:

```
String[][] x1 = {{"a","b"},{"x","y"}};
String[][] x2 = {{"a","b"},{"x","y"}};
String[][] y =  {{"a","b"},{"y","y"}};

System.out.println(Objects.equals(x1, x2));        //prints: false
System.out.println(Objects.equals(x1, y));         //prints: false
System.out.println(Objects.deepEquals(x1, x2));    //prints: true
System.out.println(Objects.deepEquals(x1, y));     //prints: false
```

The `Objects.hash()` method handles null values too. One important thing to remember is that the list of the properties compared in the `equals()` method has to match the list of the properties passed into `Objects.hash()` as parameters. Otherwise, two equal `Person` objects will have different hash values, which makes hash-based collections work incorrectly.

Another thing worth noticing is that there is another hash-related `Objects.hashCode()` method that accepts only one parameter. But the value it generates is not equal to the value generated by `Objects.hash()` with only one parameter. Observe the following, for example:

```
System.out.println(Objects.hash(42) == Objects.hashCode(42));
                                               //prints: false
System.out.println(Objects.hash("abc") == Objects.hashCode("abc"));
                                               //prints: false
```

To avoid this caveat, always use `Objects.hash()`.

Another potential source of confusion is demonstrated by the following code:

```
System.out.println(Objects.hash(null));        //prints: 0
System.out.println(Objects.hashCode(null));    //prints: 0
System.out.println(Objects.hash(0));           //prints: 31
System.out.println(Objects.hashCode(0));       //prints: 0
```

As you can see, the `Objects.hashCode()` method generates the same hash value for `null` and 0, which can be problematic for some algorithms based on the hash value.

`static <T> int compare (T a, T b, Comparator<T> c)` is another popular method that returns 0 (if the arguments are equal), otherwise the result of `c.compare(a, b)`. It is very useful for implementing the `Comparable` interface (establishing a natural order for a custom object sorting). Observe the following, for example:

```
class Person implements Comparable<Person> {
    private int age;
    private String name;
    public Person(int age, String name) {
        this.age = age;
        this.name = name;
    }
    public int getAge(){ return this.age; }
    public String getName(){ return this.name; }
    @Override
    public int compareTo(Person p){
        int result = Objects.compare(name, p.getName(),
                                    Comparator.naturalOrder());
        if (result != 0) {
            return result;
        }
        return Objects.compare(age, p.getAge(),
                                    Comparator.naturalOrder());
    }
}
```

This way, you can easily change the sorting algorithm by setting the `Comparator.reverseOrder()` value, or by adding `Comparator.nullFirst()` or `Comparator.nullLast()`.

Also, the `Comparator` implementation we used in the previous section can be made more flexible by using the `Objects.compare()` method:

```
class ComparePersons implements Comparator<Person> {
    public int compare(Person p1, Person p2){
        int result = Objects.compare(p1.getName(), p2.getName(),
                                    Comparator.naturalOrder());
        if (result != 0) {
            return result;
        }
        return Objects.compare(p1.getAge(), p2.getAge(),
                                    Comparator.naturalOrder());
    }
}
```

Finally, the last two methods of the `Objects` class that we are going to discuss are the methods that generate a string representation of an object. They come in handy when you need to call a `toString()` method on an object but are not sure if the object reference is `null`. Observe the following, for example:

```
List<String> list = Arrays.asList("s1", null);
for(String e: list){
    //String s = e.toString();   //NullPointerException
}
```

In the preceding example, we know the exact value of each element. But imagine a scenario where the list is passed into the method as a parameter. Then, we are forced to write something as follows:

```
void someMethod(List<String> list){
    for(String e: list){
        String s = e == null ? "null" : e.toString();
    }
}
```

It seems like this is not a big deal. But after writing such code a dozen times, a programmer naturally thinks about some kind of utility method that does all of that, and that is when the following two methods of the `Objects` class help:

- `static String toString(Object o)`: The returns the result of calling `toString()` on the parameter when it is not `null`, and returns `null` when the parameter value is `null`.

- `static String toString(Object o, String nullDefault)`: This returns the result of calling `toString()` on the first parameter when it is not `null`, and returns the second `nullDefault` parameter value when the first parameter value is `null`.

The following code demonstrates these two methods:

```
List<String> list = Arrays.asList("s1", null);
for(String e: list){
    String s = Objects.toString(e);
    System.out.print(s + " ");          //prints: s1 null
}
for(String e: list){
    String s = Objects.toString(e, "element was null");
    System.out.print(s + " ");          //prints: s1 element was null
}
```

As of the time of writing, the Objects class has 17 methods. We recommend you become familiar with them so as to avoid writing your own utilities in the event that the same utility already exists.

ObjectUtils class

The last statement of the previous section applies to the org.apache.commons.lang3.ObjectUtils class of the Apache Commons library that complements the methods of the java.util.Objects class described in the preceding section. The scope of this book and its allotted size does not allow for a detailed review of all the methods under the ObjectUtils class, so we will describe them briefly in groups according to their related functionality. To use this class, you would need to add the following dependency to the Maven pom.xml configuration file:

```
<dependency>
    <groupId>org.apache.commons</groupId>
    <artifactId>commons-lang3</artifactId>
    <version>3.8.1</version>
</dependency>
```

All the methods of the ObjectUtils class can be organized into seven groups:

- Object cloning methods.
- Methods that support a comparison of two objects.
- The notEqual() method, which compares two objects for inequality, where either one or both objects may be null.
- Several identityToString() methods that generate a String representation of the provided object as if produced by the toString(), which is a default method of the Object base class and, optionally, append it to another object.
- The allNotNull() and anyNotNull() methods, which analyze an array of objects for null
- The firstNonNull() and defaultIfNull() methods, which analyze an array of objects and return the first not-null object or default value.
- The max(), min(), median(), and mode() methods, which analyze an array of objects and return one of them that corresponds to the method name.

java.time package

There are many classes in the `java.time` package and its sub-packages. They were introduced as a replacement for other (older packages) that handled date and time. The new classes are thread-safe (hence, better suited for multithreaded processing) and what is also important is that they are more consistently designed and easier to understand. Also, the new implementation follows **International Standard Organization (ISO)** standards as regards the date and time formats, but allows any other custom format to be used as well.

We will describe the main five classes, and demonstrate how to use them:

- `java.time.LocalDate`
- `java.time.LocalTime`
- `java.time.LocalDateTime`
- `java.time.Period`
- `java.time.Duration`

All these, and other classes of the `java.time` package, as well as its sub-packages, are rich in various functionality that cover all practical cases. But we are not going to discuss all of them; we will just introduce the basics and the most popular use cases.

LocalDate class

`LocalDate` class does not carry time. It represents a date in ISO 8601 format (YYYY-MM-DD):

```
System.out.println(LocalDate.now()); //prints: 2019-03-04
```

That is the current date in this location at the time of writing. The value was picked up from the computer clock. Similarly, you can get the current date in any other time zone using that static `now(ZoneId zone)` method. The `ZoneId` object can be constructed using the static `ZoneId.of(String zoneId)` method, where `String zoneId` is any of the string values returned by the `ZonId.getAvailableZoneIds()` method:

```
Set<String> zoneIds = ZoneId.getAvailableZoneIds();
for(String zoneId: zoneIds){
    System.out.println(zoneId);
}
```

The preceding code prints almost 600 time zone IDs. Here are a few of them:

```
Asia/Aden
Etc/GMT+9
Africa/Nairobi
America/Marigot
Pacific/Honolulu
Australia/Hobart
Europe/London
America/Indiana/Petersburg
Asia/Yerevan
Europe/Brussels
GMT
Chile/Continental
Pacific/Yap
CET
Etc/GMT-1
Canada/Yukon
Atlantic/St_Helena
Libya
US/Pacific-New
Cuba
Israel
GB-Eire
GB
Mexico/General
Universal
Zulu
Iran
Navajo
Egypt
Etc/UTC
SystemV/AST4ADT
Asia/Tokyo
```

Let's try to use "`Asia/Tokyo`", for example:

```
ZoneId zoneId = ZoneId.of("Asia/Tokyo");
System.out.println(LocalDate.now(zoneId)); //prints: 2019-03-05
```

An object of `LocalDate` can represent any date in the past, or in the future too, using the following methods:

- `LocalDate parse(CharSequence text)`: This constructs an object from a string in ISO 8601 format (YYYY-MM-DD).

- LocalDate parse(CharSequence text, DateTimeFormatter formatter): This constructs an object from a string in a format specified by the DateTimeFormatter object that has a rich system of patterns and many predefined formats as well; the following are a few of them:

    - BASIC_ISO_DATE, for example, 20111203
    - ISO_LOCAL_DATE ISO, for example, 2011-12-03
    - ISO_OFFSET_DATE, for example, 2011-12-03+01:00
    - ISO_DATE, for example, 2011-12-03+01:00; 2011-12-03
    - ISO_LOCAL_TIME, for example, 10:15:30
    - ISO_OFFSET_TIME, for example, 10:15:30+01:00
    - ISO_TIME, for example, 10:15:30+01:00; 10:15:30
    - ISO_LOCAL_DATE_TIME, for example, 2011-12-03T10:15:30

- LocalDate of(int year, int month, int dayOfMonth): This constructs an object from a year, month, and day.
- LocalDate of(int year, Month month, int dayOfMonth): This constructs an object from a year, month (enum constant), and day.
- LocalDate ofYearDay(int year, int dayOfYear): This constructs an object form from a year and day-of-year.

The following code demonstrates the preceding methods listed:

```
LocalDate lc1 = LocalDate.parse("2020-02-23");
System.out.println(lc1);                    //prints: 2020-02-23

LocalDate lc2 =
        LocalDate.parse("20200223", DateTimeFormatter.BASIC_ISO_DATE);
System.out.println(lc2);                    //prints: 2020-02-23

DateTimeFormatter formatter = DateTimeFormatter.ofPattern("dd/MM/yyyy");
LocalDate lc3 =  LocalDate.parse("23/02/2020", formatter);
System.out.println(lc3);                    //prints: 2020-02-23

LocalDate lc4 =  LocalDate.of(2020, 2, 23);
System.out.println(lc4);                    //prints: 2020-02-23

LocalDate lc5 =  LocalDate.of(2020, Month.FEBRUARY, 23);
System.out.println(lc5);                    //prints: 2020-02-23

LocalDate lc6 = LocalDate.ofYearDay(2020, 54);
System.out.println(lc6);                    //prints: 2020-02-23
```

A `LocalDate` object can provide various values:

```
LocalDate lc = LocalDate.parse("2020-02-23");
System.out.println(lc);                    //prints: 2020-02-23
System.out.println(lc.getYear());          //prints: 2020
System.out.println(lc.getMonth());         //prints: FEBRUARY
System.out.println(lc.getMonthValue());    //prints: 2
System.out.println(lc.getDayOfMonth());    //prints: 23
System.out.println(lc.getDayOfWeek());     //prints: SUNDAY
System.out.println(lc.isLeapYear());       //prints: true
System.out.println(lc.lengthOfMonth());    //prints: 29
System.out.println(lc.lengthOfYear());     //prints: 366
```

The `LocalDate` object can be modified as follows:

```
LocalDate lc = LocalDate.parse("2020-02-23");
System.out.println(lc.withYear(2021)); //prints: 2021-02-23
System.out.println(lc.withMonth(5));        //prints: 2020-05-23
System.out.println(lc.withDayOfMonth(5));   //prints: 2020-02-05
System.out.println(lc.withDayOfYear(53));   //prints: 2020-02-22
System.out.println(lc.plusDays(10));        //prints: 2020-03-04
System.out.println(lc.plusMonths(2));       //prints: 2020-04-23
System.out.println(lc.plusYears(2));        //prints: 2022-02-23
System.out.println(lc.minusDays(10));       //prints: 2020-02-13
System.out.println(lc.minusMonths(2));      //prints: 2019-12-23
System.out.println(lc.minusYears(2));       //prints: 2018-02-23
```

The `LocalDate` objects can be compared as follows:

```
LocalDate lc1 = LocalDate.parse("2020-02-23");
LocalDate lc2 = LocalDate.parse("2020-02-22");
System.out.println(lc1.isAfter(lc2));       //prints: true
System.out.println(lc1.isBefore(lc2));      //prints: false
```

There are many other helpful methods in the `LocalDate` class. If you have to work with dates, we recommend that you read the API of this class and other classes of the `java.time` package and its sub-packages.

LocalTime class

The `LocalTime` class contains time without a date. It has similar methods to the methods of the `LocalDate` class. Here is how an object of the `LocalTime` class can be created:

```
System.out.println(LocalTime.now());        //prints: 21:15:46.360904

ZoneId zoneId = ZoneId.of("Asia/Tokyo");
```

```
System.out.println(LocalTime.now(zoneId));     //prints: 12:15:46.364378

LocalTime lt1 =  LocalTime.parse("20:23:12");
System.out.println(lt1);                        //prints: 20:23:12

LocalTime lt2 = LocalTime.of(20, 23, 12);
System.out.println(lt2);                        //prints: 20:23:12
```

Each component of time value can be extracted from a `LocalTime` object as follows:

```
LocalTime lt2 =  LocalTime.of(20, 23, 12);
System.out.println(lt2);                        //prints: 20:23:12

System.out.println(lt2.getHour());              //prints: 20
System.out.println(lt2.getMinute());            //prints: 23
System.out.println(lt2.getSecond());            //prints: 12
System.out.println(lt2.getNano());              //prints: 0
```

The object of the `LocalTime` class can be modified as follows:

```
LocalTime lt2 = LocalTime.of(20, 23, 12);
System.out.println(lt2.withHour(3));   //prints: 03:23:12
System.out.println(lt2.withMinute(10)); //prints: 20:10:12
System.out.println(lt2.withSecond(15)); //prints: 20:23:15
System.out.println(lt2.withNano(300)); //prints: 20:23:12.000000300
System.out.println(lt2.plusHours(10));         //prints: 06:23:12
System.out.println(lt2.plusMinutes(2));        //prints: 20:25:12
System.out.println(lt2.plusSeconds(2));        //prints: 20:23:14
System.out.println(lt2.plusNanos(200));        //prints: 20:23:12.000000200
System.out.println(lt2.minusHours(10));        //prints: 10:23:12
System.out.println(lt2.minusMinutes(2));       //prints: 20:21:12
System.out.println(lt2.minusSeconds(2));       //prints: 20:23:10
System.out.println(lt2.minusNanos(200));       //prints: 20:23:11.999999800
```

And two objects of the `LocalTime` class can also be compared, as follows:

```
LocalTime lt2 =  LocalTime.of(20, 23, 12);
LocalTime lt4 =  LocalTime.parse("20:25:12");
System.out.println(lt2.isAfter(lt4));        //prints: false
System.out.println(lt2.isBefore(lt4));       //prints: true
```

There are many other helpful methods in the `LocalTime` class. If you have to work with dates, we recommend that you read the API of this class and other classes of the `java.time` package and its sub-packages.

LocalDateTime class

The `LocalDateTime` class contains both the date and time and has all the methods
the `LocalDate` and `LocalTime` classes have, so we are not going to repeat them here. We
will only show how an object of the `LocalDateTime` class can be created:

```
System.out.println(LocalDateTime.now());
                              //prints: 2019-03-04T21:59:00.142804
ZoneId zoneId = ZoneId.of("Asia/Tokyo");
System.out.println(LocalDateTime.now(zoneId));
                              //prints: 2019-03-05T12:59:00.146038
LocalDateTime ldt1 = LocalDateTime.parse("2020-02-23T20:23:12");
System.out.println(ldt1);             //prints: 2020-02-23T20:23:12
DateTimeFormatter formatter =
      DateTimeFormatter.ofPattern("dd/MM/yyyy HH:mm:ss");
LocalDateTime ldt2 =
      LocalDateTime.parse("23/02/2020 20:23:12", formatter);
System.out.println(ldt2);             //prints: 2020-02-23T20:23:12
LocalDateTime ldt3 = LocalDateTime.of(2020, 2, 23, 20, 23, 12);
System.out.println(ldt3);             //prints: 2020-02-23T20:23:12
LocalDateTime ldt4 =
      LocalDateTime.of(2020, Month.FEBRUARY, 23, 20, 23, 12);
System.out.println(ldt4);             //prints: 2020-02-23T20:23:12

LocalDate ld = LocalDate.of(2020, 2, 23);
LocalTime lt = LocalTime.of(20, 23, 12);
LocalDateTime ldt5 = LocalDateTime.of(ld, lt);
System.out.println(ldt5);             //prints: 2020-02-23T20:23:12
```

There are many other helpful methods in the `LocalDateTime` class. If you have to work
with dates, we recommend that you read the API of this class and other classes of
the `java.time` package and its sub-packages.

Period and Duration classes

The `java.time.Period` and `java.time.Duration` classes are designed to contain an
amount of time:

- A `Period` object contains an amount of time in units of years, months, and days.
- A `Duration` object contains an amount of time in hours, minutes, seconds, and
  nanoseconds.

The following code demonstrates their creation and usage using the
`LocalDateTime` class, but the same methods exist in the `LocalDate` (for `Period`) and
`LocalTime` (for `Duration`) classes:

```
LocalDateTime ldt1 = LocalDateTime.parse("2020-02-23T20:23:12");
LocalDateTime ldt2 = ldt1.plus(Period.ofYears(2));
System.out.println(ldt2);        //prints: 2022-02-23T20:23:12
```

The following methods work the same way:

```
LocalDateTime ldt = LocalDateTime.parse("2020-02-23T20:23:12");
ldt.minus(Period.ofYears(2));
ldt.plus(Period.ofMonths(2));
ldt.minus(Period.ofMonths(2));
ldt.plus(Period.ofWeeks(2));
ldt.minus(Period.ofWeeks(2));
ldt.plus(Period.ofDays(2));
ldt.minus(Period.ofDays(2));
ldt.plus(Duration.ofHours(2));
ldt.minus(Duration.ofHours(2));
ldt.plus(Duration.ofMinutes(2));
ldt.minus(Duration.ofMinutes(2));
ldt.plus(Duration.ofMillis(2));
ldt.minus(Duration.ofMillis(2));
```

Some other methods of creating and using `Period` objects are demonstrated by the
following code:

```
LocalDate ld1 =  LocalDate.parse("2020-02-23");
LocalDate ld2 =  LocalDate.parse("2020-03-25");
Period period = Period.between(ld1, ld2);
System.out.println(period.getDays());        //prints: 2
System.out.println(period.getMonths());      //prints: 1
System.out.println(period.getYears());       //prints: 0
System.out.println(period.toTotalMonths());  //prints: 1
period = Period.between(ld2, ld1);
System.out.println(period.getDays());        //prints: -2
```

`Duration` objects can be similarly created and used:

```
LocalTime lt1 =  LocalTime.parse("10:23:12");
LocalTime lt2 =  LocalTime.parse("20:23:14");
Duration duration = Duration.between(lt1, lt2);
System.out.println(duration.toDays());       //prints: 0
System.out.println(duration.toHours());      //prints: 10
System.out.println(duration.toMinutes());    //prints: 600
System.out.println(duration.toSeconds());    //prints: 36002
System.out.println(duration.getSeconds());   //prints: 36002
```

```
System.out.println(duration.toNanos());    //prints: 3600200000000
System.out.println(duration.getNano());     //prints: 0
```

There are many other helpful methods in `Period` and `Duration` classes. If you have to work with dates, we recommend that you read the API of this class and other classes of the `java.time` package and its sub-packages.

Summary

This chapter introduced the reader to the Java collections framework and its three main interfaces: `List`, `Set`, and `Map`. Each of the interfaces was discussed and its methods demonstrated with one of the implementing classes. The generics were explained and demonstrated as well. The `equals()` and `hashCode()` methods have to be implemented in order for the object to be capable of being handled by Java collections correctly.

The `Collections` and `CollectionUtils` utility classes have many useful methods for collection handling and were presented in examples, along with `Arrays`, `ArrayUtils`, `Objects`, and `ObjectUtils`.

The class methods of the `java.time` package allow time/date values to be managed, class was demonstrated in specific practical code snippets.

In the next chapter, we will overview the Java Class Library and some external libraries, including those that support testing. Specifically, we will explore the `org.junit`, `org.mockito`, `org.apache.log4j`, `org.slf4j`, and `org.apache.commons` packages and their sub-packages.

Quiz

1. What is a Java collections framework? Select all that apply:
    a. A collection of frameworks
    b. Classes and interfaces of the `java.util` package
    c. `List`, `Set`, and `Map` Interfaces
    d. Classes and interfaces that implement a collection data structure

2. What is generics in a collection? Select all that apply:

    a. A collection structure definition

    b. The element type declaration

    c. The type generalization

    d. The mechanism that provides compile-time safety

3. What are the limitations of the collection of `of()` factory methods? Select all that apply:

    a. They do not allow a `null` element.

    b. They do not allow elements to be added to the initialized collection.

    c. They do not allow elements to be added to the initialized collection.

    d. They do not allow modification of elements in relation to the initialized collection.

4. What does the implementation of the `java.lang.Iterable` interface allow? Select all that apply:

    a. It allows elements of the collection to be accessed one by one.

    b. It allows the collection to be used in `FOR` statements.

    c. It allows the collection to be used in `WHILE` statements.

    d. It allows the collection to be used in `DO...WHILE` statements.

5. What does the implementation of the `java.util.Collection` interface allow? Select all that apply:

    a. Addition to the collection of elements from another collection

    b. Removal from the collection of objects that are elements of another collection

    c. Modification of just only those elements of the collection that belong to another collection

    d. Removal from the collection of objects that do not belong to another collection

6. Select all the correct statements pertaining to the `List` interface methods:

    a. `E get(int index)`: This returns the element at the specified position in the list.

    b. `E remove(int index)`: This removes the element at the specified position in the list; it returns the removed element.

    c. `static List<E> copyOf(Collection<E> coll)`: This returns an unmodifiable `List` containing the elements of the given `Collection` and preserves their order.

    d. `int indexOf(Object o)`: This returns the position of the specified element in the list.

7. Select all the correct statements pertaining to the `Set` interface methods:

    a. `E get(int index)`: This returns the element at the specified position in the list.

    b. `E remove(int index)`: This removes the element at the specified position in the list; it returns the removed element.

    c. `static Set<E> copyOf(Collection<E> coll)`: This returns an unmodifiable `Set` containing the elements of the given `Collection`.

    d. `int indexOf(Object o)`: This returns the position of the specified element in the list.

8. Select all the correct statements pertaining to the `Map` interface methods:

    a. `int size()`: This returns the count of key-value pairs stored in the map; when the `isEmpty()` method returns `true`, this method returns 0.

    b. `V remove(Object key)`: This removes both key and value from the map; returns value, or `null` if there is no such key or the value is `null`.

    c. `default boolean remove(Object key, Object value)`: This removes the key-value pair if such a pair exists in the map; returns `true` if the value is removed.

    d. `default boolean replace(K key, V oldValue, V newValue)`: This replaces the `oldValue` value with the `newValue` provided if the key provided is currently mapped to the `oldValue`; it returns `true` if the `oldValue` was replaced; otherwise, it returns `false`.

9. Select all correct statements pertaining to the `static void sort (List<T>` `list, Comparator<T> comparator)` method of the `Collections` class:

    a. It sorts the list's natural order if the list elements implement the `Comparable` interface.

    b. It sorts the list's order according to the `Comparator` object provided.

    c. It sorts the list's order according to the `Comparator` object provided if list elements implement the `Comparable` interface.

    d. It sorts the list's order according to the provided `Comparator` object irrespective of whether the list elements implement the `Comparable` interface.

10. What is the outcome of executing the following code?

```
List<String> list1 = Arrays.asList("s1","s2", "s3");
List<String> list2 = Arrays.asList("s3", "s4");
Collections.copy(list1, list2);
System.out.println(list1);
```

    a. `[s1, s2, s3, s4]`

    b. `[s3, s4, s3]`

    c. `[s1, s2, s3, s3, s4]`

    d. `[s3, s4]`

11. What is the functionality of the `CollectionUtils` class methods? Select all that apply:

    a. It matches the functionality of the `Collections` class methods, but by handling `null`.

    b. It complements the functionality of the `Collections` class methods.

    c. It searches, processes, and compares Java collections in a way that the `Collections` class methods do not do.

    d. It duplicates the functionality of the `Collections` class methods.

12. What is the result of executing the following code?

```
Integer[][] ar1 = {{42}};
Integer[][] ar2 = {{42}};
System.out.print(Arrays.equals(ar1, ar2) + " ");
System.out.println(Arrays.deepEquals(arr3, arr4));
```

   a. false true
   b. false false
   c. true false
   d. true true

13. What is the result of executing the following code?

```
String[] arr1 = { "s1", "s2" };
String[] arr2 = { null };
String[] arr3 = null;
System.out.print(ArrayUtils.getLength(arr1) + " ");
System.out.print(ArrayUtils.getLength(arr2) + " ");
System.out.print(ArrayUtils.getLength(arr3) + " ");
System.out.print(ArrayUtils.isEmpty(arr2) + " ");
System.out.print(ArrayUtils.isEmpty(arr3));
```

   a. 1 2 0 false true
   b. 2 1 1 false true
   c. 2 1 0 false true
   d. 2 1 0 true false

14. What is the result of executing the following code?

```
String str1 = "";
String str2 = null;
System.out.print((Objects.hash(str1) ==
                Objects.hashCode(str2)) + " ");
System.out.print(Objects.hash(str1) + " ");
System.out.println(Objects.hashCode(str2) + " ");
```

   a. true 0 0
   b. Error
   c. false -1 0
   d. false 31 0

15. What is the result of executing the following code?

```
String[] arr = {"c", "x", "a"};
System.out.print(ObjectUtils.min(arr) + " ");
System.out.print(ObjectUtils.median(arr) + " ");
System.out.println(ObjectUtils.max(arr));
```

    a. c x a

    b. a c x

    c. x c a

    d. a x c

16. What is the result of executing the following code?

```
LocalDate lc = LocalDate.parse("1900-02-23");
System.out.println(lc.withYear(21));
```

    a. 1921-02-23

    b. 21-02-23

    c. 0021-02-23

    d. Error

17. What is the result of executing the following code?

```
LocalTime lt2 = LocalTime.of(20, 23, 12);
System.out.println(lt2.withNano(300));
```

    a. 20:23:12.000000300

    b. 20:23:12.300

    c. 20:23:12:300

    d. Error

18. What is the result of executing the following code?

```
LocalDate ld = LocalDate.of(2020, 2, 23);
LocalTime lt = LocalTime.of(20, 23, 12);
LocalDateTime ldt = LocalDateTime.of(ld, lt);
System.out.println(ldt);
```

    a. 2020-02-23 20:23:12

    b. 2020-02-23T20:23:12

    c. 2020-02-23:20:23:12

    d. Error

19. What is the result of executing the following code?

```
LocalDateTime ldt = LocalDateTime.parse("2020-02-23T20:23:12");
System.out.print(ldt.minus(Period.ofYears(2)) + " ");
System.out.print(ldt.plus(Duration.ofMinutes(12)) + " ");
System.out.println(ldt);
```

  a. `2020-02-23T20:23:12 2020-02-23T20:23:12`
   `2020-02-23T20:23:12`

  b. `2020-02-23T20:23:12 2020-02-23T20:35:12`
   `2020-02-23T20:35:12`

  c. `2018-02-23T20:23:12 2020-02-23T20:35:12`
   `2020-02-23T20:23:12`

  d. `2018-02-23T20:23:12 2020-02-23T20:35:12`
   `2018-02-23T20:35:12`

7
Java Standard and External Libraries

It is not possible to write a Java program without using the standard libraries, also called **Java Class Library** (**JCL**). That is why a solid familiarity with such libraries is as vital for successful programming as the knowledge of the language itself.

There are also *non-standard* libraries, called **external libraries** or **third-party libraries** because they are not included in **Java Development Kit** (**JDK**) distribution. Some of them have long become a permanent fixture of any programmer toolkit.

To keep track of all the functionality available in these libraries is not easy. That is because an **Integrated Development Environment** (**IDE**) gives you a hint about the language possibilities, but it cannot advise about the functionality of a package not imported yet. The only package that is imported automatically is `java.lang`.

This purpose of this chapter is to provide you with an overview of the functionality of the most popular packages of JCL and external libraries.

The topics we are discussing in this chapter are as follows:

- Java Class Library (JCL)
- `java.lang`
- `java.util`
- `java.time`
- `java.io` and `java.nio`
- `java.sql` and `javax.sql`
- `java.net`
- `java.lang.math` and `java.math`
- `java.awt`, `javax.swing`, and `javafx`

- External libraries
- `org.junit`
- `org.mockito`
- `org.apache.log4j` and `org.slf4j`
- `org.apache.commons`

Java Class Library

JCL is a collection of packages that implement the language. In simpler terms, it is a collection of the `.class` files included in the JDK and ready to be used. Once you have installed Java, you get them as part of the installation and can start building your application code up using the classes of JCL as building blocks that take care of a lot of low-level plumbing. The JCL richness and ease of usage has substantially contributed to Java popularity.

In order to use a JCL package, one can import it without adding a new dependency to the `pom.xml` file. Maven adds JCL to the classpath automatically. And that is what separates standard library and external libraries; if you need to add a library (typically, a `.jar` file) as a dependency in the Maven `pom.xml` configuration file, this library is an external one. Otherwise, it is a standard library or JCL.

Some JCL package names start with `java`. Traditionally, they are called **core Java packages**, while those that start with `javax` used to be called "extensions." It was done so probably because the extensions were thought to be optional and maybe even released independently of JDK. There was also an attempt to promote the former extension library to become a core package. But that would require changing the package name from "java" to "javax," which would break the already existing applications that used the `javax` package. Therefore, the idea was abandoned, so the distinction between core and extensions gradually disappeared.

That is why, if you look at the official Java API on the Oracle website, you will see listed as standard not only `java` and `javax` packages, but also `jdk`, `com.sun`, `org.xml`, and some other packages too. These extra packages are primarily used by the tools or other specialized applications. In our book, we will concentrate mostly on the mainstream Java programming and talk only about the `java` and `javax` packages.

java.lang

This package is so fundamental that is not required to be imported in order to use it. The JVM authors decided to import it automatically. It contains the most often used classes of JCL:

- `Object` class: The base class of any other Java class
- `Class` class: Carries metadata of every loaded class at runtime
- `String`, `StringBuffer`, and `StringBuilder` classes: Support operations with type `String`
- The wrapper classes of all primitive types: `Byte`, `Boolean`, `Short`, `Character`, `Integer`, `Long`, `Float`, and `Double`
- `Number` class: The base class for the wrapper classes of the numeric primitive types—all the previously listed, except `Boolean`
- `System` class: Provides access to important system operations and the standard input and output (we have used the `System.out` object in every code example in this book)
- `Runtime` class: Provides access to the execution environment
- The `Thread` and `Runnable` interfaces: Fundamental for creating Java threads
- The `Iterable` interface: Used by the iteration statements
- `Math` class: Provides methods for basic numeric operations
- `Throwable` class: The base class for all exceptions
- `Error` class: An exception class; all its children are used to communicate system errors that can't be caught by an application
- `Exception` class: This class and its direct children represent checked exceptions
- `RuntimeException` class: This class and its children that represent unchecked exceptions, also called runtime exceptions
- `ClassLoader` class: Reads the `.class` files and puts (loads) them into memory; it also can be used to build a customized class-loader
- `Process` and `ProcessBuilder` classes: Allow creating other JVM processes
- Many other useful classes and interfaces

java.util

Most of the content of the `java.util` package is dedicated to supporting Java collections:

- The `Collection` interface: The base interface of many other interfaces of collections, it declares all the basic methods necessary to manage collection elements: `size()`, `add()`, `remove()`, `contains()`, `stream()`, and others; it also extends the `java.lang.Iterable` interface and inherits its methods, including `iterator()` and `forEach()`, which means that any implementation of the `Collection` interface or any of its children—List, Set, Queue, Deque, and others—can be used in iteration statements too: `ArrayList`, `LinkedList`, `HashSet`, `AbstractQueue`, `ArrayDeque`, and others.
- The `Map` interface and classes that implement it: `HashMap`, `TreeMap`, and others.
- The `Collections` class: Provides many static methods to analyze, manipulate, and convert collections.
- Many other collection interfaces, classes, and related utilities.

We have talked about Java collections and saw examples of their usage in Chapter 6, *Data Structures, Generics, and Popular Utilities*.

The `java.util` package also includes several other useful classes:

- `Objects`: Provides various object-related utility methods, some of which we have over-viewed in Chapter 6, *Data Structures, Generics, and Popular Utilities*.
- `Arrays`: Contains some 160 static methods to manipulate arrays, some of which we have over-viewed in Chapter 6, *Data Structures, Generics, and Popular Utilities*.
- `Formatter`: Allows formatting of any primitive type, `String`, `Date`, and other types; we have demonstrated the examples of its usage in Chapter 6, *Data Structures, Generics, and Popular Utilities*.
- `Optional`, `OptionalInt`, `OptionalLong`, and `OptionalDouble`: These classes help to avoid `NullPointerException` by wrapping the actual value that can be `null` or not.
- `Properties`: Helps to read, and create key-value pairs used for application configuration, and similar purposes.
- `Random`: Complements `java.lang.Math.random()` method by generating streams of pseudo-random numbers.
- `StringTokenizer`: Breaks the `String` object into the tokens separated by the specified delimiter.

- `StringJoiner`: Constructs a sequence of characters separated by the specified delimiter, and optionally surrounded by the specified prefix and suffix.
- Many other useful utility classes including the classes that support internationalization and base64-encoding and -decoding.

java.time

The `java.time` package contains classes for managing dates, time, periods, and durations. The package includes the following:

- The `Month` enum
- The `DayOfWeek` enum
- The `Clock` class that returns current instant, date, and time using a time zone
- The `Duration` and `Period` classes represent and compare amounts of time in different time units
- The `LocalDate`, `LocalTime`, and `LocalDateTime` classes represent dates and times without a time zone
- The `ZonedDateTime` class represents date and time with a time zone
- The `ZoneId` class identifies a time zone such as *America/Chicago*
- The `java.time.format.DateTimeFormatter` class allows presenting date and time in accordance with the **International Standards Organization (ISO)** formats, like the pattern *YYYY-MM-DD* and similar
- some other classes that support date and time manipulation

We discussed most of these classes in Chapter 6, *Data Structures, Generics, and Popular Utilities.*

java.io and java.nio

The `java.io` and `java.nio` packages contain classes and interfaces that support reading and writing data using streams, serialization, and file systems. The difference between these two packages is as follows:

- The `java.io` package classes allow reading/writing data as they come without caching (we discussed it in Chapter 5, *Strings, Input/Output, and Files*), while classes of the `java.nio` package create a buffer that allows moving back and forth along the populated buffer.

- The java.io package classes block the stream until all the data is read or written, while classes of the java.nio package are implemented in a non-blocking style (we will talk about non-blocking style in Chapter 15, *Reactive Programming*).

java.sql and javax.sql

These two packages compose the **Java Database Connectivity (JDBC)** API that allows accessing and processing data stored in a data source, typically a relational database. The javax.sql package complements the java.sql package by providing support for the following:

- The DataSource interface as an alternative to the DriverManager class
- Connections and statements pooling
- Distributed transactions
- Rowsets

We will talk about these packages and see code examples in Chapter 10, *Manage Data in a Database*.

java.net

The java.net package contains classes that support applications networking on the following two levels:

- **Low-level networking,** based on:
    - IP addresses
    - Sockets, which are basic bidirectional data communication mechanisms
    - Various network interfaces
- **High-level networking,** based on:
    - **Universal Resource Identifier (URI)**
    - **Universal Resource Locator (URL)**
    - Connections to the resource pointed to by URLs

We will talk about this package and see code examples in Chapter 11, *Network Programming*.

java.lang.math and java.math

The `java.lang.math` package contains methods for performing basic numeric operations, such as calculating the minimum and maximum of two numeric values, the absolute value, the elementary exponential, logarithms, square roots, trigonometric functions, and many other mathematical operations.

The `java.math` package complements Java primitive types and wrapper classes of the `java.lang` package by allowing working with much bigger numbers using the `BigDecimal` and `BigInteger` classes.

java.awt, javax.swing, and javafx

The first Java library that supported building a **Graphical User Interface** (**GUI**) for desktop applications was the **Abstract Window Toolkit** (**AWT**) in the `java.awt` package. It provided an interface to the native system of the executing platform that allowed creating and managing windows, layouts, and events. It also had the basic GUI widgets (like text fields, buttons, and menus), provided access to the system tray, and allowed to launch a web browser and email a client from the Java code. Its heavy dependence on the native code made AWT-based GUI look different on different platforms.

In 1997, Sun Microsystems and Netscape Communication Corporation introduced Java **Foundation Classes**, later called **Swing** and placed them in the `javax.swing` package. The GUI components built with Swing were able to emulate the look and feel of some native platforms but also allowed you to plug in a look and feel that did not depend on the platform it was running on. It expanded the list of widgets the GUI could have by adding tabbed panels, scroll panes, tables, and lists. Swing components are called lightweight because they do not depend on the native code and are fully implemented in Java.

In 2007, Sun Microsystems announced the creation of JavaFX, which has eventually become a software platform for creating and delivering desktop applications across many different devices. It was intended to replace Swing as the standard GUI library for Java SE. The JavaFX framework is located in the packages that start with `javafx` and supports all major desktop **Operating Systems** (**OS**) and multiple mobile OSes, including Symbian OS, Windows Mobile, and some proprietary real-time OSes.

JavaFX adds the support of smooth animation, web views, audio and video playback, and styles to the arsenal of a GUI developer, based on **Cascading Style Sheets (CSS)**. However, Swing has more components and the third-party libraries, so using JavaFX may require creating custom components and plumbing that was implemented in Swing a long time ago. That's why, although JavaFX is recommended as the first choice for desktop GUI implementation, Swing will remain part of Java *for the foreseeable future*, according to the official response on the Oracle website (`http://www.oracle.com/technetwork/java/javafx/overview/faq-1446554.html#6`). So, it is possible to continue using Swing, but, if possible, better to switch to JavaFX.

We will talk about JavaFX and see code examples in `Chapter 12`, *Java GUI Programming*.

External libraries

Different lists of the most used third-party non-JCL libraries include between 20 and 100 libraries. In this section, we are going to discuss those that are included in the majority of such lists. All of them are open source projects.

org.junit

The `org.junit` package is the root package of an open source testing framework JUnit. It can be added to the project as the following `pom.xml` dependency:

```
<dependency>
    <groupId>junit</groupId>
    <artifactId>junit</artifactId>
    <version>4.12</version>
    <scope>test</scope>
</dependency>
```

The `scope` value in the preceding `dependency` tag tells Maven to include the library `.jar` file only when the test code is going to be run, not into the production `.jar` file of the application. With the dependency in place, you can now create a test. You can write code yourself or let the IDE do it for you using the following steps:

1. Right-click on the class name you would like to test.
2. Select **Go To**.
3. Select **Test**.
4. Click **Create New Test**.

5. Click the checkbox for the methods of the class you would like to test.

6. Write code for the generated test methods with the @Test annotation.

7. Add methods with the @Before and @After annotations if necessary.

Let's assume we have the following class:

```
public class SomeClass {
    public int multiplyByTwo(int i){
        return i * 2;
    }
}
```

If you follow the preceding steps listed, the following test class will be created under the test source tree:

```
import org.junit.Test;
public class SomeClassTest {
    @Test
    public void multiplyByTwo() {
    }
}
```

Now you can implement the void multiplyByTwo() method as follows:

```
@Test
public void multiplyByTwo() {
    SomeClass someClass = new SomeClass();
    int result = someClass.multiplyByTwo(2);
    Assert.assertEquals(4, result);
}
```

A **unit** is a minimal piece of code that can be tested, thus the name. The best testing practices consider a method as a minimal testable unit. That's why a unit test usually tests a method.

org.mockito

One of the problems a unit test often faces is the need to test a method that uses a third-party library, a data source, or a method of another class. While testing, you want to control all the inputs, so you can predict the expected result of the tested code. That is where the technique of simulating or mocking the behavior of the objects the tested code interacts with comes in handy.

An open source framework Mockito (`org.mockito` root package name) allows the accomplishment of exactly that—the creation of **mock objects**. To use it is quite easy and straightforward. Here is one simple case. Let's assume we need to test another `SomeClass` method:

```
public class SomeClass {
    public int multiplyByTwoTheValueFromSomeOtherClass(SomeOtherClass
                                                      someOtherClass){
        return someOtherClass.getValue() * 2;
    }
}
```

To test this method, we need to make sure the `getValue()` method returns a certain value, so we are going to mock this method. In order to do it, follow these steps:

1. Add a dependency to the Maven `pom.xml` configuration file:

```
<dependency>
    <groupId>org.mockito</groupId>
    <artifactId>mockito-core</artifactId>
    <version>2.23.4</version>
    <scope>test</scope>
</dependency>
```

2. Call the `Mockito.mock()` method for the class you need to simulate:

```
SomeOtherClass mo = Mockito.mock(SomeOtherClass.class);
```

3. Set the value you need to be returned from a method:

```
Mockito.when(mo.getValue()).thenReturn(5);
```

4. Now you can pass the mocked object as a parameter into the method you are testing that calls the mocked method:

```
SomeClass someClass = new SomeClass();
int result = someClass.multiplyByTwoTheValueFromSomeOtherClass(mo);
```

5. The mocked method returns the result you have predefined:

```
Assert.assertEquals(10, result);
```

6. After the preceding steps have been performed, here's how the test method looks:

```
@Test
public void multiplyByTwoTheValueFromSomeOtherClass() {
    SomeOtherClass mo = Mockito.mock(SomeOtherClass.class);
    Mockito.when(mo.getValue()).thenReturn(5);

    SomeClass someClass = new SomeClass();
    int result =
            someClass.multiplyByTwoTheValueFromSomeOtherClass(mo);
    Assert.assertEquals(10, result);
}
```

Mockito has certain limitations. For example, you cannot mock static methods and private methods. Otherwise, it is a great way to isolate the code you are testing by reliably predicting the results of the used third-party classes.

org.apache.log4j and org.slf4j

Throughout this book, we used `System.out` to display the results. In the real-life application, one can do it too and redirect the output to a file, for example, for later analysis. After doing it for some time, you will notice that you need more details about each output: the date and time of each statement, and the class name where the logging statement was generated, for example. As the code base grows, you will find that it would be nice to send output from different subsystems or packages to different files or turn off some messages, when everything works as expected, and turn them back on when an issue has been detected and more detailed information about code behavior is needed. And you don't want the size of the log file to grow uncontrollably.

It is possible to write your own code that accomplishes all that. But there are several frameworks that do it based on the settings in a configuration file, which you can change every time you need to change the logging behavior. The two most popular frameworks used for that are called `log4j` (pronounced as *LOG-FOUR-JAY*) and `slf4j` (pronounced as *S-L- F-FOUR-JAY*).

In fact, these two frameworks are not rivals. The `slf4j` framework is a facade that provides unified access to an underlying actual logging framework, one of them can be `log4j` too. Such a facade is especially helpful during a library development when programmers do not know in advance what kind of a logging framework will be used by the application that uses the library. By writing code using `slf4j`, the programmers allow later to configure it to use any logging system.

So, if your code is going to be used only by the application your team develops, using just log4j is quite enough. Otherwise, consider using slf4j.

And, as in the case of any third-party library, before you can use the log4j framework, you have to add a corresponding dependency to the Maven pom.xml configuration file:

```
<dependency>
    <groupId>org.apache.logging.log4j</groupId>
    <artifactId>log4j-api</artifactId>
    <version>2.11.1</version>
</dependency>
<dependency>
    <groupId>org.apache.logging.log4j</groupId>
    <artifactId>log4j-core</artifactId>
    <version>2.11.1</version>
</dependency>
```

For example, here's how the framework can be used:

```
import org.apache.logging.log4j.LogManager;
import org.apache.logging.log4j.Logger;
public class SomeClass {
    static final Logger logger =
                        LogManager.getLogger(SomeClass.class.getName());
    public int multiplyByTwoTheValueFromSomeOtherClass(SomeOtherClass
                                                        someOtherClass){
        if(someOtherClass == null){
            logger.error("The parameter should not be null");
            System.exit(1);
        }
        return someOtherClass.getValue() * 2;
    }
    public static void main(String... args){
        new SomeClass().multiplyByTwoTheValueFromSomeOtherClass(null);
    }
}
```

If we run the preceding main() method, the result will be the following:

```
18:34:07.672 [main] ERROR SomeClass - The parameter should not be null
Process finished with exit code 1
```

As you can see, if no `log4j`-specific configuration file is added to the project, `log4j` will provide a default configuration in the `DefaultConfiguration` class. The default configuration is as follows:

1. The log message will go to a console.
2. The pattern of the message is going to be `"%d{HH:mm:ss.SSS} [%t] %-5level %logger{36} - %msg%n"`.
3. The level of `logging` will be `Level.ERROR` (other levels are `OFF`, `FATAL`, `WARN`, `INFO`, `DEBUG`, `TRACE`, and `ALL`).

The same result is achieved by adding the `log4j2.xml` file to the `resources` folder (which Maven places on the classpath) with the following content:

```xml
<?xml version="1.0" encoding="UTF-8"?>
<Configuration status="WARN">
    <Appenders>
        <Console name="Console" target="SYSTEM_OUT">
            <PatternLayout pattern="%d{HH:mm:ss.SSS} [%t] %-5level
                                    %logger{36} - %msg%n"/>
        </Console>
    </Appenders>
    <Loggers>
        <Root level="error">
            <AppenderRef ref="Console"/>
        </Root>
    </Loggers>
</Configuration>
```

If that is not good enough for you, it is possible to change the configuration to log messages of different levels, to different files, and so on. Read the `log4J` documentation (`https://logging.apache.org`).

org.apache.commons

The `org.apache.commons` package is another popular library developed as a project called **Apache Commons**. It is maintained by an open source community of programmers called **Apache Software Foundation**. This organization was formed from the Apache Group in 1999. The Apache Group has grown around the development of the Apache HTTP Server since 1993. The Apache HTTP Server is an open source cross-platform web server that has remained the most popular web server since April 1996.

The Apache Commons project has the following three parts:

- **Commons Sandbox**: A workspace for Java component development; you can contribute to the open source working there.
- **Commons Dormant**: A repository of components that are currently inactive; you can use the code there, but have to build the components yourself since these components will probably not be released in the near future.
- **Commons Proper**: The reusable Java components, which compose the actual `org.apache.commons` library.

We discussed the `org.apache.commons.io` package in Chapter 5, *String, Input/Output, and Files*.
In the following subsections, we will discuss only three of Commons Proper most popular packages:

- `org.apache.commons.lang3`
- `org.apache.commons.collections4`
- `org.apache.commons.codec.binary`

But there are many more packages under `org.apache.commons` that contain thousands of classes that can easily be used and can help make your code more elegant and efficient.

lang and lang3

The `org.apache.commons.lang3` package is actually the version 3 of the `org.apache.commons.lang` package. The decision to create a new package was forced by the fact that changes introduced in version 3 were backward incompatible, which means that the existing applications that use the previous version of `org.apache.commons.lang` package may stop working after the upgrade to version 3. But in the majority of mainstream programming, adding 3 to an import statement (as the way to migrate to the new version) typically does not break anything.

According to the documentation, the `org.apache.commons.lang3` package provides highly reusable static utility methods, chiefly concerned with adding value to the `java.lang` classes. Here are a few notable examples:

- The `ArrayUtils` class: Allows to search and manipulate arrays; we discussed and demonstrated it in Chapter 6, *Data Structures, Generics, and Popular Utilities*
- The `ClassUtils` class: Provides some metadata about a class

- The `ObjectUtils` class: Checks an array of objects for `null`, compares objects, and calculates the median and minimum/maximum of an array of objects in a null-safe manner; we discussed and demonstrated it in `Chapter 6`, *Data Structures, Generics, and Popular Utilities*
- The `SystemUtils` class: Provides information about the execution environment
- The `ThreadUtils` class: Finds information about currently running threads
- The `Validate` class: Validates individual values and collections, compares them, checks for `null`, matches, and performs many other validations
- The `RandomStringUtils` class: Generates `String` objects from the characters of various character sets
- The `StringUtils` class: We discussed in `Chapter 5`, *String, Input/Output, and Files*

collections4

Although on the surface the content of the `org.apache.commons.collections4` package looks quite similar to the content of the `org.apache.commons.collections` package (which is the version 3 of the package), the migration to the version 4 may not be as smooth as just adding "4" to the import statement. Version 4 removed deprecated classes, added generics and other features incompatible with the previous versions.

One has to be hard-pressed to come up with a collection type or a collection utility that is not present in this package or one of its sub-packages. The following is just a high-level list of features and utilities included:

- The `Bag` interface for collections that have several copies of each object.
- A dozen classes that implement the `Bag` interface, for example, here is how the `HashBag` class can be used:

```
Bag<String> bag = new HashBag<>();
bag.add("one", 4);
System.out.println(bag);                    //prints: [4:one]
bag.remove("one", 1);
System.out.println(bag);                    //prints: [3:one]
System.out.println(bag.getCount("one")); //prints: 3
```

- The `BagUtils` class that transforms `Bag`-based collections.

- The `BidiMap` interface for bidirectional maps that allow you to retrieve not only value by its key but also a key by its value; it has several implementations, for example:

```
BidiMap<Integer, String> bidi = new TreeBidiMap<>();
bidi.put(2, "two");
bidi.put(3, "three");
System.out.println(bidi);                  //prints: {2=two, 3=three}
System.out.println(bidi.inverseBidiMap());
                                           //prints: {three=3, two=2}
System.out.println(bidi.get(3));           //prints: three
System.out.println(bidi.getKey("three")); //prints: 3
bidi.removeValue("three");
System.out.println(bidi);                  //prints: {2=two}
```

- `MapIterator` interface to provide simple and quick iteration over maps, for example:

```
IterableMap<Integer, String> map =
                   new HashedMap<>(Map.of(1, "one", 2, "two"));
MapIterator it = map.mapIterator();
while (it.hasNext()) {
    Object key = it.next();
    Object value = it.getValue();
    System.out.print(key + ", " + value + ", ");
                                        //prints: 2, two, 1, one,
    if(((Integer)key) == 2){
        it.setValue("three");
    }
}
System.out.println("\n" + map);         //prints: {2=three, 1=one}
```

- Ordered maps and sets that keep the elements in a certain order, like `List` does, for example:

```
OrderedMap<Integer, String> map = new LinkedMap<>();
map.put(4, "four");
map.put(7, "seven");
map.put(12, "twelve");
System.out.println(map.firstKey()); //prints: 4
System.out.println(map.nextKey(2)); //prints: null
System.out.println(map.nextKey(7)); //prints: 12
System.out.println(map.nextKey(4)); //prints: 7
```

- Reference maps, their keys, and/or values can be removed by the garbage collector.
- Various implementations of the `Comparator` interface.
- Various implementations of the `Iterator` interface.
- Classes that convert array and enumerations to collections.
- Utilities that allow testing or creating a union, intersection, and closure of collections.
- The `CollectionUtils`, `ListUtils`, `MapUtils`, `MultiMapUtils`, `MultiSetUtils`, `QueueUtils`, `SetUtils` classes and many other interface-specific utility classes.

Read the package documentation (`https://commons.apache.org/proper/commons-collections`) for more details.

codec.binary

The `org.apache.commons.codec.binary` package provides support for Base64, Base32, Binary, and Hexadecimal String encoding and decoding. The encoding is necessary to make sure that the data you sent across different systems will not be changed on the way because of the restrictions on the range of characters in different protocols. Besides, some systems interpret the sent data as control characters (a modem, for example).

Here is the code snippet that demonstrates the basic encoding and decoding capabilities of the `Base64` class of this package:

```
String encodedStr =
          new String(Base64.encodeBase64("Hello, World!".getBytes()));
System.out.println(encodedStr);          //prints: SGVsbG8sIFdvcmxkIQ==

System.out.println(Base64.isBase64(encodedStr));          //prints: true

String decodedStr =
          new String(Base64.decodeBase64(encodedStr.getBytes()));
System.out.println(decodedStr);          //prints: Hello, World!
```

You can read more about this package on the Apache Commons project site (`https://commons.apache.org/proper/commons-codec`).

Summary

In this chapter, we provided an overview of the functionality of the most popular packages of JCL: `java.lang`, `java.util`, `java.time`, `java.io`, `java.nio`, `java.sql`, `javax.sql`, `java.net`, `java.lang.math`, `java.math`, `java.awt`, `javax.swing`, and `javafx`.

The most popular external libraries were represented by the `org.junit`, `org.mockito`, `org.apache.log4j`, `org.slf4j`, and `org.apache.commons` packages. It helps you to avoid writing custom code in the cases when such functionality already exists and can just be imported and used out of the box.

In the next chapter, we will talk about Java threads and demonstrate their usage. We will also explain the difference between parallel and concurrent processing. We will demonstrate how to create a thread and how to execute, monitor, and stop it. It will be very useful material not only for those who are going to write code for multi-threaded processing but also for those who would like to improve their understanding of how the JVM works, which will be the topic of the following chapter.

Quiz

1. What is the Java Class Library? Select all that apply:
   a. Collection of compiled classes.
   b. Packages that come with Java installation.
   c. A `.jar` file that Maven adds to the classpath automatically.
   d. Any library written in Java.

2. What is the Java external library? Select all that apply:

   a. A `.jar` file that is not included with Java installation.
   b. A `.jar` file that has to be added as a dependency in `pom.xml` before it can be used.
   c. Classes not written by the authors of JVM.
   d. Classes that do not belong to JCL.

3. What is the functionality included in the `java.lang` package? Select all that apply:

       a. It is the only package that contains Java language implementation.
       b. It contains the most often used classes of JCL.
       c. It contains the `Object` class that is the base class for any Java class.
       d. It contains all types listed in the Java Language Specification.

4. What is the functionality included in the `java.util` package? Select all that apply:

       a. All implementations of Java collection interfaces
       b. All interfaces of the Java collections framework
       c. All utilities of JCL
       d. Classes `Arrays, Objects, Properties`

5. What is the functionality included in the `java.time` package? Select all that apply:

       a. Classes that manage date.
       b. It is the only package that manages time.
       c. Classes that represent date and time.
       d. It is the only package that manages date.

6. What is the functionality included in the `java.io` package? Select all that apply:

       a. Processing of streams of binary data
       b. Processing of streams of characters
       c. Processing of streams of bytes
       d. Processing of streams of numbers

7. What is the functionality included in the `java.sql` package? Select all that apply:

       a. Supports database connection pooling
       b. Supports database statement execution
       c. Provides the capability to read/write data from/to a database
       d. Supports database transactions

8. What is the functionality included in the `java.net` package? Select all that apply:

   a. Supports .NET programming
   b. Supports sockets communication
   c. Supports URL-based communication
   d. Supports RMI-based communication

9. What is the functionality included in the `java.math` package? Select all that apply:

   a. Supports minimum and maximum calculations
   b. Supports big numbers
   c. Supports logarithms
   d. Supports square root calculations

10. What is the functionality included in the `javafx` package? Select all that apply:

    a. Supports fax-message sending
    b. Supports fax-message receiving
    c. Supports graphic user interface programming
    d. Supports animation

11. What is the functionality included in the `org.junit` package? Select all that apply:

    a. Supports testing of Java classes
    b. Supports Java units of measure
    c. Supports unit testing
    d. Supports organizational unity

12. What is the functionality included in the `org.mockito` package? Select all that apply:

    a. Supports Mockito protocol
    b. Allows simulating a method behavior
    c. Supports a static method simulation
    d. Generates objects that behave like third-party classes

13. What is the functionality included in the `org.apache.log4j` package? Select all that apply:

    a. Supports writing messages to a file
    b. Supports reading messages from a file
    c. Supports the LOG4 protocol for Java
    d. Supports control of the number and size of log files

14. What is the functionality included in the `org.apache.commons.lang3` package? Select all that apply:

    a. Supports Java language version 3
    b. Complements the `java.lang` classes
    c. Contains the `ArrayUtils, ObjectUtils, StringUtils` classes
    d. Contains the `SystemUtils` class

15. What is the functionality included in the `org.apache.commons.collections4` package? Select all that apply:

    a. Various implementations of Java collections framework interfaces
    b. Various utilities for Java collections framework implementations
    c. The `Vault` interface and its implementations
    d. Contains the `CollectionUtils` class

16. What is the functionality included in the `org.apache.commons.codec.binary` package? Select all that apply:

    a. Supports sending binary data across the network
    b. Allows encoding and decoding of data
    c. Supports data encryption
    d. Contains the `StringUtils` class

8
Multithreading and Concurrent Processing

In this chapter, we will discuss the ways to increase Java application performance by using the workers (threads) that process data concurrently. We will explain the concept of Java threads and demonstrate their usage. We will also talk about the difference between parallel and concurrent processing and how to avoid unpredictable results caused by the concurrent modification of the shared resource.

The following topics will be covered in this chapter:

- Thread vs process
- User thread vs daemon
- Extending class Thread
- Implementing interface `Runnable`
- Extending Thread vs implementing `Runnable`
- Using pool of threads
- Getting results from a thread
- Parallel vs concurrent processing
- Concurrent modification of the same resource

Thread versus process

Java has two units of execution—process and thread. A **process** usually represents the whole JVM, although an application can create another process using `java.lang.ProcessBuilder`. But, since the multi-process case is outside the scope of this book, we will focus on the second unit of execution, that is, a **thread**, which is similar to a process but less isolated from other threads and requires fewer resources for execution.

A process can have many threads running and at least one thread called the **main thread**—the one that starts the application—which we use it in every example. Threads can share resources, including memory and open files, which allows for better efficiency. But it comes with a price of higher risk of unintended mutual interference and even blocking of the execution. That is where programming skills and an understanding of the concurrency techniques are required.

User thread versus daemon

There is a particular kind of thread called a daemon.

 The word daemon has an ancient Greek origin meaning *a divinity or supernatural being of a nature between gods and humans* and *an inner or attendant spirit or inspiring force.*

In computer science, the term **daemon** has more mundane usage and is applied to *a computer program that runs as a background process, rather than being under the direct control of an interactive user.* That is why there are the following two types of threads in Java:

- User thread (default), initiated by an application (main thread is one such an example)
- Daemon thread that works in the background in support of user thread activity

That is why all daemon threads exit immediately after the last user thread exits or are terminated by JVM after an unhandled exception.

Extending class thread

One way to create a thread is to extend the `java.lang.Thread` class and override its `run()` method. For example:

```
class MyThread extends Thread {
    private String parameter;
    public MyThread(String parameter) {
        this.parameter = parameter;
    }
    public void run() {
        while(!"exit".equals(parameter)){
            System.out.println((isDaemon() ? "daemon" : "  user") +
                " thread " + this.getName() + "(id=" + this.getId() +
```

```
                                                ") parameter: " + parameter);
                pauseOneSecond();
            }
            System.out.println((isDaemon() ? "daemon" : "  user") +
                " thread " + this.getName() + "(id=" + this.getId() +
                                        ") parameter: " + parameter);
        }
        public void setParameter(String parameter) {
            this.parameter = parameter;
        }
    }
}
```

If the run() method is not overridden, the thread does nothing. In our example, the thread prints its name and other properties every second as long as the parameter is not equal to string "exit"; otherwise it exits. The pauseOneSecond() method looks as follows:

```
private static void pauseOneSecond(){
    try {
        TimeUnit.SECONDS.sleep(1);
    } catch (InterruptedException e) {
        e.printStackTrace();
    }
}
```

We can now use the MyThread class to run two threads—one user thread and one daemon thread:

```
public static void main(String... args) {
    MyThread thr1 = new MyThread("One");
    thr1.start();
    MyThread thr2 = new MyThread("Two");
    thr2.setDaemon(true);
    thr2.start();
    pauseOneSecond();
    thr1.setParameter("exit");
    pauseOneSecond();
    System.out.println("Main thread exists");
}
```

As you can see, the main thread creates two other threads, pauses for one second, sets parameter exit on the user thread, pauses another second, and, finally, exits. (The main() method completes its execution.)

If we run the preceding code, we see something like the following screenshot (The thread
`id` may be different in different operating systems.):

```
daemon thread Thread-1(id=14) parameter: Two
  user thread Thread-0(id=13) parameter: One
daemon thread Thread-1(id=14) parameter: Two
  user thread Thread-0(id=13) parameter: exit
Main thread exists
```

The preceding screenshot shows that the daemon thread exits automatically as soon as the
last user thread (main thread in our example) exits.

Implementing interface Runnable

The second way to create a thread is to use a class that implements `java.lang.Runnable`.
Here is an example of such a class that has almost exactly the same functionality
as `MyThread` class:

```java
class MyRunnable implements Runnable {
    private String parameter, name;
    public MyRunnable(String name) {
        this.name = name;
    }
    public void run() {
        while(!"exit".equals(parameter)){
            System.out.println("thread " + this.name +
                            ", parameter: " + parameter);
            pauseOneSecond();
        }
        System.out.println("thread " + this.name +
                        ", parameter: " + parameter);
    }
    public void setParameter(String parameter) {
        this.parameter = parameter;
    }
}
```

The difference is that there is no `isDaemon()` method, `getId()`, or any other out-of-
box method. The `MyRunnable` class can be any class that implements the `Runnable`
interface, so we cannot print whether the thread is daemon or not. That is why we have
added the `name` property, so we can identify the thread.

We can use the MyRunnable class to create threads similar to how we have used the MyThread class:

```
public static void main(String... args) {
    MyRunnable myRunnable1 = new MyRunnable("One");
    MyRunnable myRunnable2 = new MyRunnable("Two");

    Thread thr1 = new Thread(myRunnable1);
    thr1.start();
    Thread thr2 = new Thread(myRunnable2);
    thr2.setDaemon(true);
    thr2.start();
    pauseOneSecond();
    myRunnable1.setParameter("exit");
    pauseOneSecond();
    System.out.println("Main thread exists");
}
```

The following screenshot proves that the behavior of the MyRunnable class is similar to the behavior of the MyThread class:

```
thread Two, parameter: null
thread One, parameter: null
thread Two, parameter: null
thread One, parameter: exit
Main thread exists
```

The daemon thread (named Two) exits after the last user thread exists—exactly how it happened with the MyThread class.

Extending thread vs implementing Runnable

Implementation of Runnable has the advantage (and in some cases the only possible option) of allowing the implementation to extend another class. It is particularly helpful when you would like to add thread-like behavior to an existing class. Implementing Runnable allows more flexibility in usage. But otherwise, there is no difference in functionality comparing to the extending of the Thread class.

Thread class has several constructors that allow setting the thread name and the group it belongs to. Grouping of threads helps to manage them in the case of many threads running in parallel. Thread class has also several methods that provide information about the thread's status, its properties, and allows to control its behavior.

As you have seen, the thread's ID is generated automatically. It cannot be changed but can be reused after the thread is terminated. Several threads, on the other hand, can be set with the same name.

The execution priority can also be set programmatically with a value between `Thread.MIN_PRIORITY` and `Thread.MAX_PRIORITY`. The smaller the value, the more time the thread is allowed to run, which means it has higher priority. If not set, the priority value defaults to `Thread.NORM_PRIORITY`.

The state of a thread can have one of the following values:

- `NEW`: When a thread has not yet started
- `RUNNABLE`: When a thread is being executed
- `BLOCKED`: When a thread is blocked and is waiting for a monitor lock
- `WAITING`: When a thread is waiting indefinitely for another thread to perform a particular action
- `TIMED_WAITING`: When a thread is waiting for another thread to perform an action for up to a specified waiting time
- `TERMINATED`: When a thread has exited

Threads and any objects for that matter also can *talk to each other* using the methods `wait()`, `notify()`, and `notifyAll()` of the `java.lang.Object` base class. But this aspect of threads' behavior is outside the scope of this book.

Using pool of threads

Each thread requires resources—**CPU** and **memory**. It means the number of threads must be controlled, and one way to do it is to create a fixed number of them—a pool. Besides, creating an object incurs an overhead that may be significant for some applications.

In this section, we will look into the `Executor` interfaces and their implementations provided in the `java.util.concurrent` package. They encapsulate thread management and minimize the time an application developer spends on writing the code related to threads' life cycles.

There are three `Executor` interfaces defined in the `java.util.concurrent` package:

- The base `Executor` interface: It has only one `void execute(Runnable r)` method in it.
- The `ExecutorService` interface: It extends `Executor` and adds four groups of methods that manage the life cycle of the worker threads and of the executor itself:
    - `submit()` methods that place a `Runnable` or `Callable` object in the queue for the execution (`Callable` allows the worker thread to return a value), return an object of `Future` interface, which can be used to access the value returned by the `Callable`, and to manage the status of the worker thread
    - `invokeAll()` methods that place a collection of objects of `Callable` interface in the queue for the execution which then returns `List` of `Future` objects when all the worker threads are complete (there is also an overloaded `invokeAll()` method with a timeout)
    - `invokeAny()` methods that place a collection of interface `Callable` objects in the queue for the execution; return one `Future` object of any of the worker threads, which has completed (there is also an overloaded `invokeAny()` method with a timeout)
    - Methods that manage the worker threads' status and the service itself as follows:
        - `shutdown()`: Prevents new worker threads from being submitted to the service.
        - `shutdownNow()`: Interrupts each worker thread that is not completed. A worker thread should be written so that it checks its own status periodically (using `Thread.currentThread().isInterrupted()`, for example) and gracefully shuts down on its own; otherwise, it will continue running even after `shutdownNow()` was called.
        - `isShutdown()`: Checks whether the shutdown of the executor was initiated.

- `awaitTermination(long timeout, TimeUnit timeUnit)`: Waits until all worker threads have completed execution after a shutdown request, or the timeout occurs, or the current thread is interrupted, whichever happens first.
- `isTerminated()`: Checks whether all the worker threads have completed after the shutdown was initiated. It never returns `true` unless either `shutdown()` or `shutdownNow()` was called first.

- The `ScheduledExecutorService` interface: It extends `ExecutorService` and adds methods that allow scheduling of the execution (one-time and periodic one) of the worker threads.

A pool-based implementation of `ExecutorService` can be created using the `java.util.concurrent.ThreadPoolExecutor` or `java.util.concurrent.ScheduledThreadPoolExecutor` class. There is also a `java.util.concurrent.Executors` factory class that covers most of the practical cases. So, before writing custom code for worker threads' pool creation, we highly recommend looking into using the following factory methods of the `java.util.concurrent.Executors` class:

- `newCachedThreadPool()` that creates a thread pool that adds a new thread as needed, unless there is an idle thread created before; threads that have been idle for 60 seconds are removed from the pool
- `newSingleThreadExecutor()` that creates an `ExecutorService` (pool) instance that executes worker threads sequentially
- `newSingleThreadScheduledExecutor()` that creates a single-threaded executor that can be scheduled to run after a given delay, or to execute periodically
- `newFixedThreadPool(int nThreads)` that creates a thread pool that reuses a fixed number of worker threads; if a new task is submitted when all the worker threads are still executing, it will be placed into the queue until a worker thread is available
- `newScheduledThreadPool(int nThreads)` that creates a thread pool of a fixed size that can be scheduled to run after a given delay, or to execute periodically

- `newWorkStealingThreadPool(int nThreads)` that creates a thread pool that uses the *work-stealing* algorithm used by `ForkJoinPool`, which is particularly useful in case the worker threads generate other threads, such as in a recursive algorithm; it also adapts to the specified number of CPUs, which you may set higher or lower than the actual CPUs count on your computer

Work-stealing algorithm

A work-stealing algorithm allows threads that have finished their assigned tasks to help other tasks that are still busy with their assignments. As an example, see the description of Fork/Join implementation in the official Oracle Java documentation (https://docs. oracle.com/javase/tutorial/essential/concurrency/forkjoin.html).

Each of these methods has an overloaded version that allows passing in a `ThreadFactory` that is used to create a new thread when needed. Let's see how it all works in a code sample. First, we run another version of `MyRunnable` class:

```
class MyRunnable implements Runnable {
    private String name;
    public MyRunnable(String name) {
        this.name = name;
    }
    public void run() {
        try {
            while (true) {
                System.out.println(this.name + " is working...");
                TimeUnit.SECONDS.sleep(1);
            }
        } catch (InterruptedException e) {
            System.out.println(this.name + " was interrupted\n" +
                this.name + " Thread.currentThread().isInterrupted()="
                    + Thread.currentThread().isInterrupted());
        }
    }
}
```

We cannot use `parameter` property anymore to tell the thread to stop executing because the thread life cycle is now going to be controlled by the `ExecutorService`, and the way it does it is by calling the `interrupt()` thread method. Also, notice that the thread we created has an infinite loop, so it will never stop executing until forced to (by calling the `interrupt()` method).

Let's write the code that does the following:

1. Creates a pool of three threads
2. Makes sure the pool does not accept more threads
3. Waits for a fixed period of time to let all the threads finish what they do
4. Stops (interrupts) the threads that did not finish what they do
5. Exits

The following code performs all the actions described in the preceding list:

```
ExecutorService pool = Executors.newCachedThreadPool();
String[] names = {"One", "Two", "Three"};
for (int i = 0; i < names.length; i++) {
    pool.execute(new MyRunnable(names[i]));
}
System.out.println("Before shutdown: isShutdown()=" + pool.isShutdown()
                        + ", isTerminated()=" + pool.isTerminated());
pool.shutdown(); // New threads cannot be added to the pool
//pool.execute(new MyRunnable("Four"));      //RejectedExecutionException
System.out.println("After shutdown: isShutdown()=" + pool.isShutdown()
                        + ", isTerminated()=" + pool.isTerminated());
try {
    long timeout = 100;
    TimeUnit timeUnit = TimeUnit.MILLISECONDS;
    System.out.println("Waiting all threads completion for "
                            + timeout + " " + timeUnit + "...");
    // Blocks until timeout, or all threads complete execution,
    // or the current thread is interrupted, whichever happens first.
    boolean isTerminated = pool.awaitTermination(timeout, timeUnit);
    System.out.println("isTerminated()=" + isTerminated);
    if (!isTerminated) {
        System.out.println("Calling shutdownNow()...");
        List<Runnable> list = pool.shutdownNow();
        System.out.println(list.size() + " threads running");
        isTerminated = pool.awaitTermination(timeout, timeUnit);
        if (!isTerminated) {
            System.out.println("Some threads are still running");
        }
        System.out.println("Exiting");
    }
} catch (InterruptedException ex) {
    ex.printStackTrace();
}
```

The attempt to add another thread to the pool after `pool.shutdown()` is called, **generates** `java.util.concurrent.RejectedExecutionException`.

The execution of the preceding code produces the following results:

```
One is working...
Three is working...
Two is working...
Before shutdown: isShutdown()=false, isTerminated()=false
After shutdown: isShutdown()=true, isTerminated()=false
Waiting all threads completion for 100 MILLISECONDS...
isTerminated()=false
Calling shutdownNow()...
0 threads running
One was interrupted
One Thread.currentThread().isInterrupted()=false
Two was interrupted
Two Thread.currentThread().isInterrupted()=false
Three was interrupted
Three Thread.currentThread().isInterrupted()=false
Exiting
```

Notice the `Thread.currentThread().isInterrupted()=false` message in the
preceding screenshot. The thread was interrupted. We know it because the thread got the
`InterruptedException`. Why then does the `isInterrupted()` method return `false`?
That is because the thread state was cleared immediately after receiving the interrupt
message. We mention it now because it is a source of some programmer mistakes. For
example, if the main thread watches the `MyRunnable` thread and calls `isInterrupted()`
on it, the return value is going to be `false`, which may be misleading after the thread was
interrupted.

So, in the case where another thread may be monitoring the `MyRunnable` thread, the
implementation of `MyRunnable` has to be changed to the following. Note how
the `interrupt()` method is called in the `catch` block:

```java
class MyRunnable implements Runnable {
    private String name;
    public MyRunnable(String name) {
        this.name = name;
    }
    public void run() {
        try {
            while (true) {
                System.out.println(this.name + " is working...");
                TimeUnit.SECONDS.sleep(1);
            }
        } catch (InterruptedException e) {
            Thread.currentThread().interrupt();
            System.out.println(this.name + " was interrupted\n" +
```

```
                    this.name + " Thread.currentThread().isInterrupted()="
                            + Thread.currentThread().isInterrupted());
        }
    }
}
```

Now, if we run this thread using the same `ExecutorService` pool again, the result will be:

```
Two is working...
Three is working...
One is working...
Before shutdown: isShutdown()=false, isTerminated()=false
After shutdown: isShutdown()=true, isTerminated()=false
Waiting all threads completion for 100 MILLISECONDS...
isTerminated()=false
Calling shutdownNow()...
0 threads running
Two was interrupted
Two Thread.currentThread().isInterrupted()=true
One was interrupted
One Thread.currentThread().isInterrupted()=true
Three was interrupted
Three Thread.currentThread().isInterrupted()=true
Exiting
```

As you can see, now the value returned by the `isInterrupted()` method is `true` and corresponds to what has happened. To be fair, in many applications, once the thread is interrupted, its status is not checked again. But setting the correct state is a good practice, especially in those cases where you are not the author of the higher level code that creates the thread.

In our example, we have used a cached thread pool that creates a new thread as needed or, if available, reuses the thread already used, but which completed its job and returned to the pool for a new assignment. We did not worry about too many threads created because our demo application had three worker threads at the most and they were quite short lived.

But, in the case where an application does not have a fixed limit of the worker threads it might need or there is no good way to predict how much memory a thread may take or how long it can execute, setting a ceiling on the worker thread count prevents an unexpected degradation of the application performance, running out of memory, or depletion of any other resources the worker threads use. If the thread behavior is extremely unpredictable, a single thread pool might be the only solution, with an option of using a custom thread pool executor. But, in the majority of the cases, a fixed-size thread pool executor is a good practical compromise between the application needs and the code complexity (earlier in this section, we listed all possible pool types created by `Executors` factory class).

Setting the size of the pool too low may deprive the application of the chance to utilize the available resources effectively. So, before selecting the pool size, it is advisable to spend some time monitoring the application with the goal of identifying the idiosyncrasy of the application behavior. In fact, the cycle deploy-monitor-adjust has to be repeated throughout the application's lifecycle in order to accommodate and take advantage of the changes that happened in the code or the executing environment.

The first characteristic you take into account is the number of CPUs in your system, so the thread pool size can be at least as big as the CPU's count. Then, you can monitor the application and see how much time each thread engages the CPU and how much of the time it uses other resources (such as I/O operations). If the time spent not using the CPU is comparable with the total executing time of the thread, then you can increase the pool size by the following ratio: the time CPU was not used divided by the total executing time. But, that is in the case where another resource (disk or database) is not a subject of contention between the threads. If the latter is the case, then you can use that resource instead of the CPU as the delineating factor.

Assuming the worker threads of your application are not too big or too long executing, and belong to the mainstream population of the typical working threads that complete their job in a reasonably short period of time, you can increase the pool size by adding the (rounded up) ratio of the desired response time and the time a thread uses the CPU or another most contentious resource. This means that, with the same desired response time, the less a thread uses the CPU or another concurrently accessed resource, the bigger the pool size should be. If the contentious resource has its own ability to improve concurrent access (like a connection pool in the database), consider utilizing that feature first.

If the required number of threads running at the same time changes at runtime under the different circumstances, you can make the pool size dynamic and create a new pool with a new size (shutting down the old pool after all its threads have completed). The recalculation of the size of a new pool might also be necessary after you add or remove the available resources. You can use `Runtime.getRuntime().availableProcessors()` to programmatically adjust the pool size based on the current count of the available CPUs, for example.

If none of the ready-to-use thread pool executor implementations that come with the JDK suit the needs of a particular application, before writing the thread managing code from scratch, try to use the `java.util.concurrent.ThreadPoolExecutor` class first. It has several overloaded constructors.

To give you an idea of its capabilities, here is the constructor with the biggest number of options:

```
ThreadPoolExecutor (int corePoolSize,
                    int maximumPoolSize,
                    long keepAliveTime,
                    TimeUnit unit,
                    BlockingQueue<Runnable> workQueue,
                    ThreadFactory threadFactory,
                    RejectedExecutionHandler handler)
```

The parameters of the preceding constructor are as follows:

- `corePoolSize` is the number of threads to keep in the pool, even if they are idle, unless the `allowCoreThreadTimeOut(boolean value)` method is called with `true` value.
- `maximumPoolSize` is the maximum number of threads to allow in the pool.
- `keepAliveTime`: When the number of threads is greater than the core, this is the maximum time that excess idle threads will wait for new tasks before terminating.
- `unit` is the time unit for the `keepAliveTime` argument.
- `workQueue` is the queue to use for holding tasks before they are executed; this queue will hold only the `Runnable` objects submitted by the `execute()` method.
- `threadFactory` is the factory to use when the executor creates a new thread.
- `handler` is the handler to use when the execution is blocked because the thread bounds and queue capacities are reached.

Each of the previous constructor parameters except the `workQueue` can also be set via the corresponding setter after the object of the `ThreadPoolExecutor` class has been created, thus allowing more flexibility and dynamic adjustment of the existing pool characteristics.

Getting results from thread

In our examples, so far, we used the `execute()` method of the `ExecutorService` interface to start a thread. In fact, this method comes from the `Executor` base interface. Meanwhile, the `ExecutorService` interface has other methods (listed in the previous *Using pool of threads* section) that can start threads and get back the results of thread execution.

The object that brings back the result of the thread execution is of type `Future`—an interface that has the following methods:

- `V get()`: Blocks until the thread finishes; returns the result (*if available*)
- `V get(long timeout, TimeUnit unit)`: Blocks until the thread finishes or the provided timeout is up; returns the result (if available)
- `boolean isDone()`: Returns `true` if the thread has finished
- `boolean cancel(boolean mayInterruptIfRunning)`: Tries to cancel execution of the thread; returns `true` if successful; returns `false` also in the case the thread had finished normally by the time the method was called
- `boolean isCancelled()`: Returns `true` if the thread execution was canceled before it has finished normally

The remark *if available* in the description of the `get()` method means that the result is not always available in principle, even when the `get()` method without parameters is called. It all depends on the method used to produce the `Future` object. Here is a list of all the methods of `ExecutorService` that return `Future` object(s):

- `Future<?> submit(Runnable task)`: Submits the thread (task) for execution; returns a `Future` representing the task; the `get()` method of the returned `Future` object returns `null`; for example, let's use `MyRunnable` class that works only 100 milliseconds:

```
class MyRunnable implements Runnable {
    private String name;
    public MyRunnable(String name) {
        this.name = name;
    }
    public void run() {
        try {
            System.out.println(this.name + " is working...");
            TimeUnit.MILLISECONDS.sleep(100);
            System.out.println(this.name + " is done");
        } catch (InterruptedException e) {
            Thread.currentThread().interrupt();
            System.out.println(this.name + " was interrupted\n" +
                this.name + " Thread.currentThread().isInterrupted()=" 
                    + Thread.currentThread().isInterrupted());
        }
    }
}
```

And, based on the code examples of the previous section, let's create a method that shuts down the pool and terminates all the threads, if necessary:

```
void shutdownAndTerminate(ExecutorService pool){
   try {
      long timeout = 100;
      TimeUnit timeUnit = TimeUnit.MILLISECONDS;
      System.out.println("Waiting all threads completion for "
                               + timeout + " " + timeUnit + "...");
      //Blocks until timeout or all threads complete execution,
      //  or the current thread is interrupted,
      //  whichever happens first.
      boolean isTerminated =
                     pool.awaitTermination(timeout, timeUnit);
      System.out.println("isTerminated()=" + isTerminated);
      if (!isTerminated) {
         System.out.println("Calling shutdownNow()...");
         List<Runnable> list = pool.shutdownNow();
         System.out.println(list.size() + " threads running");
         isTerminated = pool.awaitTermination(timeout, timeUnit);
         if (!isTerminated) {
            System.out.println("Some threads are still running");
         }
         System.out.println("Exiting");
      }
   } catch (InterruptedException ex) {
      ex.printStackTrace();
   }
}
```

We will use the preceding `shutdownAndTerminate()` method in a `finally` block to make sure no running threads were left behind. And here is the code we are going to execute:

```
ExecutorService pool = Executors.newSingleThreadExecutor();

Future future = pool.submit(new MyRunnable("One"));
System.out.println(future.isDone());          //prints: false
System.out.println(future.isCancelled());     //prints: false
try{
    System.out.println(future.get());          //prints: null
    System.out.println(future.isDone());       //prints: true
    System.out.println(future.isCancelled());//prints: false
} catch (Exception ex){
    ex.printStackTrace();
} finally {
    shutdownAndTerminate(pool);
}
```

The output of this code you can see on this screenshot:

```
false
false
One is working...
One is done
null
true
false
Waiting all threads completion for 100 MILLISECONDS...
isTerminated()=false
Calling shutdownNow()...
0 threads running
Exiting
```

As expected, the `get()` method of the `Future` object returns `null`, because `run()` method of `Runnable` does not return anything. All we can get back from the returned `Future` is the information that the task was completed, or not.

- `Future<T> submit(Runnable task, T result)`: Submits the thread (task) for execution; returns a `Future` representing the task with the provided `result` in it; for example, we will use the following class as the result:

```java
class Result {
    private String name;
    private double result;
    public Result(String name, double result) {
        this.name = name;
        this.result = result;
    }
    @Override
    public String toString() {
        return "Result{name=" + name +
                ", result=" + result + "}";
    }
}
```

The following code demonstrates how the default result is returned by the `Future` returned by the `submit()` method:

```java
ExecutorService pool = Executors.newSingleThreadExecutor();
Future<Result> future = pool.submit(new MyRunnable("Two"),
                                    new Result("Two", 42.));
System.out.println(future.isDone());        //prints: false
System.out.println(future.isCancelled());   //prints: false
```

```
try{
    System.out.println(future.get());          //prints: null
    System.out.println(future.isDone());        //prints: true
    System.out.println(future.isCancelled()); //prints: false
} catch (Exception ex){
    ex.printStackTrace();
} finally {
    shutdownAndTerminate(pool);
}
```

If we execute the preceding code, the output is going to be as follows:

```
false
false
Two is working...
Two is done
Result{name=Two, result=42.0}
true
false
Waiting all threads completion for 100 MILLISECONDS...
isTerminated()=false
Calling shutdownNow()...
0 threads running
Exiting
```

As was expected, the `get()` method of `Future` returns the object passed in as a parameter.

- `Future<T> submit(Callable<T> task)`: Submits the thread (task) for execution; returns a `Future` representing the task with the result produced and returned by the `V call()` method of the `Callable` interface; that is the only `Callable` method the interface has. For example:

```
class MyCallable implements Callable {
    private String name;
    public MyCallable(String name) {
        this.name = name;
    }
    public Result call() {
        try {
            System.out.println(this.name + " is working...");
            TimeUnit.MILLISECONDS.sleep(100);
            System.out.println(this.name + " is done");
            return new Result(name, 42.42);
        } catch (InterruptedException e) {
            Thread.currentThread().interrupt();
            System.out.println(this.name + " was interrupted\n" +
```

```
                    this.name + " Thread.currentThread().isInterrupted()="
                        + Thread.currentThread().isInterrupted());
        }
        return null;
    }
```

The result of the preceding code is as follows:

```
false
false
Three is working...
Three is done
Result{name=Three, result=42.42}
true
false
Waiting all threads completion for 100 MILLISECONDS...
isTerminated()=false
Calling shutdownNow()...
0 threads running
Exiting
```

As you can see, the get() method of the Future returns the value produced by the call() method of the MyCallable class

- List<Future<T>> invokeAll(Collection<Callable<T>> tasks):
  Executes all the Callable tasks of the provided collection; returns a list of
  Futures with the results produced by the executed Callable objects

- List<Future<T>> invokeAll(Collection<Callable<T>>: Executes all
  the Callable tasks of the provided collection; returns a list of Futures with the
  results produced by the executed Callable objects or the timeout expires,
  whichever happens first

- T invokeAny(Collection<Callable<T>> tasks): Executes all
  the Callable tasks of the provided collection; returns the result of one that has
  completed successfully (meaning, without throwing an exception), if any do

- T invokeAny(Collection<Callable<T>> tasks, long timeout,
  TimeUnit unit): Executes all the Callable tasks of the provided collection;
  returns the result of one that has completed successfully (meaning, without
  throwing an exception), if such is available before the provided timeout expires

As you can see, there are many ways to get the results from a thread. The method you
choose depends on the particular needs of your application.

Parallel vs concurrent processing

When we hear about working threads executing at the same time, we automatically assume that they literally do what they are programmed to do in parallel. Only after we look under the hood of such a system we, do realize that such parallel processing is possible only when the threads are executed each by a different CPU. Otherwise, they time-share the same processing power. We perceive them working at the same time only because the time slots they use are very short—a fraction of the time units we have used in our everyday life. When the threads share the same resource, in computer science, we say they do it *concurrently*.

Concurrent modification of the same resource

Two or more threads modifying the same value while other threads read it is the most general description of one of the problems of concurrent access. Subtler problems include **thread interference** and **memory consistency** errors, which both produce unexpected results in seemingly benign fragments of code. In this section, we are going to demonstrate such cases and ways to avoid them.

At first glance, the solution seems quite straightforward: just allow only one thread at a time to modify/access the resource and that's it. But if the access takes a long time, it creates a bottleneck that might eliminate the advantage of having many threads working in parallel. Or, if one thread blocks access to one resource while waiting for access to another resource and the second thread blocks access to the second resource while waiting for access to the first one, it creates a problem called a **deadlock**. These are two very simple examples of the possible challenges a programmer encounters while using multiple threads.

First, we'll reproduce a problem caused by the concurrent modification of the same value. Let's create a `Calculator` interface:

```
interface Calculator {
    String getDescription();
    double calculate(int i);
}
```

We will use `getDescription()` method to capture the description of the implementation. Here is the first implementation:

```
class CalculatorNoSync implements Calculator{
    private double prop;
    private String description = "Without synchronization";
    public String getDescription(){ return description; }
    public double calculate(int i){
        try {
            this.prop = 2.0 * i;
            TimeUnit.MILLISECONDS.sleep(i);
            return Math.sqrt(this.prop);
        } catch (InterruptedException e) {
            Thread.currentThread().interrupt();
            System.out.println("Calculator was interrupted");
        }
        return 0.0;
    }
}
```

As you can see, the `calculate()` method assigns a new value to the `prop` property, then does something else (we simulate it by calling the `sleep()` method), and then calculates the square root of the value assigned to the `prop` property. The `"Without synchronization"` description depicts the fact that the value of the `prop` property is changing every time the `calculate()` method is called—without any coordination or **synchronization**, as it is called in the case of the coordination between threads when they concurrently modify the same resource.

We are now going to share this object between two threads, which means that the `prop` property is going to be updated and used concurrently. So some kind of thread synchronization around the `prop` property is necessary, but we have decided that our first implementation does not do it.

The following is the method we are going to use while executing every `Calculator` implementation we are going to create:

```
void invokeAllCallables(Calculator c){
    System.out.println("\n" + c.getDescription() + ":");
    ExecutorService pool = Executors.newFixedThreadPool(2);
    List<Callable<Result>> tasks = List.of(new MyCallable("One", c),
                                           new MyCallable("Two", c));
    try{
        List<Future<Result>> futures = pool.invokeAll(tasks);
        List<Result> results = new ArrayList<>();
        while (results.size() < futures.size()){
            TimeUnit.MILLISECONDS.sleep(5);
```

```
                    for(Future future: futures){
                        if(future.isDone()){
                            results.add((Result)future.get());
                        }
                    }
                }
                for(Result result: results){
                    System.out.println(result);
                }
            } catch (Exception ex){
                ex.printStackTrace();
            } finally {
                shutdownAndTerminate(pool);
            }
        }
    }
```

As you can see, the preceding method does the following:

- Prints description of the passed-in `Calculator` implementation.
- Creates a fixed-size pool for two threads.
- Creates a list of two `Callable` tasks—the objects of the following `MyCallable` class:

```
class MyCallable implements Callable<Result> {
    private String name;
    private Calculator calculator;
    public MyCallable(String name, Calculator calculator) {
        this.name = name;
        this.calculator = calculator;
    }
    public Result call() {
        double sum = 0.0;
        for(int i = 1; i < 20; i++){
            sum += calculator.calculate(i);
        }
        return new Result(name, sum);
    }
}
```

- The list of tasks is passed into the `invokeAll()` method of the pool, where each of the tasks is executed by invoking the `call()` method; each `call()` method applies the `calculate()` method of the passed-in `Calculator` object to every one of the 19 numbers from 1 to 20 and sums up the results; the resulting sum is returned inside the `Result` object along with the name of the `MyCallable` object.

- Each `Result` object is eventually returned inside a `Future` object.
- The `invokeAllCallables()` method then iterates over the list of `Future` objects and checks each of them if the task is completed; when a task is completed, the result is added to the `List<Result> results`.
- After all the tasks are completed, the `invokeAllCallables()` method then prints all the elements of the `List<Result> results` and terminates the pool.

Here is the result we got from one of our runs of `invokeAllCallables(new CalculatorNoSync())`:

```
Without synchronization:
Result{name=One, result=81.43661054438327}
Result{name=Two, result=83.13051012374216}
Waiting all threads completion for 100 MILLISECONDS...
isTerminated()=false
Calling shutdownNow()...
0 threads running
Exiting
```

The actual numbers are slightly different every time we run the preceding code, but the result of task `One` never equals the result of task `Two`. That is because, in the period between setting the value of the `prop` field and returning its square root in the `calculate()` method, the other thread managed to assign a different value to `prop`. This is a case of thread interference.

There are several ways to address this problem. We start with an atomic variable as the way to achieve thread-safe concurrent access to a property. Then we will also demonstrate two methods of thread synchronization.

Atomic variable

An **atomic variable** is a that can be updated only when its current value matches the expected one. In our case, it means that a `prop` value should not be used if it has been changed by another thread.

The `java.util.concurrent.atomic` package has a dozen classes that support this logic: `AtomicBoolean`, `AtomicInteger`, `AtomicReference`, and `AtomicIntegerArray`, to name a few. Each of these classes has many methods that can be used for different synchronization needs. Check the online API documentation for each of these classes (`https://docs.oracle.com/en/java/javase/12/docs/api/java.base/java/util/concurrent/atomic/package-summary.html`). For the demonstration, we will use only two methods present in all of them:

- `V get()`: Returns the current value
- `boolean compareAndSet(V expectedValue, V newValue)`: Sets the value to `newValue` if the current value equals via operator (`==`) the `expectedValue`; returns `true` if successful or `false` if the actual value was not equal to the expected value

Here is how the `AtomicReference` class can be used to solve the problem of threads' interference while accessing the `prop` property of the `Calculator` object concurrently using these two methods:

```
class CalculatorAtomicRef implements Calculator {
    private AtomicReference<Double> prop = new AtomicReference<>(0.0);
    private String description = "Using AtomicReference";
    public String getDescription(){ return description; }
    public double calculate(int i){
        try {
            Double currentValue = prop.get();
            TimeUnit.MILLISECONDS.sleep(i);
            boolean b = this.prop.compareAndSet(currentValue, 2.0 * i);
            //System.out.println(b);    //prints: true for one thread
                                        //and false for another thread
            return Math.sqrt(this.prop.get());
        } catch (InterruptedException e) {
            Thread.currentThread().interrupt();
            System.out.println("Calculator was interrupted");
        }
        return 0.0;
    }
}
```

As you can see, the preceding code makes sure that the `currentValue` of the `prop` property does not change while the thread was sleeping. The following is the screenshot of the messages produced when we run `invokeAllCallables(new CalculatorAtomicRef())`:

```
Using AtomicReference:
Result{name=One, result=80.88430683757149}
Result{name=Two, result=80.88430683757149}
Waiting all threads completion for 100 MILLISECONDS...
isTerminated()=false
Calling shutdownNow()...
0 threads running
Exiting
```

Now the results produced by the threads are the same.

The following classes of the `java.util.concurrent` package provide synchronization support too:

- `Semaphore`: Restricts the number of threads that can access a resource
- `CountDownLatch`: Allows one or more threads to wait until a set of operations being performed in other threads are completed
- `CyclicBarrier`: Allows a set of threads to wait for each other to reach a common barrier point
- `Phaser`: Provides a more flexible form of barrier that may be used to control phased computation among multiple threads
- `Exchanger`: Allows two threads to exchange objects at a rendezvous point and is useful in several pipeline designs

Synchronized method

Another way to solve the problem is to use a synchronized method. Here is another implementation of the `Calculator` interface that uses this method of solving threads interference:

```java
class CalculatorSyncMethod implements Calculator {
    private double prop;
    private String description = "Using synchronized method";
    public String getDescription(){ return description; }
    synchronized public double calculate(int i){
        try {
```

```
            this.prop = 2.0 * i;
            TimeUnit.MILLISECONDS.sleep(i);
            return Math.sqrt(this.prop);
        } catch (InterruptedException e) {
            Thread.currentThread().interrupt();
            System.out.println("Calculator was interrupted");
        }
        return 0.0;
    }
}
```

We have just added the synchronized keyword in front of the calculate() method. Now, if we run invokeAllCallables(new CalculatorSyncMethod()), the results of both threads are always going to be the same:

```
Using synchronized method:
Result{name=One, result=80.88430683757149}
Result{name=Two, result=80.88430683757149}
Waiting all threads completion for 100 MILLISECONDS...
isTerminated()=false
Calling shutdownNow()...
0 threads running
Exiting
```

This is because another thread cannot enter the synchronized method until the current thread (the one that has entered the method already) has exited it. This is probably the simplest solution, but this approach may cause performance degradation if the method takes a long time to execute. In such cases, a synchronized block can be used, which wraps only several lines of code in an atomic operation.

Synchronized block

Here is an example of a synchronized block used to solve the problem of the threads' interference:

```
class CalculatorSyncBlock implements Calculator {
    private double prop;
    private String description = "Using synchronized block";
    public String getDescription(){
        return description;
    }
    public double calculate(int i){
        try {
```

```
        //there may be some other code here
        synchronized (this) {
            this.prop = 2.0 * i;
            TimeUnit.MILLISECONDS.sleep(i);
            return Math.sqrt(this.prop);
        }
    } catch (InterruptedException e) {
        Thread.currentThread().interrupt();
        System.out.println("Calculator was interrupted");
    }
    return 0.0;
    }
}
```

As you can see, the synchronized block acquires a lock on `this` object, which is shared by both threads, and releases it only after the threads exit the block. In our demo code, the block covers all the code of the method, so there is no difference in performance. But imagine there is more code in the method (we commented the location as there may be some other code here). If that is the case, the synchronized section of the code is smaller, thus having fewer chances to become a bottleneck.

If we run `invokeAllCallables(new CalculatorSyncBlock())`, the results are as follows:

```
Using synchronized block:
Result{name=One, result=80.88430683757149}
Result{name=Two, result=80.88430683757149}
Waiting all threads completion for 100 MILLISECONDS...
isTerminated()=false
Calling shutdownNow()...
0 threads running
Exiting
```

As you can see, the results are exactly the same as in the previous two examples. Different types of locks for different needs and with different behavior are assembled in the `java.util.concurrent.locks` package.

Each object in Java inherits the `wait()`, `notify()`, and `notifyAll()` methods from the base object. These methods can also be used to control the threads' behavior and their access to the locks.

Concurrent collections

Another way to address concurrency is to use a thread-safe collection from the `java.util.concurrent` package. Before you select which collection to use, read the *Javadoc* (`https://docs.oracle.com/en/java/javase/12/docs/api/index.html`) to see whether the limitations of the collection are acceptable for your application. Here is the list of these collections and some recommendations:

- `ConcurrentHashMap<K,V>`: Supports full concurrency of retrievals and high-expected concurrency for updates; use it when the concurrency requirements are very demanding and you need to allow locking on the write operation but do not need to lock the element.

- `ConcurrentLinkedQueue<E>`: A thread-safe queue based on linked nodes; employs an efficient non-blocking algorithm.

- `ConcurrentLinkedDeque<E>`: A concurrent queue based on linked nodes; both `ConcurrentLinkedQueque` and `ConcurrentLinkedDeque` are an appropriate choice when many threads share access to a common collection.

- `ConcurrentSkipListMap<K,V>`: A concurrent `ConcurrentNavigableMap` interface implementation.

- `ConcurrentSkipListSet<E>`: A concurrent `NavigableSet` implementation based on a `ConcurrentSkipListMap`.
  The `ConcurrentSkipListSet` and `ConcurrentSkipListMap` classes, as per the *Javadoc, provide expected average log(n) time cost for the contains, add, and remove operations and their variants. Ascending ordered views and their iterators are faster than descending ones*; use them when you need to iterate quickly through the elements in a certain order.

- `CopyOnWriteArrayList<E>`: A thread-safe variant of `ArrayList` in which all mutative operations (add, set, and so on) are implemented by making a fresh copy of the underlying array; as per the *Javadoc*, the `CopyOnWriteArrayList` class *is ordinarily too costly, but may be more efficient than alternatives when traversal operations vastly outnumber mutations, and is useful when you cannot or don't want to synchronize traversals, yet need to preclude interference among concurrent threads*; use it when you do not need to add new elements at different positions and do not require sorting; otherwise, use `ConcurrentSkipListSet`.

- `CopyOnWriteArraySet<E>`: A set that uses an internal `CopyOnWriteArrayList` for all of its operations.

- `PriorityBlockingQueue`: It is a better choice when a natural order is acceptable and you need fast adding of elements to the tail and fast removing of elements from the head of the queue; **blocking** means that the queue waits to become non-empty when retrieving an element and waits for space to become available in the queue when storing an element.
- `ArrayBlockingQueue`, `LinkedBlockingQueue`, and `LinkedBlockingDeque` have a fixed size (bounded); the other queues are unbounded.

Use these and similar characteristics and recommendations as with the guidelines, but execute comprehensive testing and performance-measuring before and after implementing your functionality. To demonstrate some of these collections capabilities, let's use `CopyOnWriteArrayList<E>`. First, let's look at how an `ArrayList` behaves when we try to modify it concurrently:

```
List<String> list = Arrays.asList("One", "Two");
System.out.println(list);
try {
    for (String e : list) {
        System.out.println(e);   //prints: One
        list.add("Three");       //UnsupportedOperationException
    }
} catch (Exception ex) {
    ex.printStackTrace();
}
System.out.println(list);        //prints: [One, Two]
```

As expected, the attempt to modify a list while iterating on it generates an exception and the list remains unmodified.

Now, let's use `CopyOnWriteArrayList<E>` in the same circumstances:

```
List<String> list =
            new CopyOnWriteArrayList<>(Arrays.asList("One", "Two"));
System.out.println(list);
try {
    for (String e : list) {
        System.out.print(e + " "); //prints: One Two
        list.add("Three");         //adds element Three
    }
} catch (Exception ex) {
    ex.printStackTrace();
}
System.out.println("\n" + list);   //prints: [One, Two, Three, Three]
```

The output this code produces looks as follows:

```
[One, Two]
One Two
[One, Two, Three, Three]
```

As you can see, the list was modified without an exception, but not the currently iterated copy. That is the behavior you can use if needed.

Addressing memory consistency error

Memory consistency errors can have many forms and causes in a multithreaded environment. They are well discussed in the *Javadoc* of the `java.util.concurrent` package. Here, we will mention only the most common case, which is caused by a lack of visibility.

When one thread changes a property value, the other might not see the change immediately, and you cannot use the synchronized keyword for a primitive type. In such a situation, consider using the `volatile` keyword for the property; it guarantees its read/write visibility between different threads.

Concurrency problems are not easy to solve. That is why it is not surprising that more and more developers now take a more radical approach. Instead of managing an object state, they prefer processing data in a set of stateless operations. We will see examples of such code in `Chapter 13`, *Functional Programming* and `Chapter 14`, *Java Standard Streams*. It seems that Java and many modern languages and computer systems are evolving in this direction.

Summary

In this chapter, we talked about multithreaded processing, the ways to organize it and avoid unpredictable results caused by the concurrent modification of the shared resource. We have shown readers how to create threads and execute them using pools of threads. We have also demonstrated how the results can be extracted from the threads that have completed successfully and discussed the difference between parallel and concurrent processing.

In the next chapter, we will provide readers with a deeper understanding of JVM, its structure and processes, and we'll discuss in detail the garbage collection process that keeps memory from being overflown. By the end of the chapter, the readers will know what constitutes Java application execution, Java processes inside JVM, garbage collection, and how JVM works in general.

Quiz

1. Select all correct statements:
   a. JVM process can have main threads.
   b. Main thread is the main process.
   c. A process can launch another process.
   d. A thread may launch another thread.

2. Select all correct statements:
   a. A daemon is a user thread.
   b. A daemon thread exits after the first user thread completes.
   c. A daemon thread exits after the last user thread completes.
   d. Main thread is a user thread.

3. Select all correct statements:
   a. All threads have `java.lang.Thread` as a base class.
   b. All threads extend `java.lang.Thread`.
   c. All threads implement `java.lang.Thread`.
   d. Daemon thread does not extend `java.lang.Thread`.

4. Select all correct statements:
   a. Any class can implement the `Runnable` interface.
   b. The `Runnable` interface implementation is a thread.
   c. The `Runnable` interface implementation is used by a thread.
   d. The `Runnable` interface has only one method.

5. Select all correct statements:
    a. A thread name has to be unique.
    b. A thread ID is generated automatically.
    c. A thread name can be set.
    d. A thread priority can be set.

6. Select all correct statements:
    a. A thread pool executes threads.
    b. A thread pool reuses threads.
    c. Some thread pools can have a fixed count of threads.
    d. Some thread pools can have an unlimited count of threads.

7. Select all correct statements:
    a. A `Future` object is the only way to get the result from a thread.
    b. A `Callable` object is the only way to get the result from a thread.
    c. A `Callable` object allows getting the result from a thread.
    d. A `Future` object represents a thread.

8. Select all correct statements:
    a. Concurrent processing can be done in parallel.
    b. Parallel processing is possible only with several CPUs or cores available on the computer.
    c. Parallel processing is concurrent processing.
    d. Without multiple CPU, concurrent processing is impossible.

9. Select all correct statements:
    a. Concurrent modification always leads to incorrect results.
    b. An atomic variable protects the property from concurrent modification.
    c. An atomic variable protects the property from the thread interference.
    d. An atomic variable is the only way to protect the property from concurrent modification.

10. Select all correct statements:
    a. A `synchronized` method is the best way to avoid thread interference.
    b. A `synchronized` keyword can be applied to any method.
    c. A `synchronized` method can create a processing bottleneck.
    d. A `synchronized` method is easy to implement.

11. Select all correct statements:
    a. A `synchronized` block makes sense only when it is smaller than the method.
    b. A `synchronized` block requires shared lock.
    c. Every Java object can provide a lock.
    d. A `synchronized` block is the best way to avoid thread interference.

12. Select all correct statements:
    a. Using a concurrent collection is preferred rather than using a non-concurrent one.
    b. Using a concurrent collection incurs some overhead.
    c. Not every concurrent collection fits every concurrent processing scenario.
    d. One can create a concurrent collection by calling `Collections.makeConcurrent()` method.

13. Select all correct statements:
    a. The only way to avoid memory consistency error is to declare the `volatile` variable.
    b. Using the `volatile` keyword guarantees visibility of the value change across all threads.
    c. One of the ways to avoid concurrency is to avoid any state management.
    d. Stateless utility methods cannot have concurrency issues.

JVM Structure and Garbage Collection

9

This chapter provides readers with an overview of **Java Virtual Machine (JVM)** structure and behavior, which are more complex than you might expect.

A JVM is just an executor of instructions according to the coded logic. It also finds and loads into the memory the `.class` files requested by the application, verifies them, interprets the bytecodes (that is, it translates them into platform-specific binary code), and passes the resulting binary code to the central processor (or processors) for execution. It uses several service threads in addition to the application threads. One of the service threads, called **garbage collection (GC)**, performs an important mission of releasing the memory from unused objects.

After reading this chapter, readers will better understand what constitutes Java application execution, Java processes inside JVM, GC, and how JVM works in general.

The following topics will be covered in this chapter:

- Java application execution
- Java processes
- JVM structure
- Garbage collection

Java application execution

Before getting deeper into how the JVM works, let's review how to run an application, bearing in mind that the following statements are used as synonyms:

- Run/execute/start the main class
- Run/execute/start the main method
- Run/execute/start/launch an application
- Run/execute/start/launch JVM or a Java process

There are also several ways to do it. In Chapter 1, *Getting Started with Java 12*, we showed you how to run the main(String[]) method using IntelliJ IDEA. In this chapter, we will just repeat some of what has been said already and add other variations that might be helpful for you.

Using an IDE

Any IDE allows for running the main() method. In IntelliJ IDEA, it can be done in three ways:

1. Click the green triangle next to the main() method name:

```
package com.packt.learnjava.ch09_jvm;

public class MyApplication {
    public static void main(String... args){
        System.out.println("Hello, world!");
    }
}
```

2. After you have executed the main() method using the green triangle at least once, the name of the class will be added to the drop-down menu (on the top line, to the left of the green triangle):

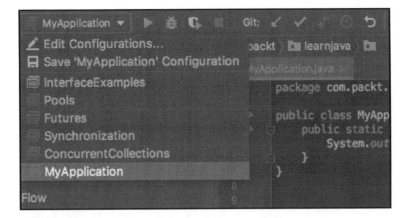

Select it and click the green triangle to the right of the menu:

- Open the **Run** menu and select the name of the class. There are several different options to select:

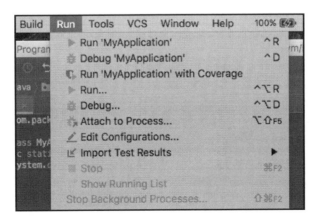

In the previous screenshot, you can also see the option to **Edit Configurations....**
It can be used for setting the **Program arguments** that are passed to the main()
method at the start, and some other options:

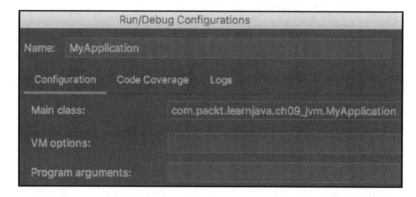

The **VM options** field allows for setting java command options. For example, if you input
the -Xlog:gc, the IDE will form the following java command:

```
java -Xlog:gc -cp . com.packt.learnjava.ch09_jvm.MyApplication
```

The -Xlog:gc option requires the GC log to be displayed. We will use this option in the
next section to demonstrate how GC works. The -cp . option (**cp** stands for **classpath**)
indicates that the class is located in a folder on the file tree that starts from the current
directory (the one where the command is entered). In our case, the .class file is located in
the com/packt/learnjava/ch09_jvm folder, where com is the subfolder of the current
directory. The classpath can include many locations where JVM has to look for the .class
files necessary for the application's execution.

For this demonstration, let's set **VM options** as follows:

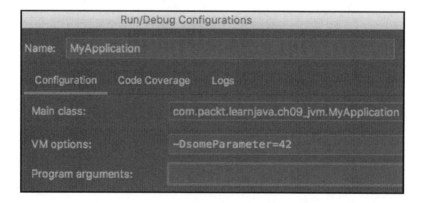

The **Program arguments** field allows for setting a parameter in the `java` command. For example, let's set `one two three` in this field:

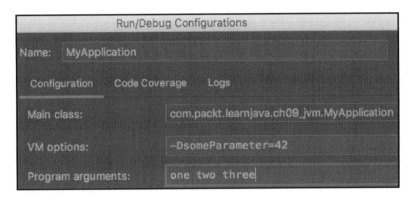

This setting will result in the following `java` command:

```
java -DsomeParameter=42 -cp . \
     com.packt.learnjava.ch09_jvm.MyApplication one two three
```

We can read these parameters in the `main()` method:

```java
public static void main(String... args){
    System.out.println("Hello, world!"); //prints: Hello, world!
    for(String arg: args){
        System.out.print(arg + " ");      //prints: one two three
    }
    String p = System.getProperty("someParameter");
    System.out.println("\n" + p);         //prints: 42
}
```

Another possible setting on the **Edit Configurations** screen is in the **Environment variables** field:

That is the way to set environment variables that can be accessed from the application using `System.getenv()`. For example, let's set the environment variables x and y , as follows:

If done as shown in the preceding screenshot, the values of x and y can be read not only in the `main()` method, but anywhere in the application using the `System.getenv("varName")` method. In our case, the values of x and y can be retrieved as follows:

```
String p = System.getenv("x");
System.out.println(p);                  //prints: 42
p = System.getenv("y");
System.out.println(p);                  //prints: 43
```

There are other parameters of the `java` command that can be set on the **Edit Configurations** screen, too. We encourage you to spend some time on that screen and view the possible options.

Using the command line with classes

Now, let's run `MyApplication` from a command line. To remind you, the main class looks as follows:

```
package com.packt.learnjava.ch09_jvm;
public class MyApplication {
    public static void main(String... args){
        System.out.println("Hello, world!"); //prints: Hello, world!
        for(String arg: args){
            System.out.print(arg + " ");      //prints all arguments
        }
        String p = System.getProperty("someParameter");
        System.out.println("\n" + p);      //prints someParameter set
                                           // as VM option -D
    }
}
```

First, it has to be compiled using the `javac` command. The command line looks as follows (provided you open the Terminal window in the root of the project, in the folder where `pom.xml` resides):

```
javac src/main/java/com/packt/learnjava/ch09_jvm/MyApplication.java
```

That is for Linux-type platforms. On Windows, the command looks similar:

```
javac src\main\java\com\packt\learnjava\ch09_jvm\MyApplication.java
```

The compiled `MyApplication.class` file is placed in the same folder with `MyApplication.java`. Now we can execute the compiled class with the `java` command:

```
java -DsomeParameter=42 -cp src/main/java \
         com.packt.learnjava.ch09_jvm.MyApplication one two three
```

Notice that `-cp` points to the folder `src/main/java` (the path is relative to the current folder), where the package of the main class starts. The result is as follows:

If the application uses other `.class` files located in different folders, all the paths to these folders (relative to the current folder) can be listed after the `-cp` option, separated by a colon (`:`). For example, consider the following:

```
java -cp src/main/java:someOtherFolder/folder \
                com.packt.learnjava.ch09_jvm.MyApplication
```

Notice that the folders listed with the `-cp` option can contain any number of `.class` files. This way, JVM can find what it needs. For example, let's create a sub-package, `example`, in the `com.packt.learnjava.ch09_jvm` package with the `ExampleClass` class in it:

```
package com.packt.learnjava.ch09_jvm.example;
public class ExampleClass {
    public static int multiplyByTwo(int i){
        return 2 * i;
    }
}
```

Now let's use it in the MyApplication class:

```
package com.packt.learnjava.ch09_jvm;
import com.packt.learnjava.ch09_jvm.example.ExampleClass;
public class MyApplication {
    public static void main(String... args){
        System.out.println("Hello, world!"); //prints: Hello, world!
        for(String arg: args){
            System.out.print(arg + " ");
        }
        String p = System.getProperty("someParameter");
        System.out.println("\n" + p);   //prints someParameter value
        int i = ExampleClass.multiplyByTwo(2);
        System.out.println(i);
    }
}
```

We are going to compile the MyApplication class using the same javac command as before:

```
javac src/main/java/com/packt/learnjava/ch09_jvm/MyApplication.java
```

The result is the following error:

```
import com.packt.learnjava.ch09_jvm.example.ExampleClass;
                                          ^

src/main/java/com/packt/learnjava/ch09_jvm/MyApplication.java:14: error: cannot find symbol
        int i = ExampleClass.multiplyByTwo(2);
                ^

  symbol:   variable ExampleClass
  location: class MyApplication
2 errors
```

It means that the compiler cannot find the ExampleClass.class file. We need to compile it and put on the classpath:

```
javac src/main/java/com/packt/learnjava/ch09_jvm/example/ExampleClass.java
javac -cp src/main/java \
            src/main/java/com/packt/learnjava/ch09_jvm/MyApplication.java
```

As you can see, we have added the location of ExampleClass.class, which is src/main/java, to the classpath. Now, we can execute MyApplication.class:

```
java -cp src/main/java com.packt.learnjava.ch09_jvm.MyApplication
```

The result is as follows:

There is no need to list folders that contain classes from the **Java Class Library** (JCL). The JVM knows where to find them.

Using the command line with JAR files

Keeping the compiled files in a folder as `.class` files is not always convenient, especially when many compiled files of the same framework belong to different packages and are distributed as a single library. In such cases, the compiled `.class` files are usually archived together in a `.jar` file. The format of such an archive is the same as the format of a `.zip` file. The only difference is that a `.jar` file also includes a manifest file that contains metadata describing the archive (we will talk more about the manifest in the next section).

To demonstrate how to use it, let's create a `.jar` file with the `ExampleClass.class` file and another `.jar` file with `MyApplication.class` in it, using the following commands:

```
cd src/main/java
jar -cf myapp.jar com/packt/learnjava/ch09_jvm/MyApplication.class
jar -cf example.jar \
        com/packt/learnjava/ch09_jvm/example/ExampleClass.class
```

Notice that we need to run the `jar` command in the folder where the package of `.class` file begins.

Now we can run the application as follows:

```
java -cp myapp.jar:example.jar \
    com.packt.learnjava.ch09_jvm.MyApplication
```

The `.jar` files are in the current folder. If we would like to execute the application from another folder (let's go back to the root directory, `cd ../../..`), the command should look like this:

```
java -cp src/main/java/myapp.jar:src/main/java/example.jar \
                com.packt.learnjava.ch09_jvm.MyApplication
```

Notice that every .jar file has to be listed on the classpath individually. To specify just a folder where all the .jar files reside (as is the case with .class files) is not good enough. If the folder contains only .jar files, all these files can be included in the classpath as follows:

```
java -cp src/main/java/* com.packt.learnjava.ch09_jvm.MyApplication
```

As you can see, the wildcard symbol has to be added after the folder name.

Using the command line with an executable JAR file

It is possible to avoid specifying the main class in the command line. Instead, we can create an executable .jar file. It can be accomplished by placing the name of the main class – the one you need to run and that contains the main() method – into the manifest file. Here are the steps:

1. Create a text file, manifest.txt (the name actually does not matter, but this name makes the intent clear), which contains the following line:

   ```
   Main-Class: com.packt.learnjava.ch09_jvm.MyApplication
   ```

   There has to be a space after the colon (:), and there has to be an invisible newline symbol at the end, so make sure you have pressed the *Enter* key and the cursor has jumped to the beginning of the next line.

2. Execute the command:

   ```
   cd src/main/java
   jar -cfm myapp.jar manifest.txt
   com/packt/learnjava/ch09_jvm/*.class \
   com/packt/learnjava/ch09_jvm/example/*.class
   ```

   Notice the sequence of jar command options (fm) and the sequence of the following files: myapp.jar manifest.txt. They have to be the same because f stands for the file the jar command is going to create, and m stands for the manifest source. If you include options as mf, then the files have to be listed as manifest.txt myapp.jar.

3. Now we can run the application using the following command:

   ```
   java -jar myapp.jar
   ```

Another way to create an executable `.jar` file is much easier:

```
jar cfe myjar.jar com.packt.learnjava.ch09_jvm.MyApplication \
                  com/packt/learnjava/ch09_jvm/*.class         \
                  com/packt/learnjava/ch09_jvm/example/*.class
```

This command generates a manifest with the specified main class name automatically: option c stands for **create a new archive**, option f means **archive file name**, and option e indicates an **application entry point**.

Java processes

As you may have guessed already, JVM does not know anything about the Java language and source code. It only knows how to read bytecode. It reads the bytecodes and other information from `.class` files, transforms (interprets) the bytecodes into a sequence of binary code instructions specific to the current platform (where JVM is running), and passes the resulting binary code to the microprocessor that executes it. When talking about this transformation, programmers often refer to it as a **Java process**, or just **process**.

The JVM is often referred to as a **JVM instance**. That is because every time a `java` command is executed, a new instance of JVM is launched, dedicated to running the particular application as a separate process with its own allocated memory (the size of the memory is set as a default value or passed in as a command option). Inside this Java process, multiple threads are running, each with its own allocated memory. Some are service threads created by the JVM; others are application threads created and controlled by the application.

That is the big picture of JVM executing the compiled code. But if you look closer and read the JVM specification, you will discover that the word *process*, in relation to JVM, is used to describe the JVM internal processes too. The JVM specification identifies several other processes running inside the JVM that usually are not mentioned by programmers, except maybe the **class loading process**.

That is so because most of the time, we can successfully write and execute Java programs without knowing anything about the internal JVM processes. But once in a while, some general understanding of JVM's internal workings helps to identify the root cause of certain issues. That is why in this section, we will provide a short overview of all the processes that happen inside the JVM. Then, in the following sections, we will discuss in more detail JVM's memory structure and other aspects of its functionality that may be useful to a programmer.

There are two subsystems that run all the JVM internal processes:

- **The classloader**: This reads `.class` file and populates a method area in JVM memory with the class-related data:
    - Static fields
    - Method bytecodes
    - Class metadata that describes the class
- **The execution engine**: This executes the bytecodes using the following:
    - A heap area for object instantiation
    - Java and native method stacks for keeping track of the methods called
    - A garbage collection process that reclaims the memory

Processes that run inside the main JVM process include the following:

- Processes performed by the classloader include the following:
    - Class loading
    - Class linking
    - Class initialization
- Processes performed by the execution engine include the following:
    - Class instantiation
    - Method execution
    - Garbage collection
    - Application termination

The JVM architecture

JVM architecture can be described as having two subsystems: the **classloader** and the **execution engine**, which run the service processes and application threads using runtime data memory areas such as method area, heap, and application thread stacks. **Threads** are lightweight processes that require less resource allocation than the JVM execution process.

The list may give you the impression that these processes are executed sequentially. To some degree, this is true, if we're talking about one class only. It is not possible to do anything with a class before loading. The execution of a method can begin only after all the previous processes are completed. However, the GC, for example, does not happen immediately after the use of an object is stopped (see the *Garbage collection* section). Also, an application can exit any time when an unhandled exception or some other error happens.

Only the classloader processes are regulated by the JVM specification. The execution engine implementation is largely at the discretion of each vendor. It is based on the language semantics and the performance goals set by the implementation authors.

Processes of the execution engine are in a realm not regulated by the JVM specification. There is common sense, tradition, known and proven solutions, and a Java language specification that can guide a JVM vendor's implementation decision. But there is no single regulatory document. The good news is that the most popular JVMs use similar solutions or, at least, that's how it looks at a high level.

With this in mind, let's discuss each of the seven processes listed previously in more detail.

Class loading

According to the JVM specification, the loading phase includes finding the `.class` file by its name (in the locations listed on a classpath) and creating its representation in the memory.

The first class to be loaded is the one passed in the command line, with the `main(String[])` method in it. The classloader reads the `.class` file, parses it, and populates the method area with static fields and method bytecodes. It also creates an instance of `java.lang.Class` that describes the class. Then the classloader links the class (see the *Class linking* section), initializes it (see the *Class initialization* section), and then passes it to the execution engine for running its bytecodes.

The `main(String[])` method is an entrance door into the application. If it calls a method of another class, that class has to be found on the classpath, loaded, initialized, and only then can its method be executed too. If this - just loaded - method calls a method of another class, that class has to be found, loaded, and initialized too, and so on. That is how a Java application starts and gets going.

The `main(String[])` **method**

 Every class can have a `main(String[])` method, and often does. Such a method is used to run the class independently as a standalone application for testing or demo purposes. The presence of such a method does not make the class `main`. The class becomes `main` only if identified as such in a `java` command line or in a `.jar` file manifest.

That being said, let's continue with the discussion of the loading process.

If you look in the API of java.lang.Class, you will not see a public constructor there. The classloader creates its instance automatically and, by the way, it is the same instance that is returned by getClass() method you can invoke on any Java object.

It does not carry the class static data (that is maintained in the method area), nor state values (they are in an object, created during the execution). It does not contain method bytecodes either (they are stored in the method area, too). Instead, the Class instance provides metadata that describes the class – its name, package, fields, constructors, method signatures, and so on. The metadata is useful not only for JVM but also for the application.

 All the data created by the classloader in the memory and maintained by the execution engine is called a **binary representation of the type**.

If the .class file has errors or does not adhere to a certain format, the process is terminated. This means that some validation of the loaded class format and its bytecodes is performed by the loading process already. More verification follows at the beginning of the next process, called **class linking**.

Here is the high-level description of the loading process. It performs three tasks:

- Finding and reading the .class file
- Parsing it according to the internal data structure into the method area
- Creating an instance of java.lang.Class with the class metadata

Class linking

According to the JVM specification, the linking resolves the references of the loaded class, so the methods of the class can be executed.

Here is a high-level description of the linking process. It performs three tasks:

1. **Verification of the binary representation of a class or an interface**:

    Although JVM can reasonably expect that the .class file was produced by the Java compiler and all the instructions satisfy the constraints and requirements of the language, there is no guarantee that the loaded file was produced by the known compiler implementation, or a compiler at all.

That's why the first step of the linking process is verification. This makes sure that the binary representation of the class is structurally correct, which means the following:

- The arguments of each method invocation are compatible with the method descriptor.
- The return instruction matches the return type of its method.
- Some other checks and verification that vary depending on the JVM vendor.

2. **Preparation of static fields in the method area**:

After verification is successfully completed, the interface or class (static) variables are created in the method area and initialized to the default values of their types. The other kinds of initialization, such as the explicit assignments specified by a programmer and static initialization blocks, are deferred to the process called **class initialization** (see the *Class initialization* section).

3. **Resolution of symbolic references into concrete references that point to the method area**:

If the loaded bytecodes refer to other methods, interfaces, or classes, the symbolic references are resolved into concrete references that point to the method area, which is done by the resolution process. If the referred interfaces and classes are not loaded yet, the classloader finds them and loads as needed.

Class initialization

According to the JVM specification, the initialization is accomplished by executing the class initialization methods. That is when the programmer-defined initialization (in static blocks and static assignments) is performed, unless the class was already initialized at the request of another class.

The last part of this statement is an important one because the class may be requested several times by different (already loaded) methods, and also because JVM processes are executed by different threads and may access the same class concurrently. So, **coordination** (called also **synchronization**) between different threads is required, which substantially complicates the JVM implementation.

Class instantiation

This step may never happen. Technically, an instantiation process, triggered by the new operator, is the first step of the execution. If the `main(String[])` method (which is static) uses only static methods of other classes, the instantiation never happens. That's why it is reasonable to identify this process as separate from the execution.

Besides, this activity has very specific tasks:

- Allocating memory for the object (its state) in the heap area
- Initialization of the instance fields to the default values
- Creating thread stacks for Java and native methods

Execution starts when the first method (not a constructor) is ready to be executed. For every application thread, a dedicated runtime stack is created, where every method call is captured in a stack frame. For example, if an exception happens, we get data from the current stack frames when we call the `printStackTrace()` method.

Method execution

The first application thread (called **main thread**) is created when the `main(String[])` method starts executing. It can create other application threads.

The execution engine reads the bytecodes, interprets them, and sends the binary code to the microprocessor for execution. It also maintains a count of how many times and how often each method was called. If the count exceeds a certain threshold, the execution engine uses a compiler, called the **Just-In-Time (JIT)** compiler, which compiles the method bytecodes into native code. This way, the next time the method is called, it will be ready without needing an interpretation. It substantially improves code performance.

The instruction currently being executed and the address of the next instruction are maintained in the **program counter (PC)** registers. Each thread has its own dedicated PC registers. It also improves performance and keeps track of the execution.

Garbage collection

The **garbage collector (GC)** runs the process that identifies the objects that are not referenced anymore and can be removed from the memory.

There is a Java static method, `System.gc()`, which can be used programmatically to trigger the GC, but the immediate execution is not guaranteed. Every GC cycle affects the application performance, so the JVM has to maintain a balance between memory availability and the ability to execute the bytecodes quickly enough.

Application termination

There are several ways an application can be terminated (and the JVM stopped or exited) programmatically:

- Normal termination, without an error status code
- Abnormal termination, because of an unhandled exception
- Forced programmatic exit, with or without an error status code

If there are no exceptions and infinite loops, the `main(String[])` method completes with a return statement or after its last statement is executed. As soon as it happens, the main application thread passes the control flow to the JVM and the JVM stops executing too. That is the happy end, and many applications enjoy it in real life. Most of our examples, except those when we have demonstrated exceptions or infinite loops, have exited successfully too.

However, there are other ways a Java application can exit, some of them quite graceful too – others not so much. If the main application thread created child threads or, in other words, a programmer has written code that generates other threads, even the graceful exit may not be easy. It all depends on the kind of child threads created.

If any of them is a user thread (the default), then the JVM instance continues to run even after the main thread exits. Only after all user threads are completed does the JVM instance stop. It is possible for the main thread to request the child user thread to complete. But until it exits, the JVM continues running. And this means that the application is still running too.

But if all child threads are daemon threads, or there are no child threads running, the JVM instance stops running as soon as the main application thread exits.

How the application exits in the case of an exception depends on the code design. We have touched on this in Chapter 4, *Exception Handling*, while discussing the best practices of exception handling. If the thread captures all the exceptions in a try-catch block in `main(String[])` or a similarly high-level method, then it is up to the application (and the programmer who wrote the code) to decide how best to proceed – to try to change the input data and repeat the block of code that generated the exception, to log the error and continue, or to exit.

If, on the other hand, the exception remains unhandled and propagates into the JVM code, the thread (where the exception happened) stops executing and exits. What happens next depends on the type of the thread and some other conditions. The following are four possible options:

- If there are no other threads, the JVM stops executing and returns an error code and the stack trace.
- If the thread with an unhandled exception was not the main one, other threads (if present) continue running.
- If the main thread has thrown an unhandled exception and the child threads (if present) are daemons, they exit too.
- If there is at least one user child thread, the JVM continues running until all user threads exit.

There are also ways to programmatically force the application to stop:

- `System.exit(0);`
- `Runtime.getRuntime().exit(0);`
- `Runtime.getRuntime().halt(0);`

All of these methods force the JVM to stop executing any thread and exit with a status code passed in as the parameter (0 in our examples):

- Zero indicates normal termination.
- The nonzero value indicates abnormal termination.

If the Java command was launched by some script or another system, the value of the status code can be used for the automation of the decision-making regarding the next step. But that is already outside the application and Java code.

The first two methods have identical functionality because this is how `System.exit()` is implemented:

```
public static void exit(int status) {
    Runtime.getRuntime().exit(status);
}
```

 To see the source code in the IDE, just click on the method.

The JVM exits when some thread invokes the `exit()` method of the `Runtime` or `System` classes, or the `halt()` method of the `Runtime` class, and the exit or halt operation is permitted by the security manager. The difference between `exit()` and `halt()` is that `halt()` forces the JVM exit immediately, while `exit()` performs additional actions that can be set using the `Runtime.addShutdownHook()` method. But all these options are rarely used by mainstream programmers.

JVM structure

JVM structure can be described in terms of the runtime data structure in the memory and in terms of the two subsystems that use the runtime data – the classloader and the execution engine.

Runtime data areas

Each of the runtime data areas of JVM memory belongs to one of two categories:

- **Shared areas** that include the following:
    - **Method area**: Class metadata, static fields, and method bytecodes
    - **Heap area**: Objects (states)
- **Unshared areas** dedicated to a particular application thread, which includes the following:
    - **Java stack**: Current and caller frames, each frame keeping the state of Java (non native) method invocation:
        - Values of local variables
        - Method parameter values
        - Values of operands for intermediate calculations (operand stack)
        - Method return value (if any)
    - **PC register**: The next instruction to execute
    - **Native method stack**: The state of the native method invocations

We have already discussed that a programmer has to be careful when using reference types and not modify the object itself unless it needs to be done. In a multi-threaded application, if a reference to an object can be passed between threads, we have to be extra careful because of the possibility of the concurrent modification of the same data. On the bright side, though, such a shared a area can be and often is used as the method of communication between threads.

Classloaders

The classloader performs the following three functions:

- Reads a `.class` file
- Populates the method area
- Initializes static fields not initialized by a programmer

Execution engine

The execution engine does the following:

- Instantiates objects in the heap area
- Initializes static and instance fields, using initializers written by the programmer
- Adds/removes frames to and from the Java stack
- Updates the PC register with the next instruction to execute
- Maintains the native method stack
- Keeps count of method calls and compiles popular ones
- Finalizes objects
- Runs garbage collection
- Terminates the application

Garbage collection

Automatic memory management is an important aspect of JVM that relieves the programmer from the need to do it programmatically. In Java, the process that cleans up memory and allows it to be reused is called a **garbage collection**.

Responsiveness, throughput, and stop-the-world

The effectiveness of GC affects two major application characteristics – **responsiveness** and **throughput**:

- **Responsiveness**: This is measured by how quickly an application responds (brings the necessary data) to the request; for example, how quickly a website returns a page, or how quickly a desktop application responds to an event. The smaller the response time, the better the user experience.
- **Throughput**: This indicates the amount of work an application can do in a unit of time; for example, how many requests a web application can serve, or how many transactions the database can support. The bigger the number, the more value the application can potentially generate and the more user requests it can support.

Meanwhile, GC needs to move data around, which is impossible to accomplish while allowing data processing because the references are going to change. That's why GC needs to stop application thread execution once in a while for a period of time, called **stop-the-world**. The longer these periods are, the quicker GC does its job and the longer an application freeze lasts, which can eventually grow big enough to affect both the application's responsiveness and throughput.

Fortunately, it is possible to tune the GC behavior using Java command options, but that is outside the scope of this book. We will concentrate instead on a high-level view of the main activity of a GC inspecting objects in the heap and removing those that don't have references in any thread stack.

Object age and generations

The basic GC algorithm determines *how old* each object is. The term **age** refers to the number of collection cycles the object has survived.

When JVM starts, the heap is empty and is divided into three sections:

- The young generation
- The old or tenured generation
- Humongous regions for holding the objects that are 50% of the size of a standard region or larger

The young generation has three areas:

- An Eden space
- Survivor 0 (S0)
- Survivor 1 (S1)

The newly created objects are placed in Eden. When it is filling up, a minor GC process starts. It removes the unreferred and circular referred objects and moves the others to the S1 area. During the next minor collection, S0 and S1 switch roles. The referenced objects are moved from Eden and S1 to S0.

During each of the minor collections, the objects that have reached a certain age are moved to the old generation. As a result of this algorithm, the old generation contains objects that are older than a certain age. This area is bigger than the young generation and, because of that, the garbage collection here is more expensive and happens not as often as in the young generation. But it is checked eventually (after several minor collections). The unreferenced objects are removed and the memory is defragmented. This cleaning of the old generation is considered a major collection.

When stop-the-world is unavoidable

Some collections of the objects in the old generation are done concurrently and some are done using stop-the-world pauses. The steps include the following:

1. **Initial marking**: This marks the survivor regions (root regions) that may have references to objects in the old generation and is done using a stop-the-world pause.
2. **Scanning**: This searches survivor regions for references to the old generation, and is done concurrently while the application continues to run.
3. **Concurrent marking**: This marks live objects over the entire heap and is done concurrently while the application continues to run.
4. **Remark**: The completes the marking of live objects and is done using a stop-the-world pause.
5. **Cleanup**: The calculates the age of live objects and frees regions (using stop-the-world) and returns them to the free list. This is done concurrently.

The preceding sequence might be interspersed with the young generation evacuations because most of the objects are short-lived and it is easier to free up a lot of memory by scanning the young generation more often. There is also a mixed phase (when G1 collects the regions already marked as mostly garbage in both the young and the old generations), and humongous allocation (when large objects are moved to or evacuated from humongous regions).

To help with GC tuning, the JVM provides platform-dependent default selections for the garbage collector, heap size, and runtime compiler. But fortunately, the JVM vendors improve and tune the GC process all the time, so most of the applications work just fine with default GC behavior.

Summary

In this chapter, the reader has learned how a Java application can be executed using an IDE or the command line. Now you can write your own applications and launch them in a manner most appropriate for the given environment. Knowledge about the JVM structure and its processes – class loading, linking, initialization, execution, garbage collection, and application termination – provides you with better control over the application's execution and transparency about the performance and current state of the JVM.

In the next chapter, we will discuss and demonstrate how to manage – insert, read, update, and delete – data in a database from a Java application. We will also provide a short introduction to SQL language and basic database operations: how to connect to a database, how to create the database structure, how to write database expressions using SQL, and how to execute them.

Quiz

1. Select all of the correct statements:
   a. An IDE executes Java code without compiling it.
   b. An IDE uses the installed Java to execute the code.
   c. An IDE checks the code without using the Java installation.
   d. An IDE uses the compiler of the Java installation.

2. Select all of the correct statements:
   - a. All the classes used by the application have to be listed on the classpath.
   - b. The locations of all the classes used by the application have to be listed on the classpath.
   - c. The compiler can find a class if it is in the folder listed on the classpath.
   - d. Classes of the main package do not need to be listed on the classpath.

3. Select all of the correct statements:
   - a. All the `.jar` files used by the application have to be listed on the classpath.
   - b. The locations of all the `.jar` files used by the application have to be listed on the classpath.
   - c. The JVM can find a class only if it is in the `.jar` file listed on the classpath.
   - d. Every class can have the `main()` method.

4. Select all of the correct statements:
   - a. Every `.jar` file that has a manifest is an executable.
   - b. If the `-jar` option is used by the `java` command, the classpath option is ignored.
   - c. Every `.jar` file has a manifest.
   - d. An executable `.jar` is a ZIP file with a manifest.

5. Select all of the correct statements:
   - a. Class loading and linking can work in parallel on different classes.
   - b. Class loading moves the class to the execution area.
   - c. Class linking connects two classes.
   - d. Class linking uses memory references.

6. Select all of the correct statements:
   - a. Class initialization assigns values to the instance properties.
   - b. Class initialization happens every time the class is referred to by another class.
   - c. Class initialization assigns values to static properties.
   - d. Class initialization provides data to the instance of `java.lang.Class`.

7. Select all of the correct statements:
    a. Class instantiation may never happen.
    b. Class instantiation includes object properties initialization.
    c. Class instantiation includes memory allocation on a heap.
    d. Class instantiation includes executing a constructor code.

8. Select all of the correct statements:
    a. Method execution includes binary code generation.
    b. Method execution includes source code compilation.
    c. Method execution includes reusing the binary code produced by the JIT compiler.
    d. Method execution counts how many times every method was called.

9. Select all of the correct statements:
    a. Garbage collection starts immediately after the System.gc() method is called.
    b. The application can be terminated with or without an error code.
    c. The application exits as soon as an exception is thrown.
    d. The main thread is a user thread.

10. Select all of the correct statements:
    a. The JVM has memory areas shared across all threads.
    b. The JVM has memory areas not shared across threads.
    c. Class metadata is shared across all threads.
    d. Method parameter values are not shared across threads.

11. Select all of the correct statements:
    a. The classloader populates the method area.
    b. The classloader allocates memory on a heap.
    c. The classloader writes to the .class file.
    d. The classloader resolves method references.

12. Select all of the correct statements:
    a. The execution engine allocates memory on a heap.
    b. The execution engine terminates the application.
    c. The execution engine runs garbage collection.
    d. The execution engine initializes static fields not initialized by a programmer.

13. Select all of the correct statements:
    a. The number of transactions per second that a database can support is a throughput measure.
    b. When the garbage collector pauses the application, it is called stop-all-things.
    c. How slowly the website returns data is a responsiveness measure.
    d. The garbage collector clears the CPU queue of jobs.

14. Select all of the correct statements:
    a. Object age is measured by the number of seconds since the object was created.
    b. The older the object, the more probable it is going to be removed from the memory.
    c. Cleaning the old generation is a major collection.
    d. Moving an object from one area of the young generation to another area of the young generation is a minor collection.

15. Select all of the correct statements:
    a. The garbage collector can be tuned by setting the parameters of the `javac` command.
    b. The garbage collector can be tuned by setting the parameters of the `java` command.
    c. The garbage collector works with its own logic and cannot change its behavior based on the set parameters.
    d. Cleaning the old generation area requires a stop-the-world pause.

10
Managing Data in a Database

his chapter explains and demonstrates how to manage – that is, insert, read, update, and delete – data in a database using a Java application. It also provides a short introduction to **Structured Query Language** (**SQL**) and basic database operations, including how to connect to a database, how to create a database structure, how to write a database expression using SQL, and how to execute these expressions.

The following topics will be covered in this chapter:

- Creating a database
- Creating a database structure
- Connecting to a database
- Releasing the connection
- Create, read, update, and delete (CRUD) operations on data

Creating a database

Java Database Connectivity (**JDBC**) is a Java functionality that allows you to access and modify data in a database. It is supported by the JDBC API (which includes the `java.sql`, `javax.sql`, and `java.transaction.xa` packages), and the database-specific class that implements an interface for database access (called a **database driver**), which is provided by each database vendor.

Using JDBC means writing Java code that manages data in a database using the interfaces and classes of the JDBC API and a database-specific driver, which knows how to establish a connection with the particular database. Using this connection, an application can then issue requests written in SQL.

Naturally, we are only referring to the databases that understand SQL here. They are called relational or tabular **Database Management Systems (DBMSes)** and make up the vast majority of the currently-used DBMSes – although some alternatives (for example, a navigational database and NoSQL) are used too.

The `java.sql` and `javax.sql` packages are included in the **Java Platform Standard Edition (Java SE)**. The `javax.sql` package contains the `DataSource` interface that supports the statement's pooling, distributed transactions, and rowsets.

Creating a database involves the following eight steps:

1. Install the database by following the vendor instructions.
2. Create a database user, a database, a schema, tables, views, stored procedures, and anything else that is necessary to support the data model of the application.
3. Add to this application the dependency on a `.jar` file with the database-specific driver.
4. Connect to the database from the application.
5. Construct the SQL statement.
6. Execute the SQL statement.
7. Use the result of the execution as your application requires.
8. Release (that is, close) the database connection and any other resources that were opened in the process.

Steps 1 to 3 are performed only once during the database setup and before the application is run. Steps 4 to 8 are performed by the application repeatedly as needed. In fact, steps 5 to 7 can be repeated multiple times with the same database connection.

For our example, we are going to use the PostgreSQL database. You will first need to perform steps 1 to 3 by yourself using the database-specific instructions. To create the database for our demonstration, we use the following commands:

```
create user student SUPERUSER;
create database learnjava owner student;
```

These commands create a `student` user that can manage all aspects of the `SUPERUSER` database, and make the `student` user an owner of the `learnjava` database. We will use the `student` user to access and manage data from the Java code. In practice, for security considerations, an application is not allowed to create or change database tables and other aspects of the database structure.

Additionally, it is a good practice to create another logical layer called schema that can have its own set of users and permissions. This way, several schemas in the same database can be isolated, and each user (one of them is your application) can only access certain schemas. On an enterprise level, the common practice is to create synonyms for the database schema so that no application can access the original structure directly. However, we do not do this in this book for the sake of simplicity.

Creating a database structure

After the database is created, the following three SQL statements will allow you to create and change the database structure. This is done through database entities, such as a table, function, or constraint:

- The CREATE statement creates the database entity
- The ALTER statement changes the database entity
- The DROP statement deletes the database entity

There are also various SQL statements that allow you to inquire about each database entity. Such statements are database-specific and, typically, they are only used in a database console. For example, in the PostgreSQL console, \d <table> can be used to describe a table, while \dt lists all the tables. Refer to your database documentation for more details.

To create a table, you can execute the following SQL statement:

```
CREATE TABLE tablename ( column1 type1, column2 type2, ... );
```

The limitations for a table name, column names, and types of values that can be used depends on the particular database. Here is an example of a command that creates the person table in PostgreSQL:

```
CREATE table person (
    id SERIAL PRIMARY KEY,
    first_name VARCHAR NOT NULL,
    last_name VARCHAR NOT NULL,
    dob DATE NOT NULL );
```

The SERIAL keyword indicates that this field is a sequential integer number that is generated by the database every time a new record is created. Additional options for generating sequential integers are SMALLSERIAL and BIGSERIAL; they differ by size and the range of possible values:

```
SMALLSERIAL: 2 bytes, range from 1 to 32,767
SERIAL: 4 bytes, range from 1 to 2,147,483,647
BIGSERIAL: 8 bytes, range from 1 to 922,337,2036,854,775,807
```

The PRIMARY_KEY keyword indicates that this is going to be the unique identifier of the record, and will most probably be used in a search. The database creates an index for each primary key to make the search process faster. An index is a data structure that helps to accelerate data search in the table without having to check every table record. An index can include one or more columns of a table. If you request the description of the table, you will see all the existing indices.

Alternatively, we can make a composite PRIMARY KEY keyword using a combination of first_name, last_name, and dob:

```
CREATE table person (
    first_name VARCHAR NOT NULL,
    last_name VARCHAR NOT NULL,
    dob DATE NOT NULL,
    PRIMARY KEY (first_name, last_name, dob) );
```

However, there is a chance that there are two persons who will have the same name and were born on the same day.

The NOT NULL keyword imposes a constraint on the field: it cannot be empty. The database will raise an error for every attempt to create a new record with an empty field or delete the value from the existing record. We did not set the size of the columns of type VARCHAR, thus allowing these columns to store string values of any length.

The Java object that matches such a record may be represented by the following Person class:

```
public class Person {
    private int id;
    private LocalDate dob;
    private String firstName, lastName;
    public Person(String firstName, String lastName, LocalDate dob) {
        if (dob == null) {
            throw new RuntimeException("Date of birth cannot be null");
        }
        this.dob = dob;
        this.firstName = firstName == null ? "" : firstName;
```

```
            this.lastName = lastName == null ? "" : lastName;
        }
        public Person(int id, String firstName,
                      String lastName, LocalDate dob) {
            this(firstName, lastName, dob);
            this.id = id;
        }
        public int getId() { return id; }
        public LocalDate getDob() { return dob; }
        public String getFirstName() { return firstName; }
        public String getLastName() { return lastName; }
    }
```

As you may have noticed, there are two constructors in the Person class: with and without id. We will use the constructor that accepts id to construct an object based on the existing record, while the other constructor will be used to create an object before inserting a new record.

Once created, the table can be deleted using the DROP command:

```
DROP table person;
```

The existing table can also be changed using the ALTER SQL command; for example, we can add a column address:

```
ALTER table person add column address VARCHAR;
```

If you are not sure whether such a column exists already, you can add IF EXISTS or IF NOT EXISTS:

```
ALTER table person add column IF NOT EXISTS address VARCHAR;
```

However, this possibility exists only with PostgreSQL 9.6 and later versions.

Another important consideration to take note of during database table creation is whether another index (in addition to PRIMARY KEY) has to be added. For example, we can allow a case-insensitive search of first and last names by adding the following index:

```
CREATE index idx_names on person ((lower(first_name), lower(last_name));
```

If the search speed improves, we leave the index in place; if not, it can be removed as follows:

```
DROP index idx_names;
```

We remove it because an index has an overhead of additional writes and storage space.

We also can remove a column from a table if we need to, as follows:

```
ALTER table person DROP column address;
```

In our examples, we follow the naming convention of PostgreSQL. If you use a different database, we suggest that you look up its naming convention and follow it, so that the names you create align with those that are created automatically.

Connecting to a database

So far, we have used a console to execute SQL statements. The same statements can be executed from the Java code using the JDBC API too. But tables are created only once, so there is not much sense in writing a program for a one-time execution.

Data management, however, is another matter. So, from now on, we will use Java code to manipulate the data in a database. In order to do this, we first need to add the following dependency to the pom.xml file:

```
<dependency>
    <groupId>org.postgresql</groupId>
    <artifactId>postgresql</artifactId>
    <version>42.2.2</version>
</dependency>
```

This matches the PostgreSQL version 9.6 that we have installed. Now we can create a database connection from the Java code, as follows:

```
String URL = "jdbc:postgresql://localhost/learnjava";
Properties prop = new Properties();
prop.put( "user", "student" );
// prop.put( "password", "secretPass123" );
try {
    Connection conn = DriverManager.getConnection(URL, prop);
} catch (SQLException ex) {
    ex.printStackTrace();
}
```

The preceding code is just an example of how to create a connection using the `java.sql.DriverManger` class. The `prop.put( "password", "secretPass123" )` statement demonstrates how to provide a password for the connection using the `java.util.Properties` class. However, we did not set a password when we created the `student` user, so we do not need it.

Many other values can be passed to `DriverManager` that configure the connection behavior. The name of the keys for the passed-in properties are the same for all major databases, but some of them are database-specific. So, read your database vendor documentation for more details.

Alternatively, for passing `user` and `password` only, we could use an overloaded `DriverManager.getConnection(String url, String user, String password)` version. It is a good practice to keep the password encrypted. We are not going to demonstrate how to do it, but there are plenty of guides available on the internet that you can refer to.

Another way of connecting to a database is to use the `javax.sql.DataSource` interface. Its implementation is included in the same `.jar` file as the database driver. In the case of `PostgreSQL`, there are two classes that implement the `DataSource` interface:

- `org.postgresql.ds.PGSimpleDataSource`
- `org.postgresq l.ds.PGConnectionPoolDataSource`

We can use these classes instead of `DriverManager`. The following code is an example of creating a database connection using the `PGSimpleDataSource` class:

```
PGSimpleDataSource source = new PGSimpleDataSource();
source.setServerName("localhost");
source.setDatabaseName("learnjava");
source.setUser("student");
//source.setPassword("password");
source.setLoginTimeout(10);
try {
    Connection conn = source.getConnection();
} catch (SQLException ex) {
    ex.printStackTrace();
}
```

Using the `PGConnectionPoolDataSource` class allows you to create a pool of `Connection` objects in memory, as follows:

```
PGConnectionPoolDataSource source = new PGConnectionPoolDataSource();
source.setServerName("localhost");
source.setDatabaseName("learnjava");
source.setUser("student");
//source.setPassword("password");
source.setLoginTimeout(10);
try {
    PooledConnection conn = source.getPooledConnection();
    Set<Connection> pool = new HashSet<>();
    for(int i = 0; i < 10; i++){
        pool.add(conn.getConnection())
    }
} catch (SQLException ex) {
    ex.printStackTrace();
}
```

This is a preferred method because creating a `Connection` object takes time. Pooling allows you to do it up front and then reuse the created objects when they are needed. After the connection is no longer required, it can be returned to the pool and reused. The pool size and other parameters can be set in a configuration file (such as `postgresql.conf` for PostgreSQL).

However, you do not need to manage the connection pool yourself. There are several mature frameworks that can do it for you, such as HikariCP (`https://brettwooldridge.github.io/HikariCP`), Vibur (`http://www.vibur.org`), and Commons DBCP (`https://commons.apache.org/proper/commons-dbcp`) – they are reliable and easy to use.

Whatever method of creating a database connection we choose, we are going to hide it inside the `getConnection()` method and use it in all our code examples in the same way. With the object of the `Connection` class acquired, we can now access the database to add, read, delete, or modify the stored data.

Releasing the connection

Keeping the database connection alive requires a significant number of resources – such as memory and CPU – so, it is a good idea to close the connection and release the allocated resources as soon as you no longer need them. In the case of pooling, the `Connection` object, when closed, is returned to the pool and consumes fewer resources.

Before Java 7, a connection was closed by invoking the `close()` method in a `finally` block:

```
try {
    Connection conn = getConnection();
    //use object conn here
} finally {
    if(conn != null){
        try {
            conn.close();
        } catch (SQLException e) {
            e.printStackTrace();
        }
    }
}
```

The code inside the `finally` block is always executed, whether the exception inside the `try` block is thrown or not. However, since Java 7, the try-with-resources construct also does the job on any object that implements the `java.lang.AutoCloseable` or `java.io.Closeable` interface. Since the `java.sql.Connection` object does implement the `AutoCloseable` interface, we can rewrite the previous code snippet as follows:

```
try (Connection conn = getConnection()) {
    //use object conn here
} catch(SQLException ex) {
    ex.printStackTrace();
}
```

The `catch` clause is necessary because the `AutoCloseable` resource throws `java.sql.SQLException`.

CRUD data

There are four kinds of SQL statements that read or manipulate data in the database:

- The `INSERT` statement adds data to the database
- The `SELECT` statement reads data from the database
- The `UPDATE` statement changes data in the database
- The `DELETE` statement deletes data from the database

Either one or several different clauses can be added to the preceding statements to identify the data that is requested (such as the WHERE clause) and the order in which the results have to be returned (such as the ORDER clause).

The JDBC connection is represented by java.sql.Connection. This, among others, has the methods required to create three types of objects that allow you to execute SQL statements that provide different functionality to the database side:

- java.sql.Statement: This simply sends the statement to the database server for execution
- java.sql.PreparedStatement: This caches the statement with a certain execution path on the database server by allowing it to be executed multiple times with different parameters in an efficient manner
- java.sql.CallableStatement: This executes the stored procedure in the database

In this section, we are going to review how to do it in Java code. The best practice is to test the SQL statement in the database console before using it programmatically.

The INSERT statement

The INSERT statement creates (populates) data in the database, and has the following format:

```
INSERT into table_name (column1, column2, column3,...)
             values (value1, value2, value3,...);
```

Alternatively, when several records need to be added, you can use the following format:

```
INSERT into table_name (column1, column2, column3,...)
             values (value1, value2, value3,... ),
                    (value21, value22, value23,...),
                    ...;
```

The SELECT statement

The SELECT statement has the following format:

```
SELECT column_name, column_name FROM table_name
                            WHERE some_column = some_value;
```

Alternatively, when all the columns need to be selected, you can use the following format:

```
SELECT * from table_name WHERE some_column=some_value;
```

A more general definition of the WHERE clause is as follows:

```
WHERE column_name operator value
Operator:
= Equal
<> Not equal. In some versions of SQL, !=
> Greater than
< Less than
>= Greater than or equal
<= Less than or equal IN Specifies multiple possible values for a column
LIKE Specifies the search pattern
BETWEEN Specifies the inclusive range of values in a column
```

The construct's column_name operator value can be combined using the AND and OR logical operators, and grouped by brackets, ( ).

For example, the following method brings all the first name values (separated by a whitespace character) from the person table:

```
String selectAllFirstNames() {
    String result = "";
    Connection conn = getConnection();
    try (conn; Statement st = conn.createStatement()) {
        ResultSet rs = st.executeQuery("select first_name from person");
        while (rs.next()) {
            result += rs.getString(1) + " ";
        }
    } catch (SQLException ex) {
        ex.printStackTrace();
    }
    return result;
}
```

The getString(int position) method of the ResultSet interface extracts the String value from position 1 (the first in the list of columns in the SELECT statement). There are similar getters for all primitive types: getInt(int position), getByte(int position), and more.

It is also possible to extract the value from the ResultSet object using the column name. In our case, it will be getString("first_name"). This method of getting values is especially useful when the SELECT statement is as follows:

```
select * from person;
```

However, bear in mind that extracting values from the `ResultSet` object using the column name is less efficient. But the difference in performance, though, is very small and only becomes important when the operation takes place many times. Only the actual measuring and testing processes can tell if the difference is significant to your application or not. Extracting values by the column name is especially attractive because it provides better code readability, which pays off in the long run during application maintenance.

There are many other useful methods in the `ResultSet` interface. If your application reads data from a database, we highly recommend that you read the official documentation of the `SELECT` statement and the `ResultSet` interface.

The UPDATE statement

The data can be changed by the UPDATE statement, as follows:

```
UPDATE table_name SET column1=value1,column2=value2,... WHERE clause;
```

We can use this statement to change the first name in one of the records from the original value, John, to a new value, Jim:

```
update person set first_name = 'Jim' where last_name = 'Adams';
```

Without the WHERE clause, all the records of the table will be affected.

The DELETE statement

To remove records from a table, use the DELETE statement, as follows:

```
DELETE FROM table_name WHERE clause;
```

Without the WHERE clause, all the records of the table are deleted. In the case of the person table, we can delete all the records using the following SQL statement:

```
delete from person;
```

Additionally, this statement only deletes the records that have the first name of Jim:

```
delete from person where first_name = 'Jim';
```

Using statements

The `java.sql.Statement` interface offers the following methods for executing SQL statements:

- `boolean execute(String sql)`: This returns `true` if the executed statement returns data (inside the `java.sql.ResultSet` object) that can be retrieved using the `ResultSet getResultSet()` method of the `java.sql.Statement` interface. Alternatively, it returns `false` if the executed statement does not return data (for the INSERT statement or the UPDATE statement) and the subsequent call to the `int getUpdateCount()` method of the `java.sql.Statement` interface returns the number of the affected rows.

- `ResultSet executeQuery(String sql)`: This returns data as a `java.sql.ResultSet` object (the SQL statement used with this method is usually a SELECT statement). The `ResultSet getResultSet()` method of the `java.sql.Statement` interface does not return data, while the `int getUpdateCount()` method of the `java.sql.Statement` interface returns –1.

- `int executeUpdate(String sql)`: This returns the number of the affected rows (the executed SQL statement is expected to be the UPDATE statement or the DELETE statement). The same number is returned by the `int getUpdateCount()` method of the `java.sql.Statement` interface; the subsequent call to the `ResultSet getResultSet()` method of the `java.sql.Statement` interface returns `null`.

We will demonstrate how these three methods work on each of the statements: INSERT, SELECT, UPDATE, and DELETE.

The execute(String sql) method

Let's try executing each of the statements; we'll start with the INSERT statement:

```
String sql = "insert into person (first_name, last_name, dob) " +
                        "values ('Bill', 'Grey', '1980-01-27')";
Connection conn = getConnection();
try (conn; Statement st = conn.createStatement()) {
    System.out.println(st.execute(sql));           //prints: false
    System.out.println(st.getResultSet() == null); //prints: true
    System.out.println(st.getUpdateCount());       //prints: 1
} catch (SQLException ex) {
    ex.printStackTrace();
```

```
}
System.out.println(selectAllFirstNames());                    //prints: Bill
```

The preceding code adds a new record to the `person` table.
The returned `false` value indicates that there is no data returned by the executed
statement; this is why the `getResultSet()` method returns `null`. But
the `getUpdateCount()` method returns 1 because one record was affected (added).
The `selectAllFirstNames()` method proves that the expected record was inserted.

Now let's execute the `SELECT` statement, as follows:

```
String sql = "select first_name from person";
Connection conn = getConnection();
try (conn; Statement st = conn.createStatement()) {
    System.out.println(st.execute(sql));                      //prints: true
    ResultSet rs = st.getResultSet();
    System.out.println(rs == null);                           //prints: false
    System.out.println(st.getUpdateCount());                  //prints: -1
    while (rs.next()) {
        System.out.println(rs.getString(1) + " ");            //prints: Bill
    }
} catch (SQLException ex) {
    ex.printStackTrace();
}
```

The preceding code selects all the first names from the `person` table. The returned
`true` value indicates that there is data returned by the executed statement. That is why
the `getResultSet()` method does not return `null`, but a `ResultSet` object instead.
The `getUpdateCount()` method returns −1 because no record was affected (changed).
Since there was only one record in the `person` table, the `ResultSet` object contains only
one result, and `rs.getString(1)` returns `Bill`.

The following code uses the `UPDATE` statement to change the first name in all the records of
the `person` table to `Adam`:

```
String sql = "update person set first_name = 'Adam'";
Connection conn = getConnection();
try (conn; Statement st = conn.createStatement()) {
    System.out.println(st.execute(sql));                      //prints: false
    System.out.println(st.getResultSet() == null);            //prints: true
    System.out.println(st.getUpdateCount());                  //prints: 1
} catch (SQLException ex) {
    ex.printStackTrace();
}
System.out.println(selectAllFirstNames());                    //prints: Adam
```

In the preceding code, the returned `false` value indicates that there is no data returned by the executed statement. This is why the `getResultSet()` method returns `null`. But the `getUpdateCount()` method returns 1 because one record was affected (changed) since there was only one record in the `person` table. The `selectAllFirstNames()` method proves that the expected change was made to the record.

The following `DELETE` statement execution deletes all records from the `person` table:

```
String sql = "delete from person";
Connection conn = getConnection();
try (conn; Statement st = conn.createStatement()) {
    System.out.println(st.execute(sql));              //prints: false
    System.out.println(st.getResultSet() == null);    //prints: true
    System.out.println(st.getUpdateCount());          //prints: 1
} catch (SQLException ex) {
    ex.printStackTrace();
}
System.out.println(selectAllFirstNames());            //prints:
```

In the preceding code, the returned `false` value indicates that there is no data returned by the executed statement. That is why the `getResultSet()` method returns `null`. But the `getUpdateCount()` method returns 1 because one record was affected (deleted) since there was only one record in the `person` table. The `selectAllFirstNames()` method proves that there are no records in the `person` table.

The executeQuery(String sql) method

In this section, we will try to execute the same statements (as a query) that we used when demonstrating the `execute()` method in *The execute(String sql) method* section. We'll start with the `INSERT` statement, as follows:

```
String sql = "insert into person (first_name, last_name, dob) " +
                        "values ('Bill', 'Grey', '1980-01-27')";
Connection conn = getConnection();
try (conn; Statement st = conn.createStatement()) {
    st.executeQuery(sql);                        //PSQLException
} catch (SQLException ex) {
    ex.printStackTrace();                        //prints: stack trace
}
System.out.println(selectAllFirstNames()); //prints: Bill
```

The preceding code generates an exception with the `No results were returned by the query` message because the `executeQuery()` method expects to execute the `SELECT` statement. Nevertheless, the `selectAllFirstNames()` method proves that the expected record was inserted.

Now let's execute the `SELECT` statement, as follows:

```
String sql = "select first_name from person";
Connection conn = getConnection();
try (conn; Statement st = conn.createStatement()) {
    ResultSet rs1 = st.executeQuery(sql);
    System.out.println(rs1 == null);            //prints: false
    ResultSet rs2 = st.getResultSet();
    System.out.println(rs2 == null);            //prints: false
    System.out.println(st.getUpdateCount());    //prints: -1
    while (rs1.next()) {
        System.out.println(rs1.getString(1)); //prints: Bill
    }
    while (rs2.next()) {
        System.out.println(rs2.getString(1)); //prints:
    }
} catch (SQLException ex) {
    ex.printStackTrace();
}
```

The preceding code selects all the first names from the `person` table. The returned `false` value indicates that `executeQuery()` always returns the `ResultSet` object, even when no record exists in the `person` table. As you can see, there appear to be two ways of getting a result from the executed statement. However, the `rs2` object has no data, so, while using the `executeQuery()` method, make sure that you get the data from the `ResultSet` object.

Now let's try to execute an `UPDATE` statement as follows:

```
String sql = "update person set first_name = 'Adam'";
Connection conn = getConnection();
try (conn; Statement st = conn.createStatement()) {
    st.executeQuery(sql);                      //PSQLException
} catch (SQLException ex) {
    ex.printStackTrace();                      //prints: stack trace
}
System.out.println(selectAllFirstNames()); //prints: Adam
```

The preceding code generates an exception with the No results were returned by
the query message because the executeQuery() method expects to execute the SELECT
statement. Nevertheless, the selectAllFirstNames() method proves that the expected
change was made to the record.

We are going to get the same exception while executing the DELETE statement:

```
String sql = "delete from person";
Connection conn = getConnection();
try (conn; Statement st = conn.createStatement()) {
    st.executeQuery(sql);                        //PSQLException
} catch (SQLException ex) {
    ex.printStackTrace();                        //prints: stack trace
}
System.out.println(selectAllFirstNames()); //prints:
```

Nevertheless, the selectAllFirstNames() method proves that all the records of
the person table were deleted.

Our demonstration shows that executeQuery() should be used for SELECT statements
only. The advantage of the executeQuery() method is that, when used for SELECT
statements, it returns a not-null ResultSet object even when there is no data selected,
which simplifies the code since there is no need to check the returned value for null.

The executeUpdate(String sql) method

We'll start demonstrating the executeUpdate() method with the INSERT statement:

```
String sql = "insert into person (first_name, last_name, dob) " +
                        "values ('Bill', 'Grey', '1980-01-27')";
Connection conn = getConnection();
try (conn; Statement st = conn.createStatement()) {
    System.out.println(st.executeUpdate(sql));   //prints: 1
    System.out.println(st.getResultSet());       //prints: null
    System.out.println(st.getUpdateCount());     //prints: 1
} catch (SQLException ex) {
    ex.printStackTrace();
}
System.out.println(selectAllFirstNames());       //prints: Bill
```

As you can see, the executeUpdate() method returns the number of the affected (inserted, in this case) rows. The same number returns the int getUpdateCount() method, while the ResultSet getResultSet() method returns null. The selectAllFirstNames() method proves that the expected record was inserted.

The executeUpdate() method can't be used for executing the SELECT statement:

```
String sql = "select first_name from person";
Connection conn = getConnection();
try (conn; Statement st = conn.createStatement()) {
    st.executeUpdate(sql);      //PSQLException
} catch (SQLException ex) {
    ex.printStackTrace();       //prints: stack trace
}
```

The message of the exception is A result was returned when none was expected.

The UPDATE statement, on the other hand, is executed by the executeUpdate() method just fine:

```
String sql = "update person set first_name = 'Adam'";
Connection conn = getConnection();
try (conn; Statement st = conn.createStatement()) {
    System.out.println(st.executeUpdate(sql));   //prints: 1
    System.out.println(st.getResultSet());       //prints: null
    System.out.println(st.getUpdateCount());     //prints: 1
} catch (SQLException ex) {
    ex.printStackTrace();
}
System.out.println(selectAllFirstNames());       //prints: Adam
```

The executeUpdate() method returns the number of the affected (updated, in this case) rows. The same number returns the int getUpdateCount() method, while the ResultSet getResultSet() method returns null. The selectAllFirstNames() method proves that the expected record was updated.

The DELETE statement produces similar results:

```
String sql = "delete from person";
Connection conn = getConnection();
try (conn; Statement st = conn.createStatement()) {
    System.out.println(st.executeUpdate(sql));   //prints: 1
    System.out.println(st.getResultSet());       //prints: null
    System.out.println(st.getUpdateCount());     //prints: 1
} catch (SQLException ex) {
```

```
        ex.printStackTrace();
    }
    System.out.println(selectAllFirstNames());        //prints:
```

By now, you have probably realized that the `executeUpdate()` method is better suited for INSERT, UPDATE, and DELETE statements.

Using PreparedStatement

`PreparedStatement` is a sub-interface of the `Statement` interface. This means that it can be used anywhere that the `Statement` interface is used. The difference is that `PreparedStatement` is cached in the database instead of being compiled every time it is invoked. This way, it is efficiently executed multiple times for different input values. Similar to `Statement`, it can be created by the `prepareStatement()` method using the same `Connection` object.

Since the same SQL statement can be used for creating `Statement` and `PreparedStatement`, it is a good idea to use `PreparedStatement` for any SQL statement that is called multiple times because it performs better than the `Statement` interface on the database side. To do this, all we need to change are these two lines from the preceding code example:

```
try (conn; Statement st = conn.createStatement()) {
    ResultSet rs = st.executeQuery(sql);
```

Instead, we can use the `PreparedStatement` class as follows:

```
try (conn; PreparedStatement st = conn.prepareStatement(sql)) {
    ResultSet rs = st.executeQuery();
```

To create the `PreparedStatement` class with parameters, you can substitute the input values with the question mark symbol (?); for example, we can create the following method:

```
List<Person> selectPersonsByFirstName(String searchName) {
    List<Person> list = new ArrayList<>();
    Connection conn = getConnection();
    String sql = "select * from person where first_name = ?";
    try (conn; PreparedStatement st = conn.prepareStatement(sql)) {
        st.setString(1, searchName);
        ResultSet rs = st.executeQuery();
        while (rs.next()) {
            list.add(new Person(rs.getInt("id"),
                    rs.getString("first_name"),
```

```
                    rs.getString("last_name"),
                    rs.getDate("dob").toLocalDate()));
      }
    } catch (SQLException ex) {
        ex.printStackTrace();
    }
    return list;
}
```

The database compiles the `PreparedStatement` class as a template and stores it without executing. Then, when it is later used by the application, the parameter value is passed to the template, which can be executed immediately without the overhead of compilation, since it has been done already.

Another advantage of a prepared statement is that it is better protected from a SQL injection attack because values are passed in using a different protocol and the template is not based on the external input.

If a prepared statement is used only once, it may be slower than a regular statement, but the difference may be negligible. If in doubt, test the performance and see if it is acceptable for your application – the increased security could be worth it.

Using CallableStatement

The `CallableStatement` interface (which extends the `PreparedStatement` interface) can be used to execute a stored procedure, although some databases allow you to call a stored procedure using either a `Statement` or
`PreparedStatement` interface. A `CallableStatement` object is created by
the `prepareCall()` method and can have parameters of three types:

- `IN` for an input value
- `OUT` for the result
- `IN OUT` for either an input or an output value

The `IN` parameter can be set the same way as the parameters of `PreparedStatement`, while the `OUT` parameter must be registered by the `registerOutParameter()` method of `CallableStatement`.

It is worth noting that executing a stored procedure from Java programmatically is one of the least standardized areas. PostgreSQL, for example, does not support stored procedures directly, but they can be invoked as functions, which have been modified for this purpose by interpreting the OUT parameters as return values. Oracle, on the other hand, allows the OUT parameters as functions too.

This is why the following differences between database functions and stored procedures can serve only as general guidelines and not as formal definitions:

- A function has a return value, but it does not allow OUT parameters (except for some databases) and can be used in a SQL statement.
- A stored procedure does not have a return value (except for some databases); it allows OUT parameters (for most databases) and can be executed using the JDBC CallableStatement interface.

You can refer to the database documentation to learn how to execute a stored procedure.

Since stored procedures are compiled and stored on the database server, the execute() method of CallableStatement performs better for the same SQL statement than the corresponding method of the Statement or PreparedStatement interface. This is one of the reasons why a lot of Java code is sometimes replaced by one or several stored procedures that even include business logic. However, there is no right answer for every case and problem, so we will refrain from making specific recommendations, except to repeat the familiar mantra about the value of testing and the clarity of the code you are writing.

Let's, for example, call the replace(string origText, from substr1, to substr2) function that comes with the PostgreSQL installation. It searches the first parameter (string origText) and replaces all the substrings in it that match the second parameter (from substr1) using the string provided by the third parameter (string substr2). The following Java method executes this function using CallableStatement:

```java
String replace(String origText, String substr1, String substr2) {
    String result = "";
    String sql = "{ ? = call replace(?, ?, ? ) }";
    Connection conn = getConnection();
    try (conn; CallableStatement st = conn.prepareCall(sql)) {
        st.registerOutParameter(1, Types.VARCHAR);
        st.setString(2, origText);
        st.setString(3, substr1);
        st.setString(4, substr2);
        st.execute();
        result = st.getString(1);
    } catch (Exception ex){
```

```
            ex.printStackTrace();
        }
        return result;
    }
```

Now we can call this method as follows:

```
String result = replace("That is original text",
                            "original text", "the result");
    System.out.println(result);   //prints: That is the result
```

A stored procedure can be without any parameters at all, with IN parameters only, with OUT parameters only, or with both. The result may be one or multiple values, or a ResultSet object. You can find the syntax of the SQL for function creation in your database documentation.

Summary

In this chapter, we discussed and demonstrated how the data in a database can be populated, read, updated, and deleted from a Java application. A short introduction to the SQL language described how to create a database and its structure, how to modify it, and how to execute SQL statements, using Statement, PreparedStatement, and CallableStatement.

In the next chapter, we will describe and discuss the most popular network protocols, demonstrate how to use them, and how to implement client-server communication using the latest Java HTTP Client API. The protocols reviewed include the Java implementation of a communication protocol based on TCP, UDP, and URLs.

Quiz

1. Select all the correct statements:
     a. JDBC stands for Java Database Communication.
     b. The JDBC API includes the java.db package.
     c. The JDBC API comes with Java installation.
     d. The JDBC API includes the drivers for all major DBMSes.

2. Select all the correct statements:
   a. A database table can be created using the CREATE statement.
   b. A database table can be changed using the UPDATE statement.
   c. A database table can be removed using the DELETE statement.
   d. Each database column can have an index.

3. Select all the correct statements:
   a. To connect to a database, you can use the Connect class.
   b. Every database connection must be closed.
   c. The same database connection may be used for many operations.
   d. Database connections can be pooled.

4. Select all the correct statements:
   a. A database connection can be closed automatically using the try-with-resources construct.
   b. A database connection can be closed using the finally block construct.
   c. A database connection can be closed using the catch block.
   d. A database connection can be closed without a try block.

5. Select all the correct statements:
   a. The INSERT statement includes a table name.
   b. The INSERT statement includes column names.
   c. The INSERT statement includes values.
   d. The INSERT statement includes constraints.

6. Select all the correct statements:
   a. The SELECT statement must include a table name.
   b. The SELECT statement must include a column name.
   c. The SELECT statement must include the WHERE clause.
   d. The SELECT statement may include the ORDER clause.

7. Select all the correct statements:
   a. The UPDATE statement must include a table name.
   b. The UPDATE statement must include a column name.
   c. The UPDATE statement may include the WHERE clause.
   d. The UPDATE statement may include the ORDER clause.

8. Select all the correct statements:
   a. The DELETE statement must include a table name.
   b. The DELETE statement must include a column name.
   c. The DELETE statement may include the WHERE clause.
   d. The DELETE statement may include the ORDER clause.

9. Select all the correct statements about the execute() method of the Statement interface:
   a. It receives a SQL statement.
   b. It returns a ResultSet object.
   c. The Statement object may return data after execute() was called.
   d. The Statement object may return the number of affected records after execute() was called.

10. Select all the correct statements about the executeQuery() method of the Statement interface:
    a. It receives a SQL statement.
    b. It returns a ResultSet object.
    c. The Statement object may return data after executeQuery() was called.
    d. The Statement object may return the number of affected records after executeQuery() was called.

11. Select all correct statements about the executeUpdate() method of the interface Statement interface:
    a. It receives a SQL statement.
    b. It returns a ResultSet object.
    c. The Statement object may return data after executeUpdate() was called.
    d. The Statement object returns the number of affected records after executeUpdate() was called.

12. Select all the correct statements about the `PreparedStatement` interface:

    a. It extends `Statement`.

    b. An object of type `PreparedStatement` is created by the `prepareStatement()` method.

    c. It is always more efficient than `Statement`.

    d. It results in a template in the database being created only once.

13. Select all the correct statements about the `CallableStatement` interface:

    a. It extends `PreparedStatement`.

    b. An object of type `CallableStatement` is created by the `prepareCall()` method.

    c. It is always more efficient than `PreparedStatement`.

    d. It results in a template in the database being created only once.

11
Network Programming

In this chapter, we will describe and discuss the most popular network protocols – **User Datagram Protocol (UDP)**, **Transmission Control Protocol (TCP)**, **HyperText Transfer Protocol (HTTP)**, and **WebSocket** – and their support from the **Java Class Library (JCL)**. We will demonstrate how to use these protocols and how to implement client-server communication in Java code. We will also review **Uniform Resource Locator (URL)**-based communication and the latest **Java HTTP Client API**.

The following topics will be covered in this chapter:

- Network protocols
- UDP-based communication
- TCP-based communication
- UDP versus TCP protocols
- URL-based communication
- Using the HTTP 2 Client API

Network protocols

Network programming is a vast area. The **internet protocol (IP)** suite consists of four layers, each of which has a dozen or more protocols:

- **The link layer**: The group of protocols used when a client is physically connected to the host; three core protocols include the **Address Resolution Protocol (ARP)**, the **Reverse Address Resolution Protocol (RARP)**, and the **Neighbor Discovery Protocol (NDP)**.

- **The internet layer**: The group of inter-networking methods, protocols, and specifications used to transport network packets from the originating host to the destination host, specified by an IP address. The core protocols of this layer are **Internet Protocol version 4 (IPv4)** and **Internet Protocol version 6 (IPv6)**; IPv6 specifies a new packet format and allocates 128 bits for the dotted IP address, compared to 32 bits in IPv4. An example of an IPv4 address is `10011010.00010111.11111110.00010001`, which results in an IP address of `154.23.254.17`.

- **The transport layer**: The group of host-to-host communication services. It includes TCP, also known as TCP/IP protocol, and UDP (which we are going to discuss shortly); other protocols in this group are the **Datagram Congestion Control Protocol (DCCP)** and the **Stream Control Transmission Protocol (SCTP)**.

- **The application layer**: The group of protocols and interface methods used by hosts in a communication network. It includes **Telnet, File Transfer Protocol (FTP), Domain Name System (DNS), Simple Mail Transfer Protocol (SMTP), Lightweight Directory Access Protocol (LDAP), Hypertext Transfer Protocol (HTTP), Hypertext Transfer Protocol Secure (HTTPS)**, and **Secure Shell (SSH)**.

The link layer is the lowest layer; it is used by the internet layer that is, in turn, used by the transport layer. This transport layer is then used by the application layer in support of the protocol implementations.

For security reasons, Java does not provide access to the protocols of the link layer and the internet layer. This means that Java does not allow you to create custom transport protocols that, for example, serve as an alternative to TCP/IP. That is why, in this chapter, we will review only the protocols of the transport layer (TCP and UDP) and the application layer (HTTP). We will explain and demonstrate how Java supports them and how a Java application can take advantage of this support.

Java supports the TCP and UDP protocols with classes of the `java.net` package, while the HTTP protocol can be implemented in the Java application using the classes of the `java.net.http` package (which was introduced with Java 11).

Both the TCP and UDP protocols can be implemented in Java using *sockets*. Sockets are identified by a combination of an IP address and a port number, and they represent a connection between two applications. Since the UDP protocol is somewhat simpler than the TCP protocol, we'll start with UDP.

UDP-based communication

The UDP protocol was designed by David P. Reed in 1980. It allows applications to send messages called **datagrams** using a simple connectionless communication model with a minimal protocol mechanism, such as a checksum, for data integrity. It has no handshaking dialogs and, thus, does not guarantee message delivery or preserving the order of messages. It is suitable for those cases when dropping messages or mixing up orders are preferred to waiting for retransmission.

A datagram is represented by the `java.net.DatagramPacket` class. An object of this class can be created using one of six constructors; the following two constructors are the most commonly used:

- `DatagramPacket(byte[] buffer, int length)`: This constructor creates a datagram packet and is used to receive the packets; `buffer` holds the incoming datagram, while `length` is the number of bytes to be read.
- `DatagramPacket(byte[] buffer, int length, InetAddress address, int port)`: This creates a datagram packet and is used to send the packets; `buffer` holds the packet data, `length` is the packet data length, `address` holds the destination IP address, and `port` is the destination port number.

Once constructed, the `DatagramPacket` object exposes the following methods that can be used to extract data from the object or set/get its properties:

- `void setAddress(InetAddress iaddr)`: This sets the destination IP address.
- `InetAddress getAddress()`: This returns the destination or source IP address.
- `void setData(byte[] buf)`: This sets the data buffer.
- `void setData(byte[] buf, int offset, int length)`: This sets the data buffer, data offset, and length.
- `void setLength(int length)`: This sets the length for the packet.
- `byte[] getData()`: This returns the data buffer
- `int getLength()`: This returns the length of the packet that is to be sent or received.
- `int getOffset()`: This returns the offset of the data that is to be sent or received.
- `void setPort(int port)`: This sets the destination port number.
- `int getPort()`: This returns the port number where data is to be sent or received from.

Once a `DatagramPacket` object is created, it can be sent or received using the `DatagramSocket` class, which represents a connectionless socket for sending and receiving datagram packets. An object of this class can be created using one of six constructors; the following three constructors are the most commonly used:

- `DatagramSocket()`: This creates a datagram socket and binds it to any available port on the local host machine. It is typically used to create a sending socket because the destination address (and port) can be set inside the packet (see the preceding `DatagramPacket` constructors and methods).
- `DatagramSocket(int port)`: This creates a datagram socket and binds it to the specified port on the local host machine. It is used to create a receiving socket when any local machine address (called a **wildcard address**) is good enough.
- `DatagramSocket(int port, InetAddress address)`: This creates a datagram socket and binds it to the specified port and the specified local address; the local port must be between 0 and 65535. It is used to create a receiving socket when a particular local machine address needs to be bound.

The following two methods of the `DatagramSocket` object are the most commonly used for sending and receiving messages (or packets):

- `void send(DatagramPacket p)`: This sends the specified packet.
- `void receive(DatagramPacket p)`: This receives a packet by filling the specified `DatagramPacket` object's buffer with the data received. The specified `DatagramPacket` object also contains the sender's IP address and the port number on the sender's machine.

Let's take a look at a code example; here is the UDP message receiver that exits after the message has been received:

```
public class UdpReceiver {
    public static void main(String[] args){
        try(DatagramSocket ds = new DatagramSocket(3333)){
            DatagramPacket dp = new DatagramPacket(new byte[16], 16);
            ds.receive(dp);
            for(byte b: dp.getData()){
                System.out.print(Character.toString(b));
            }
        } catch (Exception ex){
            ex.printStackTrace();
        }
    }
}
```

As you can see, the receiver is listening for a text message (it interprets each byte as a character) on any address of the local machine on port 3333. It uses a buffer of 16 bytes only; as soon as the buffer is filled with the received data, the receiver prints its content and exits.

Here is an example of the UDP message sender:

```
public class UdpSender {
    public static void main(String[] args) {
        try(DatagramSocket ds = new DatagramSocket()){
            String msg = "Hi, there! How are you?";
            InetAddress address = InetAddress.getByName("127.0.0.1");
            DatagramPacket dp = new DatagramPacket(msg.getBytes(),
                                        msg.length(), address, 3333);
            ds.send(dp);
        } catch (Exception ex){
            ex.printStackTrace();
        }
    }
}
```

As you can see, the sender constructs a packet with the message, the local machine address, and the same port as the one that the receiver uses. After the constructed packet is sent, the sender exits.

We can run the sender now, but without the receiver running there is nobody to get the message. So, we'll start the receiver first. It listens on port 3333, but there is no message coming – so it waits. Then, we run the sender and the receiver displays the following message:

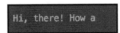

Since the buffer is smaller than the message, it was only partially received – the rest of the message is lost. We can create an infinite loop and let the receiver run indefinitely:

```
while(true){
    ds.receive(dp);
    for(byte b: dp.getData()){
        System.out.print(Character.toString(b));
    }
    System.out.println();
}
```

By doing so, we can run the sender several times; here is what the receiver prints if we run the sender three times:

```
Hi, there! How a
Hi, there! How a
Hi, there! How a
```

As you can see, all three messages are received; however, only the first 16 bytes of each message are captured by the receiver.

Now let's make the receiving buffer bigger than the message:

```
DatagramPacket dp = new DatagramPacket(new byte[30], 30);
```

If we send the same message now, the result will be as follows:

```
Hi, there! How are you?
```

To avoid processing empty buffer elements, you can use the getLength() method of the DatagramPacket class, which returns the actual number of buffer elements filled with the message:

```
int i = 1;
for(byte b: dp.getData()){
    System.out.print(Character.toString(b));
    if(i++ == dp.getLength()){
        break;
    }
}
```

The result of the preceding code will be as follows:

```
Hi, there! How are you?
```

So, this is the basic idea of the UDP protocol. The sender sends a message to a certain address and port even if there is no socket that *listens* on this address and port. It does not require establishing any kind of connection before sending the message, which makes the UDP protocol faster and more lightweight than the TCP protocol (which requires you to establish the connection first). This way, the TCP protocol takes message sending to another level of reliability – by making sure that the destination exists and that the message can be delivered.

TCP-based communication

TCP was designed by the **Defense Advanced Research Projects Agency (DARPA)** in the 1970s for use in the **Advanced Research Projects Agency Network (ARPANET)**. It complements IP and, thus, is also referred to as TCP/IP. The TCP protocol, even by its name, indicates that it provides reliable (that is, error-checked or controlled) data transmission. It allows the ordered delivery of bytes in an IP network and is widely used by the web, email, secure shell, and file transfer.

An application that uses TCP/IP is not even aware of all the handshaking that takes place between the socket and the transmission details – such as network congestion, traffic load balancing, duplication, and even loss of some IP packets. The underlying protocol implementation of the transport layer detects these problems, resends the data, reconstructs the order of the sent packets, and minimizes network congestion.

In contrast to the UDP protocol, TCP/IP-based communication is focused on accurate delivery at the expense of the delivery period. That's why it is not used for real-time applications, such as voice over IP, where reliable delivery and correct sequential ordering are required. However, if every bit needs to arrive exactly as it was sent and in the same sequence, then TCP/IP is irreplaceable.

To support such behavior, TCP/IP communication maintains a session throughout the communication. The session is identified by the client address and port. Each session is represented by an entry in a table on the server. This contains all the metadata about the session: the client IP address and port, the connection status, and the buffer parameters. But these details are usually hidden from the application developer, so we won't go into any more detail here. Instead, we will turn to the Java code.

Similar to the UDP protocol, the TCP/IP protocol implementation in Java uses sockets. But instead of the `java.net.DatagramSocket` class that implements the UDP protocol, the TCP/IP-based sockets are represented by the `java.net.ServerSocket` and `java.net.Socket` classes. They allow sending and receiving messages between two applications, one of them being a server and the other a client.

The ServerSocket and SocketClass classes perform very similar jobs. The only difference is that the ServerSocket class has the accept() method that *accepts* the request from the client. This means that the server has to be up and ready to receive the request first. Then, the connection is initiated by the client that creates its own socket that sends the connection request (from the constructor of the Socket class). The server then accepts the request and creates a local socket connected to the remote socket (on the client side).

After establishing the connection, data transmission can occur using I/O streams as described in Chapter 5, *Strings, Input/Output, and Files*. The Socket object has the getOutputStream() and getInputStream() methods that provide access to the socket's data streams. Data from the java.io.OutputStream object on the local computer appear as coming from the java.io.InputStream object on the remote machine.

Let's now take a closer look at the java.net.ServerSocket and java.net.Socket classes, and then run some examples of their usage.

The java.net.ServerSocket class

The java.net.ServerSocket class has four constructors:

- ServerSocket(): This creates a server socket object that is not bound to a particular address and port. It requires the use of the bind() method to bind the socket.
- ServerSocket(int port): This creates a server socket object bound to the provided port. The port value must be between 0 and 65535. If the port number is specified as a value of 0, this means that the port number needs to be bound automatically. By default, the maximum queue length for incoming connections is 50.
- ServerSocket(int port, int backlog): This provides the same functionality as the ServerSocket(int port) constructor, and allows you to set the maximum queue length for incoming connections by the backlog parameter.
- ServerSocket(int port, int backlog, InetAddress bindAddr): This creates a server socket object that is similar to the preceding constructor, but also bound to the provided IP address. When the bindAddr value is null, it will default to accepting connections on any or all local addresses.

The following four methods of the ServerSocket class are the most commonly used, and they are essential for establishing a socket's connection:

- void bind(SocketAddress endpoint): This binds the ServerSocket object to a specific IP address and port. If the provided address is null, then the system will pick up a port and a valid local address automatically (which can be later retrieved using the getLocalPort(), getLocalSocketAddress(), and getInetAddress() methods). Additionally, if the ServerSocket object was created by the constructor without any parameters, then this method, or the following bind() method, needs to be invoked before a connection can be established.

- void bind(SocketAddress endpoint, int backlog): This acts in a similar way to the preceding method; the backlog argument is the maximum number of pending connections on the socket (that is, the size of the queue). If the backlog value is less than or equal to 0, then an implementation-specific default will be used.

- void setSoTimeout(int timeout): This sets the value (in milliseconds) of how long the socket waits for a client after the accept() method is called. If the client has not called and the timeout expires, an java.net.SocketTimeoutException exception is thrown, but the ServerSocket object remains valid and can be reused. The timeout value of 0 is interpreted as an infinite timeout (the accept() method blocks until a client calls).

- Socket accept(): This blocks until a client calls or the timeout period (if set) expires.

Other methods of the class allow you to set or get other properties of the Socket object and they can be used for better dynamic management of the socket connection. You can refer to the online documentation of the class to understand the available options in more detail.

The following code is an example of a server implementation using the ServerSocket class:

```
public class TcpServer {
    public static void main(String[] args){
        try(Socket s = new ServerSocket(3333).accept();
            DataInputStream dis = new DataInputStream(s.getInputStream());
            DataOutputStream dout = new DataOutputStream(s.getOutputStream());
            BufferedReader console =
                        new BufferedReader(new InputStreamReader(System.in))){
            while(true){
                String msg = dis.readUTF();
```

```
            System.out.println("Client said: " + msg);
            if("end".equalsIgnoreCase(msg)){
                break;
            }
            System.out.print("Say something: ");
            msg = console.readLine();
            dout.writeUTF(msg);
            dout.flush();
            if("end".equalsIgnoreCase(msg)){
                break;
            }
        }
    } catch(Exception ex) {
        ex.printStackTrace();
    }
  }
}
```

Let's walk through the preceding code. In the try-with-resources statement, we create `Socket`, `DataInputStream`, and `DataOutputStream` objects based on our newly-created socket, and the `BufferedReader` object to read the user input from the console (we will use it to enter the data). While creating the socket, the `accept()` method blocks until a client tries to connect to port 3333 of the local server.

Then, the code enters an infinite loop. First, it reads the bytes sent by the client as a Unicode character string encoded in a modified UTF-8 format by using the `readUTF()` method of `DataInputStream`. The result is printed with the `"Client said: "` prefix. If the received message is an `"end"` string, then the code exits the loop and the server's program exits. If the message is not `"end"`, then the `"Say something: "` prompt is displayed on the console and the `readLine()` method blocks until a user types something and clicks on *Enter*.

The server takes the input from the screen and writes it as a Unicode character string to the output stream using the `writeUtf()` method. As we mentioned already, the output stream of the server is connected to the input stream of the client. If the client reads from the input stream, it receives the message sent by the server. If the sent message is `"end"`, then the sever exits the loop and the program. If not, then the loop body is executed again.

The described algorithm assumes that the client exits only when it sends or receives the `"end"` message. Otherwise, the client generates an exception if it tries to send a message to the server afterward. This demonstrates the difference between the UDP and TCP protocols that we mentioned already – TCP is based on the session that is established between the server and client sockets. If one side drops it, the other side immediately encounters an error.

Now let's review an example of a TCP-client implementation.

The java.net.Socket class

The `java.net.Socket` class should now be familiar to you since it was used in the preceding example. We used it to access the input and output streams of the connected sockets. Now we are going to review the `Socket` class systematically and explore how it can be used to create a TCP client. The `Socket` class has four constructors:

- `Socket()`: This creates an unconnected socket. It uses the `connect()` method to establish a connection of this socket with a socket on a server.
- `Socket(String host, int port)`: This creates a socket and connects it to the provided port on the `host` server. If it throws an exception, the connection to the server is not established; otherwise; you can start sending data to the server.
- `Socket(InetAddress address, int port)`: This acts in a similar way to the preceding constructor, except that the host is provided as an `InetAddress` object.
- `Socket(String host, int port, InetAddress localAddr, int localPort)`: This works in a similar way to the preceding constructor, except that it also allows you to bind the socket to the provided local address and port (if the program is run on a machine with multiple IP addresses). If the provided `localAddr` value is `null`, any local address is selected. Alternatively, if the provided `localPort` value is `null`, then the system picks up a free port in the bind operation.
- `Socket(InetAddress address, int port, InetAddress localAddr, int localPort)`: This acts in a similar way to the preceding constructor, except that the local address is provided as an `InetAddress` object.

Here are the following two methods of the `Socket` class that we have used already:

- `InputStream getInputStream()`: This returns an object that represents the source (the remote socket) and brings the data (inputs them) into the program (the local socket).

- `OutputStream getOutputStream()`: This returns an object that represents the source (the local socket) and sends the data (outputs them) to a remote socket.

Let's now examine the TCP-client code, as follows:

```
public class TcpClient {
  public static void main(String[] args) {
    try(Socket s = new Socket("localhost",3333);
      DataInputStream dis = new DataInputStream(s.getInputStream());
      DataOutputStream dout = new DataOutputStream(s.getOutputStream());
      BufferedReader console =
                new BufferedReader(new InputStreamReader(System.in))){
        String prompt = "Say something: ";
        System.out.print(prompt);
        String msg;
        while ((msg = console.readLine()) != null) {
          dout.writeUTF( msg);
          dout.flush();
          if (msg.equalsIgnoreCase("end")) {
            break;
          }
          msg = dis.readUTF();
          System.out.println("Server said: " +msg);
          if (msg.equalsIgnoreCase("end")) {
            break;
          }
          System.out.print(prompt);
        }
    } catch(Exception ex){
        ex.printStackTrace();
    }
  }
}
```

The preceding `TcpClient` code looks almost exactly the same as the `TcpServer` code we reviewed. The only principal difference is that the `new Socket("localhost", 3333)` constructor attempts to establish a connection with the `"localhost:3333"` server immediately, so it expects that the `localhost` server is up and listening on port 3333; the rest is the same as the server code.

Therefore, the only reason we need to use the `ServerSocket` class is to allow the server to run while waiting for the client to connect to it; everything else can be done using only the `Socket` class.

Other methods of the `Socket` class allow you to set or get other properties of the `socket` object, and they can be used for better dynamic management of the socket connection. You can read the online documentation of the class to understand the available options in more detail.

Running the examples

Let's now run the `TcpServer` and `TcpClient` programs. If we start `TcpClient` first, we get `java.net.ConnectException` with the **Connection refused** message. So, we launch the `TcpServer` program first. When it starts, no messages are displayed, instead, it just waits until the client connects. So, we then start `TcpClient` and see the following message on the screen:

```
Say something:
```

We type `Hello!` and click on *Enter*:

```
Say something: Hello!
```

Now let's look at the server-side screen:

```
Client said: Hello!
Say something:
```

We type `Hi!` on the server-side screen and click on *Enter*:

```
Client said: Hello!
Say something: Hi!
```

On the client-side screen, we see the following messages:

```
Say something: Hello!
Server said: Hi!
Say something:
```

We can continue this dialog indefinitely until the server or the client sends the message, end. Let's make the client do it; the client says end and then exits:

```
Say something: Hello!
Server said: Hi!
Say something: end

Process finished with exit code 0
```

Then, the server follows suit:

```
Client said: Hello!
Say something: Hi!
Client said: end

Process finished with exit code 0
```

That's all we wanted to demonstrate while discussing the TCP protocol. Now let's review the differences between the UDP and TCP protocols.

UDP versus TCP protocols

The differences between the UDP and TCP/IP protocol can be listed as follows:

- UDP simply sends data, whether the data receiver is up and running or not. That's why UDP is better suited to sending data compared to many other clients using multicast distribution. TCP, on the other hand, requires establishing the connection between the client and the server first. The TCP client sends a special control message; the server receives it and responds with a confirmation. The client then sends a message to the server that acknowledges the server confirmation. Only after this, data transmission between the client and server is possible.
- TCP guarantees message delivery or raises an error, while UDP does not, and a datagram packet may be lost.
- TCP guarantees the preservation of the order of messages on delivery, while UDP does not.
- As a result of these provided guarantees, TCP is slower than UDP.
- Additionally, protocols require headers to be sent along with the packet. The header size of a TCP packet is 20 bytes, while a datagram packet is 8 bytes. The UDP header contains Length, Source Port, Destination Port, and Checksum, while the TCP header contains Sequence Number, Ack Number, Data Offset, Reserved, Control Bit, Window, Urgent Pointer, Options, and Padding, in addition to the UDP headers.
- There are different application protocols that are based on the TCP or UDP protocols. The **TCP**-based protocols are **HTTP, HTTPS, Telnet, FTP**, and **SMTP**. The **UDP**-based protocols are **Dynamic Host Configuration Protocol (DHCP), DNS, Simple Network Management Protocol (SNMP), Trivial File Transfer Protocol (TFTP), Bootstrap Protocol (BOOTP)**, and early versions of the **Network File System (NFS)**.

We can capture the difference between UDP and TCP in one sentence: the UDP protocol is faster and more lightweight than TCP, but less reliable. As with many things in life, you have to pay a higher price for additional services. However, not all these services will be needed in all cases, so think about the task in hand and decide which protocol to use based on your application requirements.

URL-based communication

Nowadays, it seems that everybody has some notion of a URL; those who use a browser on their computers or smartphones will see URLs every day. In this section, we will briefly explain the different parts that make up a URL and demonstrate how it can be used programmatically to request data from a website (or a file) or to send (post) data to a website.

The URL syntax

Generally speaking, the URL syntax complies with the syntax of a **Uniform Resource Identifier** (**URI**) that has the following format:

```
scheme:[//authority]path[?query][#fragment]
```

The square brackets indicate that the component is optional. This means that a URI will consist of `scheme:path` at the very least. The `scheme` component can be `http`, `https`, `ftp`, `mailto`, `file`, `data`, or another value. The `path` component consists of a sequence of path segments separated by a slash (`/`). Here is an example of a URL consisting only of `scheme` and `path`:

```
file:src/main/resources/hello.txt
```

The preceding URL points to a file on a local filesystem that is relative to the directory where this URL is used. We will demonstrate how it works shortly.

The `path` component can be empty, but then the URL would seem useless. Nevertheless, an empty path is often used in conjunction with `authority`, which has the following format:

```
[userinfo@]host[:port]
```

The only required component of authority is `host`, which can be either an IP address (`137.254.120.50`, for example) or a domain name (`oracle.com`, for example).

The userinfo component is typically used with the mailto value of the scheme component, so userinfo@host represents an email address.

The port component, if omitted, assumes a default value. For example, if the scheme value is http, then the default port value is 80, and if the scheme value is https, then the default port value is 443.

An optional query component of a URL is a sequence of key-value pairs separated by a delimiter (&):

```
key1=value1&key2=value2
```

Finally, the optional fragment component is an identifier of a section of an HTML document, so that a browser can scroll this section into view.

It is necessary to mention that Oracle's online documentation uses slightly different terminology:

- protocol instead of scheme
- reference instead of fragment
- file instead of path[?query][#fragment]
- resource instead of host[:port]path[?query][#fragment]

So, from the Oracle documentation perspective, the URL is composed of protocol and resource values.

Let's now take a look at the programmatic usage of URLs in Java.

The java.net.URL class

In Java, a URL is represented by an object of the java.net.URL class that has six constructors:

- URL(String spec): This creates a URL object from the URL as a string.
- URL(String protocol, String host, String file): This creates a URL object from the provided values of protocol, host, and file (path and query), and the default port number based on the provided protocol value.
- URL(String protocol, String host, int port, String path): This creates a URL object from the provided values of protocol, host, port, and file (path and query). A port value of −1 indicates that the default port number needs to be used based on the provided protocol value.

- URL(String protocol, String host, int port, String file, URLStreamHandler handler): This acts in the same way as the preceding constructor and additionally allows you to pass in an object of the particular protocol handler; all the preceding constructors load default handlers automatically.
- URL(URL context, String spec): This creates a URL object that extends the provided URL object or overrides its components using the provided spec value, which is a string representation of a URL or some of its components. For example, if the scheme is present in both parameters, the scheme value from spec overrides the scheme value in context and many others.
- URL(URL context, String spec, URLStreamHandler handler): This acts in the same way as the preceding constructor and additionally allows you to pass in an object of the particular protocol handler.

Once created, a URL object allows you to get the values of various components of the underlying URL. The InputStream openStream() method provides access to the stream of data received from the URL. In fact, it is implemented as openConnection.getInputStream(). The URLConnection openConnection() method of the URL class returns a URLConnection object with many methods that provide details about the connection to the URL, including the getOutputStream() method that allows you to send data to the URL.

Let's take a look at the code example; we start with reading data from a hello.txt file, which is a local file that we created in Chapter 5, *Strings, Input/Output, and Files*. The file contains only one line: "*Hello!*"; here is the code that reads it:

```
try {
    URL url = new URL("file:src/main/resources/hello.txt");
    System.out.println(url.getPath());    // src/main/resources/hello.txt
    System.out.println(url.getFile());    // src/main/resources/hello.txt
    try(InputStream is = url.openStream()){
        int data = is.read();
        while(data != -1){
            System.out.print((char) data); //prints: Hello!
            data = is.read();
        }
    }
} catch (Exception e) {
    e.printStackTrace();
}
```

In the preceding code, we used the `file:src/main/resources/hello.txt` URL. It is based on the path to the file that is relative to the program's executing location. The program is executed in the root directory of our project. To begin, we demonstrate the `getPath()` and `getFile()` methods. The returned values are not different because the URL does not have a `query` component value. Otherwise, the `getFile()` method would include it too. We will see this in the following code example.

The rest of the preceding code is opening an input stream of data from a file and prints the incoming bytes as characters. The result is shown in the inline comment.

Now, let's demonstrate how Java code can read data from the URL that points to a source on the internet. Let's call the Google search engine with a `Java` keyword:

```java
try {
    URL url = new URL("https://www.google.com/search?q=Java&num=10");
    System.out.println(url.getPath()); //prints: /search
    System.out.println(url.getFile()); //prints: /search?q=Java&num=10
    URLConnection conn = url.openConnection();
    conn.setRequestProperty("Accept", "text/html");
    conn.setRequestProperty("Connection", "close");
    conn.setRequestProperty("Accept-Language", "en-US");
    conn.setRequestProperty("User-Agent", "Mozilla/5.0");
    try(InputStream is = conn.getInputStream();
     BufferedReader br = new BufferedReader(new InputStreamReader(is))){
        String line;
        while ((line = br.readLine()) != null){
            System.out.println(line);
        }
    }
} catch (Exception e) {
  e.printStackTrace();
}
```

Here, we came up with the `https://www.google.com/search?q=Java&num=10` URL and requested the properties after some research and experimentation. There is no guarantee that it will always work, so do not be surprised if it does not return the same data we describe, Besides, it is a live search, so the result may change at any time.

The preceding code also demonstrates the difference in the values returned by the `getPath()` and `getFile()` methods. You can view the inline comments in the preceding code example.

In comparison to the example of using a file URL, the Google search example used the `URLConnection` object because we need to set the request header fields:

- `Accept` tells the server what type of content the caller requests (`understands`).
- `Connection` tells the server that the connection will be closed after the response is received.
- `Accept-Language` tells the server which language the caller requests (`understands`).
- `User-Agent` tells the server information about the caller; otherwise, the Google search engine (`www.google.com`) responds with a 403 (forbidden) HTTP code.

The remaining code in the preceding example just reads from the input stream of data (HTML code) coming from the URL, and prints it, line by line. We captured the result (copied it from the screen), pasted it into the online HTML Formatter (`https://jsonformatter.org/html-pretty-print`), and ran it. The result is presented in the following screenshot:

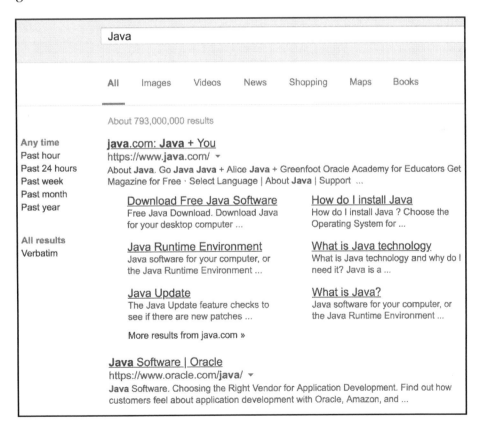

As you can see, it looks like a typical page with the search results, except there is no *Google* image in the upper-left corner with the returned HTML.

Similarly, it is possible to send (post) data to a URL; here is an example code:

```
try {
    URL url = new URL("http://localhost:3333/something");
    URLConnection conn = url.openConnection();
    //conn.setRequestProperty("Method", "POST");
    //conn.setRequestProperty("User-Agent", "Java client");
    conn.setDoOutput(true);
    OutputStreamWriter osw =
            new OutputStreamWriter(conn.getOutputStream());
    osw.write("parameter1=value1&parameter2=value2");
    osw.flush();
    osw.close();

    BufferedReader br =
        new BufferedReader(new InputStreamReader(conn.getInputStream()));
    String line;
    while ((line = br.readLine()) != null) {
        System.out.println(line);
    }
    br.close();
} catch (Exception e) {
    e.printStackTrace();
}
```

The preceding code expects a server running on the `localhost` server on port 3333 that can process the POST request with the "`/something`" path. If the server does not check the method (is it POST or any other HTTP method) and it does not check `User-Agent` value, there is no need to specify any of it. So, we comment the settings out and keep them there just to demonstrate how these, and similar, values can be set if required.

Notice we used the `setDoOutput()` method to indicate that output has to be sent; by default, it is set to `false`. Then, we let the output stream send the query parameters to the server.

Another important aspect of the preceding code is that the output stream has to be closed before the input stream is opened. Otherwise, the content of the output stream will not be sent to the server. While we did it explicitly, a better way to do it is by using the try-with-resources block that guarantees the `close()` method is called, even if an exception was raised anywhere in the block.

Here is a better version of the preceding example:

```
try {
    URL url = new URL("http://localhost:3333/something");
    URLConnection conn = url.openConnection();
    //conn.setRequestProperty("Method", "POST");
    //conn.setRequestProperty("User-Agent", "Java client");
    conn.setDoOutput(true);
    try(OutputStreamWriter osw =
                new OutputStreamWriter(conn.getOutputStream())){
        osw.write("parameter1=value1&parameter2=value2");
        osw.flush();
    }
    try(BufferedReader br =
      new BufferedReader(new InputStreamReader(conn.getInputStream()))){
        String line;
        while ((line = br.readLine()) != null) {
            System.out.println(line);
        }
    }
} catch (Exception ex) {
    ex.printStackTrace();
}
```

To demonstrate how this example works, we also created a simple server that listens on port 3333 of localhost and has a handler assigned to process all the requests that come with the "/something" path:

```
public static void main(String[] args) throws Exception {
    HttpServer server = HttpServer.create(new InetSocketAddress(3333),0);
    server.createContext("/something", new PostHandler());
    server.setExecutor(null);
    server.start();
}
static class PostHandler implements HttpHandler {
    public void handle(HttpExchange exch) {
        System.out.println(exch.getRequestURI());    //prints: /something
        System.out.println(exch.getHttpContext().getPath());///something
        try(BufferedReader in = new BufferedReader(
                new InputStreamReader(exch.getRequestBody()));
          OutputStream os = exch.getResponseBody()){
            System.out.println("Received as body:");
            in.lines().forEach(l -> System.out.println("   " + l));

            String confirm = "Got it! Thanks.";
            exch.sendResponseHeaders(200, confirm.length());
            os.write(confirm.getBytes());
        } catch (Exception ex){
```

```
                    ex.printStackTrace();
            }
        }
    }
```

To implement the server, we used the classes of the `com.sun.net.httpserver`
package that comes with the JCL. To demonstrate that the URL comes without parameters,
we print the URI and the path. They both have the same `"/something"` value; the
parameters come from the body of the request.

After the request is processed, the server sends back the message *"Got it! Thanks."* Let's see
how it works; we first run the server. It starts listening on port 3333 and blocks until the
request comes with the `"/something"` path. Then, we execute the client and observe the
following output on the server-side screen:

As you can see, the server received the parameters (or any other message for that matter)
successfully. Now it can parse them and use as needed.

If we look at the client-side screen, we will see the following output:

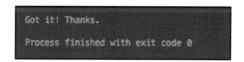

This means that the client received the message from the server and exited as expected.
Notice that the server in our example does not exit automatically and has to be closed
manually.

Other methods of the URL and URLConnection classes allow you to set/get other properties
and can be used for more dynamic management of the client-server communication. There
is also the HttpUrlConnection class (and other classes) in the `java.net` package that
simplifies and enhances URL-based communication. You can read the online
documentation of the `java.net` package to understand the available options better.

Using the HTTP 2 Client API

The HTTP Client API was introduced with Java 9 as an incubating API in the `jdk.incubator.http` package. In Java 11, it was standardized and moved to the `java.net.http` package. It is a far richer and easier-to-use alternative to the `URLConnection` API. In addition to all the basic connection-related functionality, it provides non-blocking (asynchronous) request and response using `CompletableFuture` and supports both HTTP 1.1 and HTTP 2.

HTTP 2 added the following new capabilities to the HTTP protocol:

- The ability to send data in a binary format rather than textual format; the binary format is more efficient for parsing, more compact, and less susceptible to various errors.
- It is fully multiplexed, thus allowing multiple requests and responses to be sent concurrently using just one connection.
- It uses header compression, thus reducing the overhead.
- It allows a server to push a response into the client's cache if the client indicates that it supports HTTP 2.

The package contains the following classes:

- `HttpClient`: This is used to send requests and receive responses both synchronously and asynchronously. An instance can be created using the static `newHttpClient()` method with default settings or using the `HttpClient.Builder` class (returned by the static `newBuilder()` method) that allows you to customize the client configuration. Once created, the instance is immutable and can be used multiple times.
- `HttpRequest`: This creates and represents an HTTP request with the destination URI, headers, and other related information. An instance can be created using the `HttpRequest.Builder` class (returned by the static `newBuilder()` method). Once created, the instance is immutable and can be sent multiple times.
- `HttpRequest.BodyPublisher`: This publishes a body (for the POST, PUT, and DELETE methods) from a certain source, such as a string, a file, input stream, or byte array.
- `HttpResponse`: This represents an HTTP response received by the client after an HTTP request has been sent. It contains the origin URI, headers, message body, and other related information. Once created, the instance can be queried multiple times.

- `HttpResponse.BodyHandler`: This is a functional interface that accepts the response and returns an instance of `HttpResponse.BodySubscriber` that can process the response body.
- `HttpResponse.BodySubscriber`: This receives the response body (its bytes) and transforms it into a string, a file, or a type.

The `HttpRequest.BodyPublishers`, `HttpResponse.BodyHandlers`, and `HttpResponse.BodySubscribers` classes are factory classes that create instances of the corresponding classes. For example, the `BodyHandlers.ofString()` method creates a `BodyHandler` instance that processes the response body bytes as a string, while the `BodyHandlers.ofFile()` method creates a `BodyHandler` instance that saves the response body in a file.

You can read the online documentation of the `java.net.http` package to learn more about these and other related classes and interfaces. Next, we will take a look at and discuss some examples of HTTP API usage.

Blocking HTTP requests

The following code is an example of a simple HTTP client that sends a GET request to an HTTP server:

```
HttpClient httpClient = HttpClient.newBuilder()
       .version(HttpClient.Version.HTTP_2) // default
       .build();
HttpRequest req = HttpRequest.newBuilder()
       .uri(URI.create("http://localhost:3333/something"))
       .GET()                              // default
       .build();
try {
 HttpResponse<String> resp =
           httpClient.send(req, BodyHandlers.ofString());
 System.out.println("Response: " +
              resp.statusCode() + " : " + resp.body());
} catch (Exception ex) {
   ex.printStackTrace();
}
```

We created a builder to configure an `HttpClient` instance. However, since we used default settings only, we can do it with the same result as follows:

```
HttpClient httpClient = HttpClient.newHttpClient();
```

To demonstrate the client's functionality, we will use the same `UrlServer` class that we used already. As a reminder, this is how it processes the client's request and responds with `"Got it! Thanks."`:

```
try(BufferedReader in = new BufferedReader(
            new InputStreamReader(exch.getRequestBody()));
    OutputStream os = exch.getResponseBody()){
    System.out.println("Received as body:");
    in.lines().forEach(l -> System.out.println("   " + l));

    String confirm = "Got it! Thanks.";
    exch.sendResponseHeaders(200, confirm.length());
    os.write(confirm.getBytes());
    System.out.println();
} catch (Exception ex){
    ex.printStackTrace();
}
```

If we launch this server and run the preceding client's code, the server prints the following message on its screen:

```
Received as body:
```

The client did not send a message because it used the HTTP GET method. Nevertheless, the server responds, and the client's screen shows the following message:

```
Response: 200 : Got it! Thanks.
```

The send() method of the `HttpClient` class is blocked until the response has come back from the server.

Using the HTTP POST, PUT, or DELETE methods produces similar results; let's run the following code now:

```
HttpClient httpClient = HttpClient.newBuilder()
        .version(Version.HTTP_2)   // default
        .build();
HttpRequest req = HttpRequest.newBuilder()
        .uri(URI.create("http://localhost:3333/something"))
        .POST(BodyPublishers.ofString("Hi there!"))
        .build();
try {
    HttpResponse<String> resp =
                httpClient.send(req, BodyHandlers.ofString());
```

```
        System.out.println("Response: " +
                          resp.statusCode() + " : " + resp.body());
    } catch (Exception ex) {
        ex.printStackTrace();
    }
```

As you can see, this time the client posts the message `Hi there!`, and the server's screen shows the following:

```
Received as body:
    Hi there!
```

The `send()` method of the `HttpClient` class is blocked until the same response has come back from the server:

```
Response: 200 : Got it! Thanks.
```

So far, the demonstrated functionality was not much different from the URL-based communication that we saw in the previous section. Now we are going to use the `HttpClient` methods that are not available in the URL streams.

Non-blocking (asynchronous) HTTP requests

The `sendAsync()` method of the `HttpClient` class allows you to send a message to a server without blocking. To demonstrate how it works, we will execute the following code:

```
HttpClient httpClient = HttpClient.newHttpClient();
HttpRequest req = HttpRequest.newBuilder()
        .uri(URI.create("http://localhost:3333/something"))
        .GET()    // default
        .build();
CompletableFuture<Void> cf = httpClient
        .sendAsync(req, BodyHandlers.ofString())
        .thenAccept(resp -> System.out.println("Response: " +
                          resp.statusCode() + " : " + resp.body()));
System.out.println("The request was sent asynchronously...");
try {
    System.out.println("CompletableFuture get: " +
                          cf.get(5, TimeUnit.SECONDS));
} catch (Exception ex) {
    ex.printStackTrace();
}
System.out.println("Exit the client...");
```

In comparison to the example with the `send()` method (which returns the `HttpResponse` object), the `sendAsync()` method returns an instance of the `CompletableFuture<HttpResponse>` class. If you read the documentation of the `CompletableFuture<T>` class, you will see that it implements the `java.util.concurrent.CompletionStage` interface that provides many methods that can be chained and allow you to set various functions to process the response.

To give you an idea, here is the list of the methods declared in the `CompletionStage` interface: `acceptEither, acceptEitherAsync, acceptEitherAsync, applyToEither, applyToEitherAsync, applyToEitherAsync, handle, handleAsync, handleAsync, runAfterBoth, runAfterBothAsync, runAfterBothAsync, runAfterEither, runAfterEitherAsync, runAfterEitherAsync, thenAccept, thenAcceptAsync, thenAcceptAsync, thenAcceptBoth, thenAcceptBothAsync, thenAcceptBothAsync, thenApply, thenApplyAsync, thenApplyAsync, thenCombine, thenCombineAsync, thenCombineAsync, thenCompose, thenComposeAsync, thenComposeAsync, thenRun, thenRunAsync, thenRunAsync, whenComplete, whenCompleteAsync`, and `whenCompleteAsync`.

We will talk about functions and how they can be passed as parameters in `Chapter 13`, *Functional Programming*. For now, we will just mention that the `resp -> System.out.println("Response: " + resp.statusCode() + " : " + resp.body())` construction represents the same functionality as the following method:

```
void method(HttpResponse resp){
    System.out.println("Response: " +
                    resp.statusCode() + " : " + resp.body());
}
```

The `thenAccept()` method applies the passed-in functionality to the result returned by the previous method of the chain.

After the `CompletableFuture<Void>` instance is returned, the preceding code prints **The request was sent asynchronously...** message and blocks it on the `get()` method of the `CompletableFuture<Void>` object. This method has an overloaded version `get(long timeout, TimeUnit unit)`, with two parameters – the `TimeUnit unit` and the `long timeout` that specifies the number of the units, indicating how long the method should wait for the task that is represented by the `CompletableFuture<Void>` object to complete. In our case, the task is to send a message to the server and to get back the response (and process it using the provided function). If the task is not completed in the allotted time, the `get()` method is interrupted (and the stack trace is printed in the catch block).

The `Exit the client...` message should appear on the screen either in five seconds (in our case) or after the `get()` method returns.

If we run the client, the server's screen shows the following message again with the blocking HTTP `GET` request:

```
Received as body:
```

The client's screen displays the following message:

```
The request was sent asynchronously...
Response: 200 : Got it! Thanks.
CompletableFuture get: null
Exit the client...
```

As you can see, **The request was sent asynchronously...** message appears before the response came back from the server. This is the point of an asynchronous call; the request to the server was sent and the client is free to continue to do anything else. The passed-in function will be applied to the server response. At the same time, you can pass the `CompletableFuture<Void>` object around and call it at any time to get the result. In our case, the result is `void`, so the `get()` method simply indicates that the task was completed.

We know that the server returns the message, and so we can take advantage of it by using another method of the `CompletionStage` interface. We have chosen the `thenApply()` method, which accepts a function that returns a value:

```
CompletableFuture<String> cf = httpClient
                .sendAsync(req, BodyHandlers.ofString())
                .thenApply(resp -> "Server responded: " + resp.body());
```

Now the `get()` method returns the value produced by the `resp -> "Server responded: " + resp.body()` function, so it should return the server message body; let's run this code and see the result:

```
The request was sent asynchronously...
CompletableFuture get: Server responded: Got it! Thanks.
Exit the client...
```

Now the `get()` method returns the server's message as expected, and it is presented by the function and passed as a parameter into the `thenApply()` method.

Similarly, we can use the HTTP POST, PUT, or DELETE methods for sending a message:

```
HttpClient httpClient = HttpClient.newHttpClient();
HttpRequest req = HttpRequest.newBuilder()
        .uri(URI.create("http://localhost:3333/something"))
        .POST(BodyPublishers.ofString("Hi there!"))
        .build();
CompletableFuture<String> cf = httpClient
        .sendAsync(req, BodyHandlers.ofString())
        .thenApply(resp -> "Server responded: " + resp.body());
System.out.println("The request was sent asynchronously...");
try {
    System.out.println("CompletableFuture get: " +
                            cf.get(5, TimeUnit.SECONDS));
} catch (Exception ex) {
    ex.printStackTrace();
}
System.out.println("Exit the client...");
```

The only difference from the previous example is that the server now displays the received client's message:

```
Received as body:
  Hi there!
```

The client's screen displays the same message as in the case of the GET method:

```
The request was sent asynchronously...
CompletableFuture get: Server responded: Got it! Thanks.
Exit the client...
```

The advantage of asynchronous requests is that they can be sent quickly and without needing to wait for each of them to complete. The HTTP 2 protocol supports it by multiplexing; for example, let's send three requests as follows:

```
HttpClient httpClient = HttpClient.newHttpClient();
List<CompletableFuture<String>> cfs = new ArrayList<>();
List<String> nums = List.of("1", "2", "3");
for(String num: nums){
    HttpRequest req = HttpRequest.newBuilder()
            .uri(URI.create("http://localhost:3333/something"))
            .POST(BodyPublishers.ofString("Hi! My name is " + num + "."))
            .build();
    CompletableFuture<String> cf = httpClient
            .sendAsync(req, BodyHandlers.ofString())
            .thenApply(rsp -> "Server responded to msg " + num + ": "
```

```
                                        + rsp.statusCode() + " : " + rsp.body());
        cfs.add(cf);
    }
    System.out.println("The requests were sent asynchronously...");
    try {
        for(CompletableFuture<String> cf: cfs){
            System.out.println("CompletableFuture get: " +
                                    cf.get(5, TimeUnit.SECONDS));
        }
    } catch (Exception ex) {
        ex.printStackTrace();
    }
    System.out.println("Exit the client...");
```

The server's screen shows the following messages:

```
Received as body:
  Hi! My name is 2.

Received as body:
  Hi! My name is 3.

Received as body:
  Hi! My name is 1.
```

Notice the arbitrary sequence of the incoming requests; this is because the client uses a pool of Executors.newCachedThreadPool() threads to send the messages. Each message is sent by a different thread, and the pool has its own logic of using the pool members (threads). If the number of messages is large, or if each of them consumes a significant amount of memory, it may be beneficial to limit the number of threads run concurrently.

The HttpClient.Builder class allows you to specify the pool that is used for acquiring the threads that send the messages:

```
ExecutorService pool = Executors.newFixedThreadPool(2);
HttpClient httpClient = HttpClient.newBuilder().executor(pool).build();
List<CompletableFuture<String>> cfs = new ArrayList<>();
List<String> nums = List.of("1", "2", "3");
for(String num: nums){
    HttpRequest req = HttpRequest.newBuilder()
            .uri(URI.create("http://localhost:3333/something"))
            .POST(BodyPublishers.ofString("Hi! My name is " + num + "."))
            .build();
    CompletableFuture<String> cf = httpClient
            .sendAsync(req, BodyHandlers.ofString())
            .thenApply(rsp -> "Server responded to msg " + num + ": "
                            + rsp.statusCode() + " : " + rsp.body());
```

```
        cfs.add(cf);
    }
    System.out.println("The requests were sent asynchronously...");
    try {
        for(CompletableFuture<String> cf: cfs){
            System.out.println("CompletableFuture get: " +
                                        cf.get(5, TimeUnit.SECONDS));
        }
    } catch (Exception ex) {
        ex.printStackTrace();
    }
    System.out.println("Exit the client...");
```

If we run the preceding code, the results will be the same, but the client will use only two threads to send messages. The performance may be a bit slower (in comparison to the previous example) as the number of messages grows. So, as is often the case in a software system design, you need to balance between the amount of memory used and the performance.

Similarly to the executor, several other objects can be set on the `HttpClient` object to configure the connection to handle authentication, request redirection, cookies management, and more.

Server push functionality

The second (after multiplexing) significant advantage of the HTTP 2 protocol over HTTP 1.1 is allowing the server to push the response into the client's cache if the client indicates that it supports HTTP 2. Here is the client code that takes advantage of this feature:

```
HttpClient httpClient = HttpClient.newHttpClient();
HttpRequest req = HttpRequest.newBuilder()
        .uri(URI.create("http://localhost:3333/something"))
        .GET()
        .build();
CompletableFuture cf = httpClient
        .sendAsync(req, BodyHandlers.ofString(),
                (PushPromiseHandler) HttpClientDemo::applyPushPromise);

System.out.println("The request was sent asynchronously...");
try {
    System.out.println("CompletableFuture get: " +
                                    cf.get(5, TimeUnit.SECONDS));
} catch (Exception ex) {
    ex.printStackTrace();
```

```
    }
    System.out.println("Exit the client...");
```

Notice the third parameter of the sendAsync() method. It is a function that handles the push response if one comes from the server. It is up to the client developer to decide how to implement this function; here is one possible example:

```
void applyPushPromise(HttpRequest initReq, HttpRequest pushReq,
        Function<BodyHandler, CompletableFuture<HttpResponse>> acceptor) {
    CompletableFuture<Void> cf = acceptor.apply(BodyHandlers.ofString())
        .thenAccept(resp -> System.out.println("Got pushed response "
                                                    + resp.uri()));
    try {
        System.out.println("Pushed completableFuture get: " +
                                        cf.get(1, TimeUnit.SECONDS));
    } catch (Exception ex) {
        ex.printStackTrace();
    }
    System.out.println("Exit the applyPushPromise function...");
}
```

This implementation of the function does not do much. It just prints out the URI of the push origin. But, if necessary, it can be used for receiving the resources from the server (for example, images that support the provided HTML) without requesting them. This solution saves the round-trip request-response model and shortens the time of the page loading. It also can be used for updating the information on the page.

You can find many code examples of a server that sends push requests; all major browsers support this feature too.

WebSocket support

HTTP is based on the request-response model. A client requests a resource, and the server provides a response to this request. As we demonstrated several times, the client initiates the communication. Without it, the server cannot send anything to the client. To get over this limitation, the idea was first introduced as TCP connection in the HTML5 specification and, in 2008, the first version of the WebSocket protocol was designed.

It provides a full-duplex communication channel between the client and the server. After the connection is established, the server can send a message to the client at any time. Together with JavaScript and HTML5, the WebSocket protocol support allows web applications to present a far more dynamic user interface.

The WebSocket protocol specification defines WebSocket (ws) and WebSocket Secure (wss) as two schemes, which are used for unencrypted and encrypted connections, respectively. The protocol does not support fragmentation but allows all the other URI components described in the *URL syntax* section.

All the classes that support the WebSocket protocol for a client are located in the java.net package. To create a client, we need to implement the WebSocket.Listener interface, which has the following methods:

- onText(): Invoked when textual data has been received
- onBinary(): Invoked when binary data has been received
- onPing(): Invoked when a ping message has been received
- onPong(): Invoked when a pong message has been received
- onError(): Invoked when an error has happened
- onClose(): Invoked when a close message has been received

All the methods of this interface are default. This means that you do not need to implement all of them, but only those that the client requires for a particular task:

```
class WsClient implements WebSocket.Listener {
    @Override
    public void onOpen(WebSocket webSocket) {
        System.out.println("Connection established.");
        webSocket.sendText("Some message", true);
        Listener.super.onOpen(webSocket);
    }
    @Override
    public CompletionStage onText(WebSocket webSocket,
                                   CharSequence data, boolean last) {
        System.out.println("Method onText() got data: " + data);
        if(!webSocket.isOutputClosed()) {
            webSocket.sendText("Another message", true);
        }
        return Listener.super.onText(webSocket, data, last);
    }
    @Override
    public CompletionStage onClose(WebSocket webSocket,
                                    int statusCode, String reason) {
        System.out.println("Closed with status " +
                             statusCode + ", reason: " + reason);
        return Listener.super.onClose(webSocket, statusCode, reason);
    }
}
```

A server can be implemented in a similar way, but server implementation is beyond the scope of this book. To demonstrate the preceding client code, we are going to use a WebSocket server provided by the `echo.websocket.org` website. It allows a WebSocket connection and sends the received message back; such a server is typically called an **echo server**.

We expect that our client will send the message after the connection is established. Then, it will receive (the same) message from the server, display it, and send back another message, and so on, until it is closed. The following code invokes the client that we created:

```
HttpClient httpClient = HttpClient.newHttpClient();
WebSocket webSocket = httpClient.newWebSocketBuilder()
    .buildAsync(URI.create("ws://echo.websocket.org"), new WsClient())
    .join();
System.out.println("The WebSocket was created and ran asynchronously.");
try {
    TimeUnit.MILLISECONDS.sleep(200);
} catch (InterruptedException ex) {
    ex.printStackTrace();
}
webSocket.sendClose(WebSocket.NORMAL_CLOSURE, "Normal closure")
        .thenRun(() -> System.out.println("Close is sent."));
```

The preceding code creates a `WebSocket` object using the `WebSocket.Builder` class. The `buildAsync()` method returns the `CompletableFuture` object. The `join()` method of the `CompletableFuture` class returns the result value when complete, or throws an exception. If an exception is not generated, then, as we mentioned already, the `WebSocket` communication continues until either side sends a **Close** message. That is why our client waits for 200 milliseconds, and then sends the **Close** message and exits. If we run this code, we will see the following messages:

```
Connection established.
The WebSocket was created and ran asynchronously.
Method onText() got data: Some message
Method onText() got data: Another message
Method onText() got data: Another message
Method onText() got data: Another message
Close is sent.
```

As you can see, the client behaves as expected. To finish our discussion, we would like to mention that all modern web browsers support the WebSocket protocol.

Summary

In this chapter, the reader was presented with a description of the most popular network protocols: UDP, TCP/IP, and WebSocket. The discussion was illustrated with code examples using JCL. We also reviewed URL-based communication and the latest Java HTTP 2 Client API.

The next chapter provides an overview of Java GUI technologies and demonstrates a GUI application using JavaFX, including code examples with control elements, charts, CSS, FXML, HTML, media, and various other effects. The reader will learn how to use JavaFX to create a GUI application.

Quiz

1. Name five network protocols of the application layer.
2. Name two network protocols of the transport layer.
3. Which Java package includes classes that support HTTP protocol?
4. Which protocol is based on exchanging datagrams?
5. Can a datagram be sent to the IP address where there is no server running?
6. Which Java package contains classes that support UDP and TCP protocols?
7. What does TCP stands for?
8. What is common between the TCP and TCP/IP protocols?
9. How is the TCP session identified?
10. Name one principal difference between the functionality of `ServerSocket` and `Socket`.
11. Which is faster, TCP or UDP?
12. Which is more reliable, TCP or UDP?
13. Name three TCP-based protocols.
14. Which of the following are the components of a URI? Select all that apply:

    a. Fragment
    b. Title
    c. Authority
    d. Query

15. What is the difference between `scheme` and `protocol`?
16. What is the difference between a URI and a URL?
17. What does the following code print?

```
URL url = new URL("http://www.java.com/something?par=42");
System.out.print(url.getPath());
System.out.println(url.getFile());
```

18. Name two new features that HTTP 2 has that HTTP 1.1 does not.
19. What is the fully qualified name of the `HttpClient` class?
20. What is the fully qualified name of the `WebSocket` class?
21. What is the difference between `HttpClient.newBuilder().build()` and `HttpClient.newHttpClient()`?
22. What is the fully qualified name of the `CompletableFuture` class?

12
Java GUI Programming

This chapter provides an overview of Java **Graphical User Interface** (**GUI**) technologies and demonstrates how JavaFX kit can be used to create a GUI application. The latest versions of JavaFX not only provide many helpful features but also allow preserving and embedding the legacy implementation and styles.

In a certain respect, GUI is the most important part of an application. It directly interacts in with the user. If GUI is inconvenient, unappealing to the eyes, or confusing, even the best of the backend solutions may not justify user to use this application. By contrast, well thought through intuitive and nicely designed GUI helps to retain user even if the application does not do the job as well as its competitions.

The following topics will be covered in this chapter:

- Java GUI technologies
- JavaFX fundamentals
- Hello with JavaFX
- Control elements
- Charts
- Applying CSS
- Using FXML
- Embedding HTML
- Playing media
- Adding effects

Java GUI technologies

The name **Java Foundation Classes (JFC)** may be a source of much confusion. It implies *the classes that are at the foundation of Java*, while, in fact, JFC includes only classes and interfaces related to GUI. To be precise, JFC is a collection of three frameworks: **Abstract Window Toolkit** (**AWT**), Swing, and Java 2D.

JFC is part of **Java Class Library (JCL)**, although the name JFC came into being only in 1997, while AWT was part of JCL from the very beginning. At that time, Netscape developed a GUI library called **Internet Foundation Classes (IFC)** and Microsoft created **Application Foundation Classes (AFC)** for GUI development too. So, when Sun Microsystems and Netscape decided to form a new GUI library, they *inherited* the word *Foundation* and created JFC. The framework Swing took over the Java GUI programming from AWT and was successfully used for almost two decades.

A new GUI programming toolkit JavaFX was added to JCL in Java 8. It was removed from JCL in Java 11, and, since then, has resided as an open source project supported by the company Gluon as a downloadable module in addition to the JDK. JavaFX uses a somewhat different approach to GUI programming than AWT and Swing. It presents a more consistent and simpler design and has a good chance to be a winning Java GUI-programming toolkit.

JavaFX fundamentals

Cities such as New York, London, Paris, and Moscow have many theaters, and people who live there cannot avoid hearing about new plays and productions almost weekly. It makes them inevitably familiar with theater terminology, among which the terms *stage*, *scene*, and *event* are probably used the most often. These three terms are also at the foundation of a JavaFX application structure too.

The top-level container in JavaFX that holds all other components is represented by the `javafx.stage.Stage` class. So, you can say that, in the JavaFX application, everything happens on a stage. From a user perspective, it is a display area or window where all the controls and components perform their actions (like actors in a theater). And, similar to the actors in a theater, they do it in the context of a scene, represented by the `javafx.scene.Scene` class. So, a JavaFX application, like a play in a theater, is composed of `Scene` objects presented inside the `Stage` object one at a time. Each `Scene` object contains a graph that defines the positions of the scene actors, called **nodes**, in JavaFX: controls, layouts, groups, shapes, and so on. Each of them extends the abstract class, `javafx.scene.Node`.

Some of the nodes controls are associated with events: a button clicked or a checkbox checked, for example. These events can be processed by the event handler associated with the corresponding control element.

The main class of JavaFX application has to extend the abstract class, `java.application.Application` that has several life cycle methods. We list them in the sequence of the invocation: `launch()`, `init()`, `notifyPreloader()`, `start()`, `stop()`. It looks like quite a few to remember. But, most probably, you need to implement only one method, `start()`, where the actual GUI is constructed and executed. So, we will review all the methods just for completeness:

- `static void launch(Class<? extends Application> appClass, String... args)`: Launches the application, often called by the main method; does not return until the `Platform.exit()` is called or all the application windows close; the `appClass` parameter must be a public subclass of `Application` with a public no-argument constructor

- `static void launch(String... args)`: Same as the preceding method, assuming that the public subclass of `Application` is the immediately enclosing class; this is the method most often used to launch the JavaFX application; we are going to use it in our examples too

- `void init()`: This method is called after the `Application` class is loaded; it is typically used for some kind of resource initialization; the default implementation does nothing; we are not going to use it

- `void notifyPreloader(Preloader.PreloaderNotification info)`: Can be used to show progress when the initialization takes a long time; we are not going to use it

- `abstract void start(Stage primaryStage)`: The method we are going to implement; it is called after the `init()` method returns, and after the system is ready to do the main job; the `primaryStage` parameter is the stage where the application is going to present its scenes

- `void stop()`: It is called when the application should stop; may be used to release the resources; the default implementation does nothing; we are not going to use it

The API of the JavaFX toolkit can be found online (`https://openjfx.io/javadoc/11/`). As of this writing, the latest version is 11. Oracle provides extensive documentation and code examples too (`https://docs.oracle.com/javafx/2/`). The documentation includes the description and user manual of Scene Builder (a development tool) that provides a visual layout environment and lets you quickly design a user interface for the JavaFX application without writing any code. This tool may be useful for creating a complex and intricate GUI, and many people do it all the time. In this book, though, we will concentrate on the JavaFX code-writing without using this tool.

To be able to do it, the following are the necessary steps:

1. Add the following dependency to the `pom.xml` file:

   ```
   <dependency>
       <groupId>org.openjfx</groupId>
       <artifactId>javafx-controls</artifactId>
       <version>11</version>
   </dependency>
   <dependency>
       <groupId>org.openjfx</groupId>
       <artifactId>javafx-fxml</artifactId>
       <version>11</version>
   </dependency>
   ```

2. Download JavaFX SDK for your OS from `https://gluonhq.com/products/javafx/` and unzip it in any directory.

3. Assuming you have unzipped JavaFX SDK into the `/path/JavaFX/` folder, add the following options to the Java command that is going to launch your JavaFX application on Linux platform:

   ```
   --module-path /path/JavaFX/lib -add-
   modules=javafx.controls,javafx.fxml
   ```

   On Windows, the same options look as follows:

   ```
   --module-path C:\path\JavaFX\lib -add-
   modules=javafx.controls,javafx.fxml
   ```

   The "/path/JavaFX/" and "C:\path\JavaFX\" are the placeholders which you need to substitute by the actual path to the folder that contains JavaFX SDK.

Assuming that the application's main class is `HelloWorld`, in the case of IntelliJ, enter the preceding options into the `VM options` field as follows:

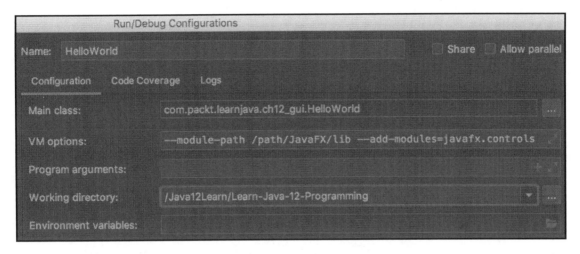

These options have to be added to the **Run/Debug Configurations** of the `HelloWorld`, `BlendEffect`, and `OtherEffects` classes of the package `ch12_gui` of the source code. If you prefer a different IDE or have a different OS, you can find recommendations on how to set it in the `openjfx.io` documentation (`https://openjfx.io/openjfx-docs/#introduction`).

To run the `HelloWorld`, `BlendEffect`, and `OtherEffects` classes from the command line, use the following commands on Linux platform in the project root directory (where the `pom.xml` file is located):

```
mvn clean package

java --module-path /path/javaFX/lib --add-
modules=javafx.controls,javafx.fxml -cp target/learnjava-1.0-
SNAPSHOT.jar:target/libs/* com.packt.learnjava.ch12_gui.HelloWorld

java --module-path /path/javaFX/lib --add-
modules=javafx.controls,javafx.fxml -cp target/learnjava-1.0-
SNAPSHOT.jar:target/libs/* com.packt.learnjava.ch12_gui.BlendEffect

java --module-path /path/javaFX/lib --add-
modules=javafx.controls,javafx.fxml -cp target/learnjava-1.0-
SNAPSHOT.jar:target/libs/* com.packt.learnjava.ch12_gui.OtherEffects
```

On Windows, the same commands look as follows:

```
mvn clean package

java --module-path C:\path\JavaFX\lib --add-
modules=javafx.controls,javafx.fxml -cp target\learnjava-1.0-
SNAPSHOT.jar;target\libs\* com.packt.learnjava.ch12_gui.HelloWorld

java --module-path C:\path\JavaFX\lib --add-
modules=javafx.controls,javafx.fxml -cp target\learnjava-1.0-
SNAPSHOT.jar;target\libs\* com.packt.learnjava.ch12_gui.BlendEffect

java --module-path C:\path\JavaFX\lib --add-
modules=javafx.controls,javafx.fxml -cp target\learnjava-1.0-
SNAPSHOT.jar;target\libs\* com.packt.learnjava.ch12_gui.OtherEffects
```

Each of the `HelloWorld`, `BlendEffect`, and `OtherEffects` classes has several `start()` methods: `start1()`, and `start2()`. After you run the class one time, rename `start()` to `start1()` and `start1()` to `start()` and run the above commands again. Then rename `start()` to `start2()` and `start2()` to `start()` and run the above commands yet again. And so on, until all the `start()` methods are executed. This way you will see the results of all the examples of this chapter.

This concludes the high-level presentation of JavaFX. With that, we move to the most exciting (for any programmer) part: writing code.

Hello with JavaFX

Here is the `HelloWorld` JavaFX application that shows the text, **Hello, World!** and **Exit**:

```
import javafx.application.Application;
import javafx.application.Platform;
import javafx.scene.control.Button;
import javafx.scene.layout.Pane;
import javafx.scene.text.Text;
import javafx.scene.Scene;
import javafx.stage.Stage;

public class HelloWorld extends Application {
    public static void main(String... args) {
        launch(args);
    }
    @Override
    public void start(Stage primaryStage) {
        Text txt = new Text("Hello, world!");
```

```
        txt.relocate(135, 40);

        Button btn = new Button("Exit");
        btn.relocate(155, 80);
        btn.setOnAction(e:> {
            System.out.println("Bye! See you later!");
            Platform.exit();
        });

        Pane pane = new Pane();
        pane.getChildren().addAll(txt, btn);

        primaryStage.setTitle("The primary stage (top-level container)");
        primaryStage.onCloseRequestProperty()
                .setValue(e:> System.out.println("Bye! See you later!"));
        primaryStage.setScene(new Scene(pane, 350, 150));
        primaryStage.show();
    }
}
```

As you can see, the application is launched by calling the static method
`Application.launch(String... args)`. The `start(Stage primaryStage)`
method creates a `Text` node with the message **Hello, World!** located at absolute position
135 (horizontally) and 40 (vertically). Then, it creates another node `Button` with the
text, **Exit,** located at absolute position 155 (horizontally) and 80 (vertically). The action,
assigned to the button (when it is clicked), prints **Bye! See you later!** on a screen and forces
the application to exit using the `Platform.exit()` method. These two nodes are added as
children to the layout pane that allows absolute positioning.

The `Stage` object assigned the **The primary stage (top-level container)** title. It also
assigned an action on clicking the close-the-window symbol (x button) in the window's
upper corner. This symbol appears on the left on the Linux system and on the right on the
Windows system.

While creating actions, we have used a lambda expression, which we are going to discuss in
`Chapter 13`, *Functional Programming*.

The created layout pane is set on a `Scene` object. The scene size set to be 350 horizontally
and 150 vertically. The scene object is placed on the stage. Then the stage is displayed by
calling the `show()` method.

If we run the preceding application, the following window will pop-up:

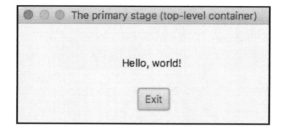

Clicking on the button or on the x button in the upper corner results in the expected message being displayed:

But, if you need to do something else after the x button is clicked and the window got closed, you can add an implementation of the stop() method to the HelloWorld class. For example, it can look as follows:

```
@Override
public void stop(){
    System.out.println("Doing what has to be done before closing");
}
```

If you did, then, after clicking the x button, the display will show this:

Bye! See you later!
Doing what has to be done before closing

This example gives you a sense of how JavaFX works. From now on, while reviewing the JavaFX capabilities, we will present only the code in the start() method.

The toolkit has a huge number of packages, each with many classes, and each class having many methods. We have no chance to discuss all of them. Instead, we are going to present an overview of all the major areas of the JavaFX functionality and present it in the most simple and straightforward way we can.

Control elements

The **control elements** are included in the `javafx.scene.control` package (`https://openjfx.io/javadoc/11/javafx.controls/javafx/scene/control/package-summary.html`). There are more than 80 of them, including a button, text field, checkbox, label, menu, progress bar, and scroll bar to name a few. As we have mentioned already, each control element is a subclass of `Node` that has more than 200 methods. So, you can imagine how rich and fine-tuned a GUI can be, built using JavaFX. However, the scope of this book allows us to cover only a few elements and their methods.

We have already seen a button. Let's now use a label and a text field to create a simple form with input fields (first name, last name, and age) and a `submit` button. We will build it in steps. All the following code snippets are sequential sections of the `start()` method.

First, let's create controls:

```
Text txt = new Text("Fill the form and click Submit");
TextField tfFirstName = new TextField();
TextField tfLastName = new TextField();
TextField tfAge = new TextField();
Button btn = new Button("Submit");
btn.setOnAction(e:> action(tfFirstName, tfLastName, tfAge));
```

As you can guess, the text will be used as the form instructions. The rest is quite straightforward and looks very similar to what we have seen in the `HelloWolrd` example. The `action()` is a function implemented as the following method:

```
void action(TextField tfFirstName,
                TextField tfLastName, TextField tfAge ) {
    String fn = tfFirstName.getText();
    String ln = tfLastName.getText();
    String age = tfAge.getText();
    int a = 42;
    try {
        a = Integer.parseInt(age);
    } catch (Exception ex){}
    fn = fn.isBlank() ? "Nick" : fn;
    ln = ln.isBlank() ? "Samoylov" : ln;
    System.out.println("Hello, " + fn + " " + ln + ", age " + a + "!");
    Platform.exit();
}
```

This function accepts three parameters (the `javafx.scene.control.TextField` objects), then gets the submitted input values and just prints them. The code makes sure that there are always some default values available for printing and that entering a non-numeric value of age does not break the application.

With the controls and action in place, we then put them into a grid layout using the class `javafx.scene.layout.GridPane`:

```
GridPane grid = new GridPane();
grid.setAlignment(Pos.CENTER);
grid.setHgap(15);
grid.setVgap(5);
grid.setPadding(new Insets(20, 20, 20, 20));
```

The `GridPane` layout pane has rows and columns that form cells in which the nodes can be set. Nodes can span columns and rows. The `setAlignment()` method sets the position of the grid to the center of a scene (default position is the top left of a scene). The `setHgap()` and `setVgap()` methods set the spacing (in pixels) between the columns (horizontally) and rows (vertically). The `setPadding()` method adds some space along the borders of the grid pane. The `Insets()` object sets the values (in pixels) in the order of top, right, bottom, and left.

Now we are going to place the created nodes in the corresponding cells (arranged in two columns):

```
int i = 0;
grid.add(txt,    1, i++, 2, 1);
GridPane.setHalignment(txt, HPos.CENTER);
grid.addRow(i++, new Label("First Name"), tfFirstName);
grid.addRow(i++, new Label("Last Name"),  tfLastName);
grid.addRow(i++, new Label("Age"), tfAge);
grid.add(btn,    1, i);
GridPane.setHalignment(btn, HPos.CENTER);
```

The `add()` method accepts either three or five parameters:

- The node, the column index, the row index
- The node, the column index, the row index, how many columns to span, how many rows to span

The columns and rows indices start from 0.

The `setHalignment()` method sets the position of the node in the cell. The enum `HPos` has values: `LEFT`, `RIGHT`, `CENTER`. The method `addRow(int i, Node... nodes)` accepts the row index and the varargs of nodes. We use it to place the `Label` and the `TextField` objects.

The rest of the `start()` method is very similar to the `HellowWorld` example (only the title and size have changed):

```
primaryStage.setTitle("Simple form example");
primaryStage.onCloseRequestProperty()
        .setValue(e -> System.out.println("Bye! See you later!"));
primaryStage.setScene(new Scene(grid, 300, 200));
primaryStage.show();
```

If we run the just-implemented `start()` method, the result will be as follows:

We can fill the data as follows, for example:

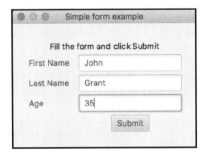

After the **Submit** button is clicked, the following message is displayed and the application exists:

```
Hello, John Grant, age 35!
```

To help visualize the layout, especially in the case of a more complex design, you can use the grid method `setGridLinesVisible(boolean v)` to make the grid lines visible. It helps to see how the cells are aligned. We can add the following line to our example:

```
grid.setGridLinesVisible(true);
```

We run it again, and the result will be as follows:

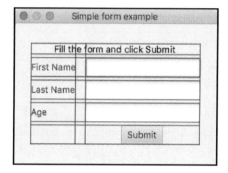

As you can see, the layout is now outlined explicitly, which helps us to visualize the design.

The `javafx.scene.layout` package includes 24 layout classes such as `Pane` (we saw it in the `HelloWorld` example), `StackPane` (allows us to overlay nodes), `FlowPane` (allows the positions of nodes to flow as the size of the window changes), `AnchorPane` (preserves the nodes' position relative to their anchor point), to name a few. The `VBox` layout will be demonstrated in the next section, *Charts*.

Charts

JavaFX provides the following chart components for data visualization in the `javafx.scene.chart` package:

- `LineChart`: Adds a line between the data points in a series; typically used to present the trends over time
- `AreaChart`: Similar to the `LineChart`, but fills the area between the line that connects the data points and the axis; typically used for comparing cumulated totals over time

- `BarChart`: Presents data as rectangular bars; used for visualization of discrete data
- `PieChart`: Presents a circle divided into segments (filled with different colors), each segment representing a value as a proportion of the total; we will demonstrate it in this section
- `BubbleChart`: Presents data as 2-dimensional oval shapes called **bubbles** that allow presenting three parameters
- `ScatterChart`: Presents the data points in a series as-is; useful to identify the presence of a clustering (data correlation)

The following example demonstrates how the result of testing can be presented as a pie chart. Each segment represents the number of tests succeeded, failed, or ignored:

```
Text txt = new Text("Test results:");

PieChart pc = new PieChart();
pc.getData().add(new PieChart.Data("Succeed", 143));
pc.getData().add(new PieChart.Data("Failed" ,  12));
pc.getData().add(new PieChart.Data("Ignored",  18));

VBox vb = new VBox(txt, pc);
vb.setAlignment(Pos.CENTER);
vb.setPadding(new Insets(10, 10, 10, 10));

primaryStage.setTitle("A chart example");
primaryStage.onCloseRequestProperty()
        .setValue(e:> System.out.println("Bye! See you later!"));
primaryStage.setScene(new Scene(vb, 300, 300));
primaryStage.show();
```

We have created two nodes—`Text` and `PieChart`—and placed them in the cells of the `VBox` layout that sets them in a column, one above another. We have added the padding of 10 pixels around the edges of the `VBox` pane. Notice that VBox extends the `Node` and `Pane` classes, as other panes do too. We have also positioned the pane in the center of the scene using the `setAlignment()` method. The rest is the same as all other previous examples, except the scene title and size.

If we run the preceding example, the result will be as follows:

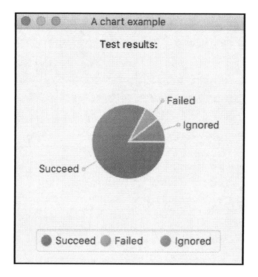

The PieChart class, as well as any other chart, has a number of other methods that can be useful for presenting more complex and dynamic data in a user-friendly manner.

Applying CSS

By default, JavaFX uses the style sheet that comes with the distribution jar file. To override the default style, you can add a style sheet to the scene using the getStylesheets() method:

```
scene.getStylesheets().add("/mystyle.css");
```

The mystyle.css file has to be placed into the src/main/resources folder. Let's do it and add the mystyle.css file with the following content to the HelloWorld example:

```
#text-hello {
  :fx-font-size: 20px;
  -fx-font-family: "Arial";
  -fx-fill: red;
}
.button {
  -fx-text-fill: white;
  -fx-background-color: slateblue;
}
```

As you can see, we would like to style the button node and the `Text` node that has an ID `text-hello` in a certain way. We also have to modify the **HelloWorld** example by adding the ID to the `Text` element and the style sheet file to the scene:

```
Text txt = new Text("Hello, world!");
txt.setId("text-hello");
txt.relocate(115, 40);

Button btn = new Button("Exit");
btn.relocate(155, 80);
btn.setOnAction(e -> {
    System.out.println("Bye! See you later!");
    Platform.exit();
});

Pane pane = new Pane();
pane.getChildren().addAll(txt, btn);

Scene scene = new Scene(pane, 350, 150);
scene.getStylesheets().add("/mystyle.css");

primaryStage.setTitle("The primary stage (top-level container)");
primaryStage.onCloseRequestProperty()
        .setValue(e -> System.out.println("\nBye! See you later!"));
primaryStage.setScene(scene);
primaryStage.show();
```

If we run this code now, the result will be as follows:

Alternatively, an inline style can be set on any node that will be used to overwrite the file style sheet, default or not. Let's add the following line to the latest version of the `HelloWorld` example:

```
btn.setStyle("-fx-text-fill: white; -fx-background-color: red;");
```

If we run the example again, the result will be as follows:

Look through the JavaFX CSS reference guide (`https://docs.oracle.com/javafx/2/api/javafx/scene/doc-files/cssref.html`) to get an idea of the variety and possible options for custom styling.

Using FXML

FXML is an XML-based language that allows building a user interface and maintaining it independently (as far as look-and-feel is concerned or other presentation-related changes) of the application (business) logic. Using FXML, you can design a user interface without even writing a line of Java code.

FXML does not have a schema, but its capabilities reflect the API of the JavaFX objects used to build a scene. This means you can use the API documentation to understand what tags and attributes are allowed in the FXML structure. Most of the times, JavaFX classes can be used as tags and their properties as attributes.

In addition to the FXML file (the view), the controller (Java class) can be used for processing the model and organizing the page flow. The model consists of domain objects managed by the view and the controller. It also allows using all the power of CSS styling and JavaScript. But, in this book, we will be able to demonstrate only the basic FXML capabilities. The rest you can find in the FXML introduction (`https://docs.oracle.com/javafx/2/api/javafx/fxml/doc-files/introduction_to_fxml.html`) and many good tutorials available online.

To demonstrate FXML usage, we are going to reproduce the simple form we have created in the *Control elements* section and then enhance it by adding the page flow. Here's how our form with first name, last name, and age can be expressed in FXML:

```
<?xml version="1.0" encoding="UTF-8"?>
<?import javafx.scene.Scene?>
<?import javafx.geometry.Insets?>
<?import javafx.scene.text.Text?>
<?import javafx.scene.control.Label?>
```

```
<?import javafx.scene.control.Button?>
<?import javafx.scene.layout.GridPane?>
<?import javafx.scene.control.TextField?>
<Scene fx:controller="com.packt.learnjava.ch12_gui.HelloWorldController"
        xmlns:fx="http://javafx.com/fxml"
        width="350" height="200">
    <GridPane alignment="center" hgap="15" vgap="5">
        <padding>
            <Insets top="20" right="20" bottom="20" left="20"/>
        </padding>
        <Text id="textFill" text="Fill the form and click Submit"
            GridPane.rowIndex="0" GridPane.columnSpan="2">
            <GridPane.halignment>center</GridPane.halignment>
        </Text>
        <Label text="First name"
            GridPane.columnIndex="0" GridPane.rowIndex="1"/>
        <TextField fx:id="tfFirstName"
            GridPane.columnIndex="1" GridPane.rowIndex="1"/>
        <Label text="Last name"
            GridPane.columnIndex="0" GridPane.rowIndex="2"/>
        <TextField fx:id="tfLastName"
            GridPane.columnIndex="1" GridPane.rowIndex="2"/>
        <Label text="Age"
            GridPane.columnIndex="0" GridPane.rowIndex="3"/>
        <TextField fx:id="tfAge"
            GridPane.columnIndex="1" GridPane.rowIndex="3"/>
        <Button text="Submit"
            GridPane.columnIndex="1" GridPane.rowIndex="4"
            onAction="#submitClicked">
            <GridPane.halignment>center</GridPane.halignment>
        </Button>
    </GridPane>
</Scene>
```

As you can see, it expresses the desired scene structure, familiar to you already, and specifies the controller class, HelloWorldController, which we are going to see shortly. As we have mentioned already, the tags match the class names we have been using to construct the same GUI with Java only. We will put the helloWorld.fxml file into the resources folder.

Now let's look at the start() method implementation of the HelloWorld class that uses the preceding FXML file:

```
try {
  FXMLLoader lder = new FXMLLoader();
  lder.setLocation(new URL("file:src/main/resources/helloWorld.fxml"));
  Scene scene = lder.load();
```

```
        primaryStage.setTitle("Simple form example");
        primaryStage.setScene(scene);
        primaryStage.onCloseRequestProperty()
                .setValue(e -> System.out.println("\nBye! See you later!"));
        primaryStage.show();
    } catch (Exception ex){
        ex.printStackTrace();
    }
```

The `start()` method just loads the `helloWorld.fxml` file and sets the stage, the latter being done exactly as in our previous examples. Now let's look at the `HelloWorldController` class. If need be, we could launch the application having only the following:

```
public class HelloWorldController {
    @FXML
    protected void submitClicked(ActionEvent e) {
    }
}
```

The form would be presented, but the button-click would do nothing. That is what we meant while talking about the user interface development independently of the application logic. Notice the @FXML annotation. It binds the method and properties to the FXML tags using their IDs. Here is how the full controller implementation looks:

```
@FXML
private TextField tfFirstName;
@FXML
private TextField tfLastName;
@FXML
private TextField tfAge;
@FXML
protected void submitClicked(ActionEvent e) {
    String fn = tfFirstName.getText();
    String ln = tfLastName.getText();
    String age = tfAge.getText();
    int a = 42;
    try {
        a = Integer.parseInt(age);
    } catch (Exception ex) {
    }
    fn = fn.isBlank() ? "Nick" : fn;
    ln = ln.isBlank() ? "Samoylov" : ln;
    System.out.println("Hello, " + fn + " " + ln + ", age " + a + "!");
    Platform.exit();
}
```

It should look very familiar to you for the most part. The only difference is that we refer to the fields and their values not directly (as previously), but using binding marked with the annotation @FXML. If we run the HelloWorld class now, the page appearance and behavior will be exactly the same as we have described it in the *Control elements* section.

Let's now add another page and modify the code so that the controller, after the Submit button is clicked, will send the submitted values to another page and close the form. To make it simple, the new page will just present the received data. Here is how its FXML will look:

```xml
<?xml version="1.0" encoding="UTF-8"?>
<?import javafx.scene.Scene?>
<?import javafx.geometry.Insets?>
<?import javafx.scene.text.Text?>
<?import javafx.scene.layout.GridPane?>

<Scene fx:controller="com.packt.lernjava.ch12_gui.HelloWorldController2"
       xmlns:fx="http://javafx.com/fxml"
       width="350" height="150">
    <GridPane alignment="center" hgap="15" vgap="5">
        <padding>
            <Insets top="20" right="20" bottom="20" left="20"/>
        </padding>
        <Text fx:id="textUser"
            GridPane.rowIndex="0" GridPane.columnSpan="2">
            <GridPane.halignment>center</GridPane.halignment>
        </Text>
        <Text id="textDo" text="Do what has to be done here"
            GridPane.rowIndex="1" GridPane.columnSpan="2">
            <GridPane.halignment>center</GridPane.halignment>
        </Text>
    </GridPane>
</Scene>
```

As you can see, the page has only two read-only Text fields. The first one (with id="textUser") will show the data passed from the previous page. The second will just show the message **Do what has to be done here**. This is not very sophisticated, but it demonstrates how the flow of data and pages can be organized.

The new page uses a different controller that looks as follows:

```java
package com.packt.learnjava.ch12_gui;
import javafx.fxml.FXML;
import javafx.scene.text.Text;
public class HelloWorldController2 {
    @FXML
```

```
        public Text textUser;
    }
```

As you might guess, the public field `textUser` has to be filled with the value by the first controller, `HelloWolrdController`. Let's do it. We modify the `submitClicked()` method as follows:

```
@FXML
protected void submitClicked(ActionEvent e) {
    String fn = tfFirstName.getText();
    String ln = tfLastName.getText();
    String age = tfAge.getText();
    int a = 42;
    try {
        a = Integer.parseInt(age);
    } catch (Exception ex) {}
    fn = fn.isBlank() ? "Nick" : fn;
    ln = ln.isBlank() ? "Samoylov" : ln;
    String user = "Hello, " + fn + " " + ln + ", age " + a + "!";
    //System.out.println("\nHello, " + fn + " " + ln + ", age " + a + "!");
    //Platform.exit();

    goToPage2(user);
    Node source = (Node) e.getSource();
    Stage stage = (Stage) source.getScene().getWindow();
    stage.close();
}
```

Instead of just printing the submitted (or default) data and exiting from the application (see the two lines commented out), we call the `goToPage2()` method and pass the submitted data as a parameter. Then, we extract from the event the reference to the stage of the current window and close it.

The `goToPage2()` method looks as follows:

```
try {
  FXMLLoader lder = new FXMLLoader();
  lder.setLocation(new URL("file:src/main/resources/helloWorld2.fxml"));
  Scene scene = lder.load();

  HelloWorldController2 c = loader.getController();
  c.textUser.setText(user);

  Stage primaryStage = new Stage();
  primaryStage.setTitle("Simple form example. Page 2.");
  primaryStage.setScene(scene);
  primaryStage.onCloseRequestProperty()
```

```
            .setValue(e -> {
                System.out.println("Bye! See you later!");
                Platform.exit();
            });
    primaryStage.show();
} catch (Exception ex) {
    ex.printStackTrace();
}
```

It loads the `helloWorld2.fxml` file, extracts from it the controller object, and sets on it the passed-in value. The rest is the same stage configuration that you have seen several times by now. The only difference is that **Page 2** is added to the title.

If we execute the `HelloWorld` class now, we will see the familiar form and fill it with data:

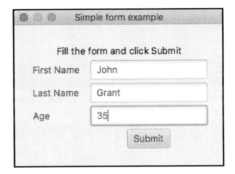

After we click the **Submit** button, this window will be closed and the new one will appear:

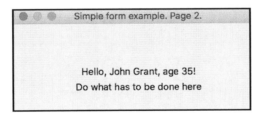

We click the x button in the upper-left corner (or in the upper-right corner on Windows) and see the same message we have seen before:

```
Bye! See you later!
Doing what has to be done before closing
```

The same stage action function and `stop()` method worked as expected.

With that, we conclude our presentation of FXML and move to the next topic of adding HTML to the JavaFX application.

Embedding HTML

To add HTML to JavaFX is easy. All you have to do is to use the `javafx.scene.web.WebView` class that provides a window where the added HTML is rendered similarly as it happens in a browser. The `WebView` class uses the open source browser engine WebKit and, thus, supports full browsing functionality.

Like all other JavaFX components, the `WebView` class extends the `Node` class and can be treated in the Java code as such. In addition, it has its own properties and methods that allow adjusting the browser window to the encompassing application, by setting the window size (max, min, and preferred height and width), font scale, zoom rate, adding CSS, enabling context (right-click) menu, and similar. The `getEngine()` method returns a `javafx.scene.web.WebEngine` object associated with it. It provides the ability to load HTML pages, to navigate them, to apply different styles to the loaded pages, to access their browsing history and the document model, and to execute JavaScript.

To start using the `javafx.scene.web` package, two steps have to be taken first:

1. Add the following dependency to the `pom.xml` file:

   ```
   <dependency>
    <groupId>org.openjfx</groupId>
    <artifactId>javafx-web</artifactId>
    <version>11.0.2</version>
   </dependency>
   ```

   The version of `javafx-web` typically stays abreast with the Java version, but at the time of this writing version 12 of `javafx-web` has not been released, so we are using the latest available version, 11.0.2.

2. Since `javafx-web` uses the packages `com.sun.*` that have been removed from Java 9 (https://docs.oracle.com/javase/9/migrate/toc.htm#JSMIG-GUID-F7696E02-A1FB-4D5A-B1F2-89E7007D4096), to access the `com.sun.*` packages from Java 9+, set the following VM options in addition to `--module-path` and `--add-modules`, described in the *JavaFX fundamentals* section in **Run/Debug Configuration** of the class `HtmlWebView` of the `ch12_gui` package:

   ```
   --add-exports javafx.graphics/com.sun.javafx.sg.prism=ALL-UNNAMED
   --add-exports javafx.graphics/com.sun.javafx.scene=ALL-UNNAMED
   ```

```
--add-exports javafx.graphics/com.sun.javafx.util=ALL-UNNAMED
--add-exports javafx.base/com.sun.javafx.logging=ALL-UNNAMED
--add-exports javafx.graphics/com.sun.prism=ALL-UNNAMED
--add-exports javafx.graphics/com.sun.glass.ui=ALL-UNNAMED
--add-exports javafx.graphics/com.sun.javafx.geom.transform=ALL-
UNNAMED
--add-exports javafx.graphics/com.sun.javafx.tk=ALL-UNNAMED
--add-exports javafx.graphics/com.sun.glass.utils=ALL-UNNAMED
--add-exports javafx.graphics/com.sun.javafx.font=ALL-UNNAMED
--add-exports javafx.graphics/com.sun.javafx.application=ALL-
UNNAMED
--add-exports javafx.controls/com.sun.javafx.scene.control=ALL-
UNNAMED
--add-exports javafx.graphics/com.sun.javafx.scene.input=ALL-
UNNAMED
--add-exports javafx.graphics/com.sun.javafx.geom=ALL-UNNAMED
--add-exports javafx.graphics/com.sun.prism.paint=ALL-UNNAMED
--add-exports javafx.graphics/com.sun.scenario.effect=ALL-UNNAMED
--add-exports javafx.graphics/com.sun.javafx.text=ALL-UNNAMED
--add-exports javafx.graphics/com.sun.javafx.iio=ALL-UNNAMED
--add-exports
javafx.graphics/com.sun.scenario.effect.impl.prism=ALL-UNNAMED
--add-exports javafx.graphics/com.sun.javafx.scene.text=ALL-UNNAMED
```

To execute the class `HtmlWebView` from the command line, use the following commands:

```
mvn clean package

java --module-path /path/javaFX/lib --add-
modules=javafx.controls,javafx.fxml --add-exports
javafx.graphics/com.sun.javafx.sg.prism=ALL-UNNAMED --add-exports
javafx.graphics/com.sun.javafx.scene=ALL-UNNAMED --add-exports
javafx.graphics/com.sun.javafx.util=ALL-UNNAMED --add-exports
javafx.base/com.sun.javafx.logging=ALL-UNNAMED --add-exports
javafx.graphics/com.sun.prism=ALL-UNNAMED --add-exports
javafx.graphics/com.sun.glass.ui=ALL-UNNAMED --add-exports
javafx.graphics/com.sun.javafx.geom.transform=ALL-UNNAMED --add-exports
javafx.graphics/com.sun.javafx.tk=ALL-UNNAMED --add-exports
javafx.graphics/com.sun.glass.utils=ALL-UNNAMED  --add-exports
javafx.graphics/com.sun.javafx.font=ALL-UNNAMED  --add-exports
javafx.graphics/com.sun.javafx.application=ALL-UNNAMED --add-exports
javafx.controls/com.sun.javafx.scene.control=ALL-UNNAMED --add-exports
javafx.graphics/com.sun.javafx.scene.input=ALL-UNNAMED --add-exports
javafx.graphics/com.sun.javafx.geom=ALL-UNNAMED  --add-exports
javafx.graphics/com.sun.prism.paint=ALL-UNNAMED  --add-exports
javafx.graphics/com.sun.scenario.effect=ALL-UNNAMED --add-exports
javafx.graphics/com.sun.javafx.text=ALL-UNNAMED --add-exports
javafx.graphics/com.sun.javafx.iio=ALL-UNNAMED --add-exports
```

```
javafx.graphics/com.sun.scenario.effect.impl.prism=ALL-UNNAMED --add-
exports javafx.graphics/com.sun.javafx.scene.text=ALL-UNNAMED  -cp
target/learnjava-1.0-SNAPSHOT.jar:target/libs/*
com.packt.learnjava.ch12_gui.HtmlWebView
```

On Windows, the same command looks as follows:

```
mvn clean package

java --module-path C:\path\JavaFX\lib --add-
modules=javafx.controls,javafx.fxml --add-exports
javafx.graphics/com.sun.javafx.sg.prism=ALL-UNNAMED --add-exports
javafx.graphics/com.sun.javafx.scene=ALL-UNNAMED --add-exports
javafx.graphics/com.sun.javafx.util=ALL-UNNAMED --add-exports
javafx.base/com.sun.javafx.logging=ALL-UNNAMED --add-exports
javafx.graphics/com.sun.prism=ALL-UNNAMED --add-exports
javafx.graphics/com.sun.glass.ui=ALL-UNNAMED --add-exports
javafx.graphics/com.sun.javafx.geom.transform=ALL-UNNAMED --add-exports
javafx.graphics/com.sun.javafx.tk=ALL-UNNAMED --add-exports
javafx.graphics/com.sun.glass.utils=ALL-UNNAMED  --add-exports
javafx.graphics/com.sun.javafx.font=ALL-UNNAMED  --add-exports
javafx.graphics/com.sun.javafx.application=ALL-UNNAMED --add-exports
javafx.controls/com.sun.javafx.scene.control=ALL-UNNAMED --add-exports
javafx.graphics/com.sun.javafx.scene.input=ALL-UNNAMED --add-exports
javafx.graphics/com.sun.javafx.geom=ALL-UNNAMED  --add-exports
javafx.graphics/com.sun.prism.paint=ALL-UNNAMED  --add-exports
javafx.graphics/com.sun.scenario.effect=ALL-UNNAMED --add-exports
javafx.graphics/com.sun.javafx.text=ALL-UNNAMED --add-exports
javafx.graphics/com.sun.javafx.iio=ALL-UNNAMED --add-exports
javafx.graphics/com.sun.scenario.effect.impl.prism=ALL-UNNAMED --add-
exports javafx.graphics/com.sun.javafx.scene.text=ALL-UNNAMED  -cp
target\learnjava-1.0-SNAPSHOT.jar;target\libs\*
com.packt.learnjava.ch12_gui.HtmlWebView
```

The class `HtmlWebView` contains several `start()` methods too. Rename and execute them one by one as described in the *JavaFX fundamentals* section.

Let's look at a few examples now. We create a new application, `HtmlWebView`, and set VM options for it with the VM options `--module-path`, `--add-modules`, and `--add-exports` we have described. Now we can write and execute a code that uses the `WebView` class.

First, here is how simple HTML can be added to the JavaFX application:

```
WebView wv = new WebView();
WebEngine we = wv.getEngine();
String html = "<html><center><h2>Hello, world!</h2></center></html>";
```

```
we.loadContent(html, "text/html");
Scene scene = new Scene(wv, 200, 60);
primaryStage.setTitle("My HTML page");
primaryStage.setScene(scene);
primaryStage.onCloseRequestProperty()
            .setValue(e -> System.out.println("Bye! See you later!"));
primaryStage.show();
```

The preceding code creates a `WebView` object, gets the `WebEngine` object from it, uses the acquired `WebEngine` object to load the HTML, sets the `WebView` object on the scene, and configures the stage. The `loadContent()` method accepts two strings: the content and its mime type. The content string can be constructed in the code or created from reading the `.html` file.

If we run the preceding example, the result will be as follows:

If necessary, you can show other JavaFX nodes along with the `WebView` object in the same window. For example, let's add a `Text` node above the embedded HTML:

```
Text txt = new Text("Below is the embedded HTML:");

WebView wv = new WebView();
WebEngine we = wv.getEngine();
String html = "<html><center><h2>Hello, world!</h2></center></html>";
we.loadContent(html, "text/html");

VBox vb = new VBox(txt, wv);
vb.setSpacing(10);
vb.setAlignment(Pos.CENTER);
vb.setPadding(new Insets(10, 10, 10, 10));

Scene scene = new Scene(vb, 300, 120);
primaryStage.setScene(scene);
primaryStage.setTitle("JavaFX with embedded HTML");
primaryStage.onCloseRequestProperty()
            .setValue(e -> System.out.println("Bye! See you later!"));
primaryStage.show();
```

As you can see, the `WebView` object is not set on the scene directly, but on the layout object instead, along with `txt` object. Then, the layout object is set on the scene. The result of the preceding code is as follows:

With a more complex HTML page, it is possible to load it from the file directly, using the `load()` method. To demonstrate this approach, let's create `form.htm` file in the `resources` folder with the following content:

```html
<!DOCTYPE html>
<html lang="en">
<head>
    <meta charset="UTF-8">
    <title>The Form</title>
</head>
<body>
<form action="http://server:port/formHandler" metrod="post">
    <table>
        <tr>
            <td><label for="firstName">Firts name:</label></td>
            <td><input type="text" id="firstName" name="firstName"></td>
        </tr>
        <tr>
            <td><label for="lastName">Last name:</label></td>
            <td><input type="text" id="lastName" name="lastName"></td>
        </tr>
        <tr>
            <td><label for="age">Age:</label></td>
            <td><input type="text" id="age" name="age"></td>
        </tr>
        <tr>
            <td></td>
            <td align="center">
                <button id="submit" name="submit">Submit</button>
            </td>
        </tr>
    </table>
</form>
</body>
</html>
```

This HTML presents a form similar to the one we have created in the Using FXML section. After the Submit button is clicked, the form data is posted to a server to the \formHandler URI. To present this form inside a JavaFX application, the following code can be used:

```
Text txt = new Text("Fill the form and click Submit");

WebView wv = new WebView();
WebEngine we = wv.getEngine();
File f = new File("src/main/resources/form.html");
we.load(f.toURI().toString());

VBox vb = new VBox(txt, wv);
vb.setSpacing(10);
vb.setAlignment(Pos.CENTER);
vb.setPadding(new Insets(10, 10, 10, 10));

Scene scene = new Scene(vb, 300, 200);

primaryStage.setScene(scene);
primaryStage.setTitle("JavaFX with embedded HTML");
primaryStage.onCloseRequestProperty()
            .setValue(e -> System.out.println("Bye! See you later!"));
primaryStage.show();
```

As you can see, the difference with our other examples is that we now use the File class and its toURI() method to access the HTML in the src/main/resources/form.html file directly, without converting the content to a string first. The result looks as follows:

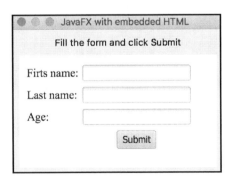

This solution is useful when you need to send a request or post data from your JavaFX application. But, when the form you would like a user to fill is already available on the server, you can just load it from the URL. For example, let's incorporate a Google search in the JavaFX application. We can do it by changing the parameter value of the `load()` method to the URL of the page we would like to load:

```
Text txt = new Text("Enjoy searching the Web!");

WebView wv = new WebView();
WebEngine we = wv.getEngine();
we.load("http://www.google.com");

VBox vb = new VBox(txt, wv);
vb.setSpacing(20);
vb.setAlignment(Pos.CENTER);
vb.setStyle("-fx-font-size: 20px;-fx-background-color: lightblue;");
vb.setPadding(new Insets(10, 10, 10, 10));

Scene scene = new Scene(vb,750,500);
primaryStage.setScene(scene);
primaryStage.setTitle("JavaFX with the window to another server");
primaryStage.onCloseRequestProperty()
        .setValue(e -> System.out.println("Bye! See you later!"));
primaryStage.show();
```

We have also added a style to the layout in order to increase the font and add color to the background, so we can see the outline of the area where the rendered HTML is embedded. When we run this example, the following window appears:

In this window, you can perform all the aspects of a search that you usually access via the browser.

And, as we have mentioned already, you can zoom into the rendered page. For example, if we add the `wv.setZoom(1.5)` line to the preceding example, the result will be as follows:

Similarly, we can set the scale for the font and even the style from a file:

```
wv.setFontScale(1.5);
we.setUserStyleSheetLocation("mystyle.css");
```

Notice, though, that we set the font scale on the `WebView` object, while we set the style in the `WebEngine` object.

We can also access (and manipulate) the DOM object of the loaded page using the `WebEngine` class method `getDocument()`:

```
Document document = we.getDocument();
```

We can also access the browsing history, get the current index, and move the history backward and forward:

```
WebHistory history = we.getHistory();
int currInd = history.getCurrentIndex();
history.go(-1);
history.go( 1);
```

And, for each entry of the history, we can extract its URL, title, or last-visited date:

```
WebHistory history = we.getHistory();
ObservableList<WebHistory.Entry> entries = history.getEntries();
for(WebHistory.Entry entry: entries){
    String url = entry.getUrl();
    String title = entry.getTitle();
    Date date = entry.getLastVisitedDate();
}
```

Read the documentation of the `WebView` and `WebEngine` classes, to get more ideas about how you can take advantage of their functionality.

Playing media

Adding an image to a scene of the JavaFX application does not require the `com.sun.*` packages, so the `--add-export` VM options listed in the *Adding HTML* section are not needed. But, it does not hurt to have them anyway, so leave the `--add-export` options in place, if you have added them already.

An image can be included in a scene using the classes `javafx.scene.image.Image` and `javafx.scene.image.ImageView`. To demonstrate how to do it, we are going to use the Packt logo `packt.png` located in the `resources` folder. Here is the code that does it:

```
Text txt = new Text("What a beautiful image!");

FileInputStream input =
                new FileInputStream("src/main/resources/packt.png");
Image image = new Image(input);
ImageView iv = new ImageView(image);

VBox vb = new VBox(txt, iv);
vb.setSpacing(20);
vb.setAlignment(Pos.CENTER);
vb.setPadding(new Insets(10, 10, 10, 10));

Scene scene = new Scene(vb, 300, 200);
primaryStage.setScene(scene);
primaryStage.setTitle("JavaFX with embedded HTML");
primaryStage.onCloseRequestProperty()
        .setValue(e -> System.out.println("Bye! See you later!"));
primaryStage.show();
```

If we run the preceding code, the result will be as follows:

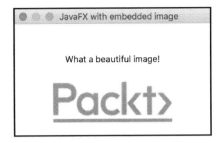

The currently supported image formats are BMP, GIF, JPEG, and PNG. Look through the API of the Image and ImageView classes (`https://openjfx.io/javadoc/11/javafx.graphics/javafx/scene/image/package-summary.html`) to learn the many ways an image can be formatted and adjusted as needed.

Now let's see how to use other media files in a JavaFX application. Playing an audio or movie file requires the `--add-export` VM options listed in the *Adding HTML* section.

The currently supported encodings are as follows:

- **AAC**: Advanced Audio Coding audio compression
- **H.264/AVC**: H.264/MPEG-4 Part 10 / **AVC** (**Advanced Video Coding**) video compression
- **MP3**: Raw MPEG-1, 2, and 2.5 audio; layers I, II, and III
- **PCM**: Uncompressed, raw audio samples

You can see a more detailed description of the supported protocols, media containers, and metadata tags in the API documentation (`https://openjfx.io/javadoc/11/javafx.media/javafx/scene/media/package-summary.html`).

The following three classes allow constructing a media player that can be added to a scene:

```
javafx.scene.media.Media;
javafx.scene.media.MediaPlayer;
javafx.scene.media.MediaView;
```

The Media class represents the source of the media. The MediaPlayer class provides all the methods that control the media playback: `play()`, `stop()`, `pause()`, `setVolume()`, and similar. You can also specify the number of times that the media should be played. The MediaView class extends the Node class and can be added to a scene. It provides a view of the media being played by the media player. It is responsible for a media appearance.

For the demonstration, let's add to the `HtmlWebView` application another version of the `start()` method that plays the `jb.mp3` file located in the `resources` folder:

```
Text txt1 = new Text("What a beautiful music!");
Text txt2 = new Text("If you don't hear music, turn up the volume.");

File f = new File("src/main/resources/jb.mp3");
Media m = new Media(f.toURI().toString());
MediaPlayer mp = new MediaPlayer(m);
MediaView mv = new MediaView(mp);

VBox vb = new VBox(txt1, txt2, mv);
vb.setSpacing(20);
vb.setAlignment(Pos.CENTER);
vb.setPadding(new Insets(10, 10, 10, 10));

Scene scene = new Scene(vb, 350, 100);
primaryStage.setScene(scene);
primaryStage.setTitle("JavaFX with embedded media player");
primaryStage.onCloseRequestProperty()
        .setValue(e -> System.out.println("Bye! See you later!"));
primaryStage.show();

mp.play();
```

Notice how a `Media` object is constructed based on the source file; then the `MediaPlayer` object is constructed based on the `Media` object and then set as a property of the `MediaView` class constructor. The `MediaView` object is set on the scene along with two `Text` objects. We use the `VBox` object to provide the layout. Finally, after the scene is set on the stage and the stage becomes visible (after the `show()` method completes), the `play()` method is invoked on the `MediaPlayer` object. By default, the media is played once.

If we execute the preceding code, the following window will appear and the `jb.m3` file will be played:

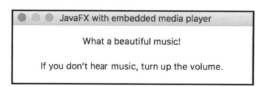

We could add controls to stop, pause, and adjust the volume, but it would require much more of the code, and that would not fit the expected size of this book. You can find a guide on how to do it in the Oracle online documentation (https://docs.oracle.com/javafx/2/media/jfxpub-media.htm).

A `sea.mp4` movie file can be played similarly:

```
Text txt = new Text("What a beautiful movie!");

File f = new File("src/main/resources/sea.mp4");
Media m = new Media(f.toURI().toString());
MediaPlayer mp = new MediaPlayer(m);
MediaView mv = new MediaView(mp);

VBox vb = new VBox(txt, mv);
vb.setSpacing(20);
vb.setAlignment(Pos.CENTER);
vb.setPadding(new Insets(10, 10, 10, 10));

Scene scene = new Scene(vb, 650, 400);
primaryStage.setScene(scene);
primaryStage.setTitle("JavaFX with embedded media player");
primaryStage.onCloseRequestProperty()
        .setValue(e -> System.out.println("Bye! See you later!"));
primaryStage.show();

mp.play();
```

The only difference is the different size of the scene needed to show the full frame of this particular clip. We figured out the necessary size after several trial-and-error adjustments. Alternatively, we could use the `MediaView` methods `autosize()`, `preserveRatioProperty()`, `setFitHeight()`, `setFitWidth()`, `fitWidthProperty()`, `fitHeightProperty()`, and similar, to adjust the size of the embedded window and to match the size of the scene automatically. If we execute the preceding example, the following window will pop-up and play the clip:

We can even combine playing both - audio and video - files in parallel, and thus provide a movie with a soundtrack:

```
Text txt1 = new Text("What a beautiful movie and sound!");
Text txt2 = new Text("If you don't hear music, turn up the volume.");

File fs = new File("src/main/resources/jb.mp3");
Media ms = new Media(fs.toURI().toString());
MediaPlayer mps = new MediaPlayer(ms);
MediaView mvs = new MediaView(mps);

File fv = new File("src/main/resources/sea.mp4");
Media mv = new Media(fv.toURI().toString());
MediaPlayer mpv = new MediaPlayer(mv);
MediaView mvv = new MediaView(mpv);

VBox vb = new VBox(txt1, txt2, mvs, mvv);
vb.setSpacing(20);
vb.setAlignment(Pos.CENTER);
vb.setPadding(new Insets(10, 10, 10, 10));

Scene scene = new Scene(vb, 650, 500);
primaryStage.setScene(scene);
primaryStage.setTitle("JavaFX with embedded media player");
primaryStage.onCloseRequestProperty()
        .setValue(e -> System.out.println("Bye! See you later!"));
primaryStage.show();

mpv.play();
mps.play();
```

It is possible to do this, because each of the players is executed by its own thread.

For more information about the `javafx.scene.media` package, read the API and the developer guide online:

- https://openjfx.io/javadoc/11/javafx.media/javafx/scene/media/package-summary.html
- https://docs.oracle.com/javafx/2/media/jfxpub-media.htm

Adding effects

The `javafx.scene.effects` package contains many classes that allow the adding of various effects to the nodes:

- `Blend`: Combines pixels from two sources (typically images) using one of the pre-defined `BlendModes`
- `Bloom`: Makes the input image brighter, so that it appears to glow
- `BoxBlur`: Adds blur to an image
- `ColorAdjust`: Allows adjustments of hue, saturation, brightness, and contrast to an image
- `ColorInput`: Renders a rectangular region that is filled with the given `Paint`
- `DisplacementMap`: Shifts each pixel by a specified distance
- `DropShadow`: Renders a shadow of the given content behind the content
- `GaussianBlur`: Adds blur using a particular (Gaussian) method
- `Glow`: Makes the input image appear to glow
- `InnerShadow`: Creates a shadow inside the frame
- `Lighting`: Simulates a light source shining on the content; makes flat objects look more realistic
- `MotionBlur`: Simulates the given content seen in motion
- `PerspectiveTransform`: Transforms the content as seen in a perspective
- `Reflection`: Renders a reflected version of the input below the actual input content
- `SepiaTone`: Produces a sepia tone effect, similar to the appearance of antique photographs
- `Shadow`: Creates a monochrome duplicate of the content with blurry edges

All effects share the parent— the abstract class `Effect`. The `Node` class has the `setEffect(Effect e)` method, which means that any of the effects can be added to any node. And that is the main way of applying effects to the nodes—the actors that produce a scene on a stage (if we recall our analogy introduced at the beginning of this chapter).

The only exception is the `Blend` effect, which makes its usage more complicated than the use of other effects. In addition to using the `setEffect(Effect e)` method, some of the `Node` class children have also the `setBlendMode(BlendMode bm)` method, which allows regulating how the images are blending into one another when they overlap. So, it is possible to set different blend effects in different ways that override one another and produce an unexpected result that may be difficult to debug. That is what makes the `Blend` effect usage more complicated, and that is why we are going to start the overview with how the `Blend` effect can be used.

There are four aspects that regulate the appearance of the area where two images overlap (we use two images in our examples to make it simpler, but, in practice, many images can overlap):

- **The value of the *opacity* property**: This defines how much can be seen through the image; the opacity value 0.0 means the image is fully transparent, while the opacity value 1.0 means nothing behind it can be seen.
- **The alpha value and strength of each color**: This defines the transparency of the color as a double value in the range 0.0-1.0 or 0-255.
- **The blending mode, defined by the** `enum BlendMode` **value:** Depending on the mode, opacity, and alpha value of each color, the result may also depend on **the sequence in which the images were added to the scene**; the first added image is called a **bottom input**, while the second of the overlapping images is called a **top input**; if the top input is completely opaque, the bottom input is hidden by the top input.

The resulting appearance of the overlapping area is calculated based on the opacity, the alpha values of the colors, the numeric values (strength) of the colors, and the blending mode, which can be one of the following:

- `ADD`: The color and alpha components from the top input are added to those from the bottom input
- `BLUE`: The blue component of the bottom input is replaced with the blue component of the top input; the other color components are unaffected
- `COLOR_BURN`: The inverse of the bottom input color components are divided by the top input color components, all of which are then inverted to produce the resulting color
- `COLOR_DODGE`: The bottom input color components are divided by the inverse of the top input color components to produce the resulting color
- `DARKEN`: The darker of the color components from the two inputs is selected to produce the resulting color

- DIFFERENCE: The darker of the color components from the two inputs is subtracted from the lighter ones to produce the resulting color
- EXCLUSION: The color components from the two inputs are multiplied and doubled, and then subtracted from the sum of the bottom input color components, to produce the resulting color
- GREEN: The green component of the bottom input is replaced with the green component of the top input; the other color components are unaffected
- HARD_LIGHT: The input color components are either multiplied or screened, depending on the top input color
- LIGHTEN: The lighter of the color components from the two inputs is selected to produce the resulting color
- MULTIPLY: The color components from the first input are multiplied with those from the second input
- OVERLAY: The input color components are either multiplied or screened, depending on the bottom input color
- RED: The red component of the bottom input is replaced with the red component of the top input; the other color components are unaffected
- SCREEN: The color components from both of the inputs are inverted, multiplied with each other, and that result is again inverted to produce the resulting color
- SOFT_LIGHT: The input color components are either darkened or lightened, depending on the top input color
- SRC_ATOP: The part of the top input lying inside of the bottom input is blended with the bottom input
- SRC_OVER: The top input is blended over the bottom input

To demonstrate the Blend effect, let's create another application called BlendEffect. It does not require the com.sun.* packages, so the --add-export VM options are not needed. Only --module-path and --add-modules options, described in the section *JavaFX fundamentals*, have to be set for compilation and execution.

The scope of this book does not allow us to demonstrate all possible combinations, so we will create a red circle and a blue square:

```
Circle createCircle(){
    Circle c = new Circle();
    c.setFill(Color.rgb(255, 0, 0, 0.5));
    c.setRadius(25);
    return c;
}
```

```
Rectangle createSquare(){
    Rectangle r = new Rectangle();
    r.setFill(Color.rgb(0, 0, 255, 1.0));
    r.setWidth(50);
    r.setHeight(50);
    return r;
}
```

We used the `Color.rgb(int red, int green, int blue, double alpha)` method to define colors of each of the figures. But there are many more ways to do it. Read the `Color` class API documentation for more details (`https://openjfx.io/javadoc/11/javafx.graphics/javafx/scene/paint/Color.html`).

To overlap the created circle and square, we will use the `Group` node:

```
Node c = createCircle();
Node s = createSquare();
Node g = new Group(s, c);
```

In the preceding code, the square is a bottom input. We will also create a group where the square is a top input:

```
Node c = createCircle();
Node s = createSquare();
Node g = new Group(c, s);
```

The distinction is important because we defined the circle as half-opaque, while the square was completely opaque. We will use the same settings throughout all our examples.

Let's compare the two modes, `MULTIPLY` and `SRC_OVER`. We will set them on the groups, using the `setEffect()` method as follows:

```
Blend blnd = new Blend();
blnd.setMode(BlendMode.MULTIPLY);
Node c = createCircle();
Node s = createSquare();
Node g = new Group(s, c);
g.setEffect(blnd);
```

For each mode, we create two groups, one with the input where circle is on top of square, another with the input where the square is on the top of the circle, and we put the four created groups in a `GridPane` layout (see the source code for details). If we run the `BlendEffect` application, the result will be this:

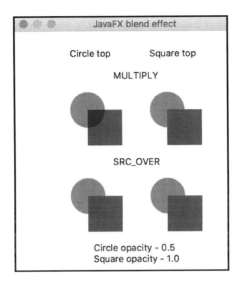

As was expected, when the square is on the top (the two images to the right), the overlapping area is completely taken by the opaque square. But, when the circle is a top input (the two images on the left), the overlapped area is somewhat visible and calculated based on the blend effect.

However, if we set the same mode directly on the group, the result will be slightly different. Let's run the same code but with the mode set on the group:

```
Node c = createCircle();
Node s = createSquare();
Node g = new Group(c, s);
g.setBlendMode(BlendMode.MULTIPLY);
```

If we run the application again, the result will look as follows:

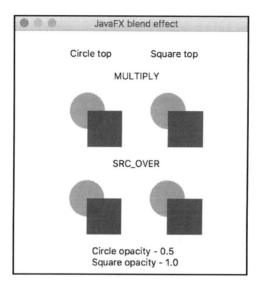

As you can see, the red color of the circle has slightly changed and there is no difference between the modes MULTIPLY and SRC_OVER. That is the issue with the sequence of adding the nodes to the scene we mentioned at the beginning of the section.

The result also changes depending on which node the effect is set. For example, instead of setting the effect on the group, let's set the blend effect on the circle only:

```
Blend blnd = new Blend();
blnd.setMode(BlendMode.MULTIPLY);
Node c = createCircle();
Node s = createSquare();
c.setEffect(blnd);
Node g = new Group(s, c);
```

We run the application and see the following:

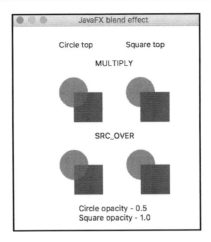

The two images on the right remain the same as in all the previous examples, but the two images on the left show the new colors of the overlapping area. Now let's set the same blend effect on the square instead of the circle as follows:

```
Blend blnd = new Blend();
blnd.setMode(BlendMode.MULTIPLY);
Node c = createCircle();
Node s = createSquare();
s.setEffect(blnd);
Node g = new Group(s, c);
```

The result will slightly change again and will look as presented on the following screenshot:

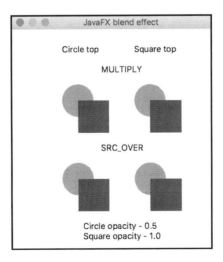

There is no difference between MULTIPLY and SRC_OVER modes, but the red color is different than it was when we set the effect on the circle.

We can change the approach again and set the blend effect mode directly on the circle only, using the following code:

```
Node c = createCircle();
Node s = createSquare();
c.setBlendMode(BlendMode.MULTIPLY);
```

The result changes again:

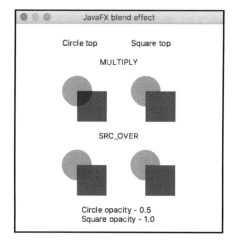

Setting the blend mode on the square only removes the difference between MULTIPLY and SRC_OVER modes again:

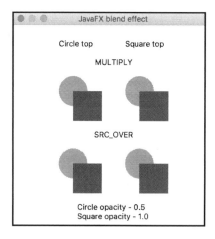

To avoid confusion and make the results of the blending more predictable, you have to watch the sequence in which the nodes are added to the scene and the consistency of the way the blend effect is applied.

In the source code provided with the book, you will find the examples for all effects included in the `javafx.scene.effects` package. They are all demonstrated by running side-by-side comparison. Here is one example:

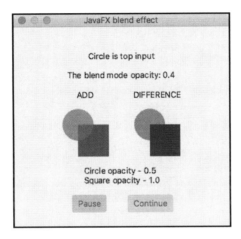

For your convenience, there are **Pause** and **Continue** buttons provided that allow you to pause the demonstration and review the result for different values of opacity set on the blend effect.

To demonstrate all other effects, we have created yet another application called `OtherEffects` that also does not require the `com.sun.*` packages, so the `--add-export` VM options are not needed. The effects demonstrated include `Bloom`, `BoxBlur`, `ColorAdjust`, `DisplacementMap`, `DropShadow`, `Glow`, `InnerShadow`, `Lighting`, `MotionBlur`, `PerspectiveTransform`, `Reflection`, `ShadowTone`, and `SepiaTone`. We have used two images to present the result of applying each of the effects, the Packt logo and a mountain lake view:

```
FileInputStream inputP =
                new FileInputStream("src/main/resources/packt.png");
Image imageP = new Image(inputP);
ImageView ivP = new ImageView(imageP);

FileInputStream inputM =
                new FileInputStream("src/main/resources/mount.jpeg");
Image imageM = new Image(inputM);
ImageView ivM = new ImageView(imageM);
```

```
ivM.setPreserveRatio(true);
ivM.setFitWidth(300);
```

We also have added two buttons that allow you to pause and continue the demonstration (it iterates over the effect and the values of their parameters):

```
Button btnP = new Button("Pause");
btnP.setOnAction(e1 -> et.pause());
btnP.setStyle("-fx-background-color: lightpink;");

Button btnC = new Button("Continue");
btnC.setOnAction(e2 -> et.cont());
btnC.setStyle("-fx-background-color: lightgreen;");
```

The et object is the object of the EffectsThread thread:

```
EffectsThread et = new EffectsThread(txt, ivM, ivP);
```

The thread goes through the list of the effects, creates a corresponding effect 10 times (with 10 different effects' parameter values) and, every time, sets the created Effect object on each of the images, then sleeps for one second to give you an opportunity to review the result:

```
public void run(){
    try {
        for(String effect: effects){
            for(int i = 0; i < 11; i++){
                double d = Math.round(i * 0.1 * 10.0) / 10.0;
                Effect e = createEffect(effect, d, txt);
                ivM.setEffect(e);
                ivP.setEffect(e);
                TimeUnit.SECONDS.sleep(1);
                if(pause){
                    while(true){
                        TimeUnit.SECONDS.sleep(1);
                        if(!pause){
                            break;
                        }
                    }
                }
            }
        }
        Platform.exit();
    } catch (Exception ex){
        ex.printStackTrace();
    }
}
```

We will show how each effect is created next, under the screenshot with the effect's result. To present the result, we have used the `GridPane` layout:

```
GridPane grid = new GridPane();
grid.setAlignment(Pos.CENTER);
grid.setVgap(25);
grid.setPadding(new Insets(10, 10, 10, 10));

int i = 0;
grid.add(txt,     0, i++, 2, 1);
GridPane.setHalignment(txt,  HPos.CENTER);
grid.add(ivP,     0, i++, 2, 1);
GridPane.setHalignment(ivP,  HPos.CENTER);
grid.add(ivM,     0, i++, 2, 1);
GridPane.setHalignment(ivM,  HPos.CENTER);
grid.addRow(i++, new Text());
HBox hb = new HBox(btnP, btnC);
hb.setAlignment(Pos.CENTER);
hb.setSpacing(25);
grid.add(hb,      0, i++, 2, 1);
GridPane.setHalignment(hb,  HPos.CENTER);
```

And, finally, the created `GridPane` object was passed to the scene, which in turn was placed on a stage familiar to you from our earlier examples:

```
Scene scene = new Scene(grid, 450, 500);
primaryStage.setScene(scene);
primaryStage.setTitle("JavaFX effect demo");
primaryStage.onCloseRequestProperty()
            .setValue(e3 -> System.out.println("Bye! See you later!"));
primaryStage.show();
```

The following screenshots depict examples of the effects of 1 of the 10 parameter values. Under each screenshot, we present the code snippet from the `createEffect(String effect, double d, Text txt)` method that created this effect:

```
//double d = 0.9;
txt.setText(effect + ".threshold: " + d);
Bloom b = new Bloom();
b.setThreshold(d);
```

```
// double d = 0.3;
int i = (int) d * 10;
int it = i / 3;
txt.setText(effect + ".iterations: " + it);
BoxBlur bb = new BoxBlur();
bb.setIterations(i);
```

```
double c = Math.round((-1.0 + d * 2) * 10.0) / 10.0;        // 0.6
txt.setText(effect + ": " + c);
ColorAdjust ca = new ColorAdjust();
ca.setContrast(c);
```

```
double h = Math.round((-1.0 + d * 2) * 10.0) / 10.0;        // 0.6
txt.setText(effect + ": " + h);
ColorAdjust ca1 = new ColorAdjust();
ca1.setHue(h);
```

```
double st = Math.round((-1.0 + d * 2) * 10.0) / 10.0;     // 0.6
txt.setText(effect + ": " + st);
ColorAdjust ca3 = new ColorAdjust();
ca3.setSaturation(st);
```

```
int w = (int)Math.round(4096 * d);   //819
int h1 = (int)Math.round(4096 * d); //819
txt.setText(effect + ": " + ": width: " + w + ", height: " + h1);
DisplacementMap dm = new DisplacementMap();
FloatMap floatMap = new FloatMap();
floatMap.setWidth(w);
floatMap.setHeight(h1);
for (int k = 0; k < w; k++) {
    double v = (Math.sin(k / 20.0 * Math.PI) - 0.5) / 40.0;
    for (int j = 0; j < h1; j++) {
        floatMap.setSamples(k, j, 0.0f, (float) v);
    }
}
dm.setMapData(floatMap);
```

```
double rd = Math.round((127.0 * d) * 10.0) / 10.0; // 127.0
System.out.println(effect + ": " + rd);
txt.setText(effect + ": " + rd);
DropShadow sh = new DropShadow();
sh.setRadius(rd);
```

```
double rad = Math.round(12.1 * d *10.0)/10.0;        // 9.7
double off = Math.round(15.0 * d *10.0)/10.0;        // 12.0
txt.setText("InnerShadow: radius: " + rad + ", offset:" + off);
InnerShadow is = new InnerShadow();
is.setColor(Color.web("0x3b596d"));
is.setOffsetX(off);
is.setOffsetY(off);
is.setRadius(rad);
```

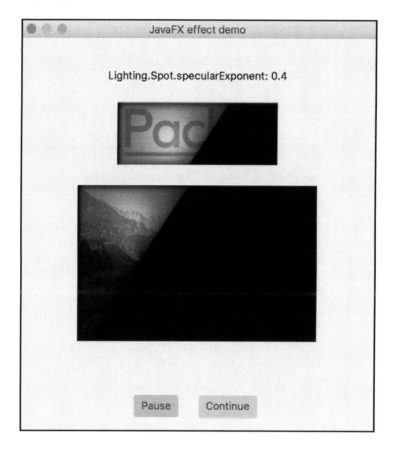

```
double sS = Math.round((d * 4)*10.0)/10.0;        // 0.4
txt.setText(effect + ": " + sS);
Light.Spot lightSs = new Light.Spot();
lightSs.setX(150);
lightSs.setY(100);
lightSs.setZ(80);
lightSs.setPointsAtX(0);
lightSs.setPointsAtY(0);
lightSs.setPointsAtZ(-50);
lightSs.setSpecularExponent(sS);
Lighting lSs = new Lighting();
lSs.setLight(lightSs);
lSs.setSurfaceScale(5.0);
```

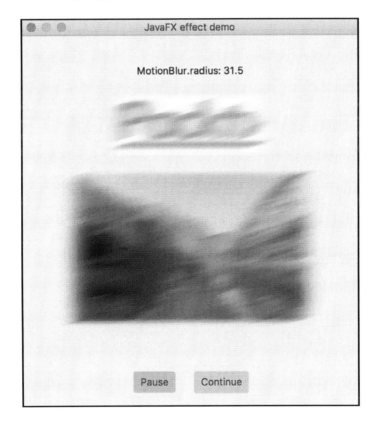

```
double r = Math.round((63.0 * d)*10.0) / 10.0;        // 31.5
txt.setText(effect + ": " + r);
MotionBlur mb1 = new MotionBlur();
mb1.setRadius(r);
mb1.setAngle(-15);
```

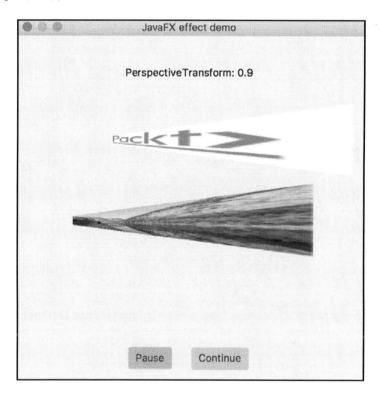

```
// double d = 0.9;
txt.setText(effect + ": " + d);
PerspectiveTransform pt =
        new PerspectiveTransform(0., 1. + 50.*d, 310., 50. - 50.*d,
                310., 50. + 50.*d + 1., 0., 100. - 50. * d + 2.);
```

```
// double d = 0.6;
txt.setText(effect + ": " + d);
Reflection ref = new Reflection();
ref.setFraction(d);
```

```
// double d = 1.0;
txt.setText(effect + ": " + d);
SepiaTone sep = new SepiaTone();
sep.setLevel(d);
```

The full source code of this demonstration is provided with the book and is available in the GitHub.

Summary

In this chapter, the reader was introduced to the JavaFX kit, its main features, and how it can be used to create a GUI application. The topics covered included an overview of Java GUI technologies, the JavaFX control elements, charts, using CSS, FXML, embedding HTML, playing media, and adding effects.

The next chapter is dedicated to functional programming. It provides an overview of functional interfaces that come with JDK, explains what a lambda expression is, and how to use a functional interface in a lambda expression. It also explains and demonstrates how to use method references.

Quiz

1. What is the top-level content container in JavaFX?
2. What is the base class of all the scene participants in JavaFX?
3. Name the base class of a JavaFX application.
4. What is one method of the JavaFX application that has to be implemented?
5. Which `Application` method has to be called by the main method to execute a JavaFX application?
6. Which two VM options are required to execute a JavaFX application?
7. Which `Application` method is called when the JavaFX application window is closed using the x button in the upper corner?
8. Which class has to be used to embed an HTML?
9. Name three classes that have to be used to play media.
10. What is the VM option required to be added in order to play media?
11. Name five JavaFX effects.

Section 3: Advanced Java

The last part of the book covers the most advanced topic of modern Java programming. It allows novices to get a solid footing in their understanding of the profession, and for those who already work in this area, to expand their skill set and expertise. The recent addition to Java of streams and functional programming makes asynchronous processing in Java almost as easy as the traditional synchronous way. It boosts the performance of Java applications, and the readers will learn how to take advantage of it and appreciate the beauty and the power of the provided solution.

The readers will also learn the new terms and the related concepts of reactive programming – asynchronous, non-blocking, and responsive – that are at the forefront of big data processing and machine learning. The building block of a reactive system is a microservice, which is demonstrated using the Vert.x toolkit.

The book concludes with an explanation of the benchmark tool, best programming practices, and the open Java projects that are currently underway. These projects, when concluded, will bring even more power to Java as a language and as a tool for creating modern, massive data processing systems. The reader will have the chance to gauge Java's future and even become a part of it.

This section contains the following chapters:

13
Functional Programming

This chapter brings the reader into the world of functional programming. It explains what a functional interface is, provides an overview of functional interfaces that come with JDK, and defines and demonstrates lambda expressions and how to use them with functional interfaces, including using **method reference**.

The following topics will be covered in this chapter:

- What is functional programming?
- Standard functional interfaces
- Functional pipelines
- Lambda expression limitations
- Method reference

What is functional programming?

We have actually used functional programming in the preceding chapters. In Chapter 6, *Data Structures, Generics and Popular Utilities*, we talked about the Iterable interface and its default void forEach (Consumer<T> function) method and provided the following example:

```
Iterable<String> list = List.of("s1", "s2", "s3");
System.out.println(list);                        //prints: [s1, s2, s3]
list.forEach(e -> System.out.print(e + " "));    //prints: s1 s2 s3
```

You can see how a Consumer e -> System.out.print(e + " ") function is passed into the forEach() method and applied to each element flowing into this method from the list. We will discuss the Consumer function shortly.

We also mentioned two methods of the `Collection` interface that accept a function as a parameter too:

- The `default boolean remove(Predicate<E> filter)` method, which attempts to remove all the elements that satisfy the given predicate from the collection; a `Predicate` function accepts an element of the collection and returns a `boolean` value

- The `default T[] toArray(IntFunction<T[]> generator)` method, which returns an array of all the elements of the collection, using the provided `IntFunction` generator function to allocate the returned array

In the same chapter, we also mentioned the following method of the `List` interface:

- `default void replaceAll(UnaryOperator<E> operator)`: Replaces each element of the list with the result of applying the provided `UnaryOperator` to that element; a `UnaryOperator` is one of the functions we are going to review in this chapter.

We described the `Map` interface, its method `default V merge(K key, V value, BiFunction<V,V,V> remappingFunction)` and how it can be used for concatenating the `String` values: `map.merge(key, value, String::concat)`.
The `BiFunction<V,V,V>` takes two parameters of the same type and returns the value of the same type as well. The `String::concat` construct is called a method reference and will be explained in the *Method reference* section.

We provided the following example of passing a `Comparator` function:

```
list.sort(Comparator.naturalOrder());
Comparator<String> cmp = (s1, s2) -> s1 == null ? -1 : s1.compareTo(s2);
list.sort(cmp);
```

It takes two `String` parameters, then compares the first one to `null`. If the first parameter is `null`, the function returns –1; otherwise, it compares the first parameter and the second one using the `compareTo()` method.

In Chapter 11, *Network Programming*, we looked at the following code:

```
HttpClient httpClient = HttpClient.newBuilder().build();
HttpRequest req = HttpRequest.newBuilder()
        .uri(URI.create("http://localhost:3333/something")).build();
try {
    HttpResponse<String> resp =
                    httpClient.send(req, BodyHandlers.ofString());
    System.out.println("Response: " +
```

```
                                    resp.statusCode() + " : " + resp.body());
} catch (Exception ex) {
    ex.printStackTrace();
}
```

The `BodyHandler` object (a function) is generated by
the `BodyHandlers.ofString()` factory method and passed into the `send()` method as a
parameter. Inside the method, the code calls its `apply()` method :

```
BodySubscriber<T> apply(ResponseInfo responseInfo)
```

Finally, in `Chapter 12`, *Java GUI Programming*, we used an `EventHandler` function as a
parameter in the following code snippet:

```
btn.setOnAction(e -> {
                     System.out.println("Bye! See you later!");
                     Platform.exit();
                 }
            );
primaryStage.onCloseRequestProperty()
       .setValue(e -> System.out.println("Bye! See you later!"));
```

The first function is `EventHanlder<ActionEvent>`. It prints a message and forces the
application to exit. The second is the `EventHandler<WindowEvent>` function. It just prints
the message.

All these examples give you a pretty good idea of how a function can be constructed and
passed around as a parameter. This ability constitutes functional programming. It is present
in many programming languages. It does not require the managing of object states. The
function is stateless. Its result depends only on the input data, no matter how many times it
was called. Such coding makes the outcome more predictable, which is the most attractive
aspect of functional programming.

The area that benefits the most from such a design is parallel data processing. Functional
programming allows for shifting the responsibility for parallelism from the client code to
the library. Before that, in order to process elements of Java collections, the client code had
to iterate over the collection and organize the processing. In Java 8, new (default) methods
were added that accept a function as a parameter and then apply it to each element of the
collection, in parallel or not, depending on the internal processing algorithm. So, it is the
library's responsibility to organize parallel processing.

What is a functional interface?

When we define a function, we, in fact, provide an implementation of an interface that has only one abstract method. That is how the Java compiler knows where to put the provided functionality. The compiler looks at the interface (`Consumer`, `Predicate`, `Comparator`, `IntFunction`, `UnaryOperator`, `BiFunction`, `BodyHandler`, and `EvenHandler` in the preceding examples), sees only one abstract method there, and uses the passed-in functionality as the method implementation. The only requirement is that the passed-in parameters must match the method signature. Otherwise, the compile-time error is generated.

That is why any interface that has only one abstract method is called a **functional interface**. Please note that the requirement of having **only one abstract method** includes the method inherited from the parent interface. For example, consider the following interfaces:

```
@FunctionalInterface
interface A {
    void method1();
    default void method2(){}
    static void method3(){}
}

@FunctionalInterface
interface B extends A {
    default void method4(){}
}

@FunctionalInterface
interface C extends B {
    void method1();
}

//@FunctionalInterface
interface D extends C {
    void method5();
}
```

The A is a functional interface because it has only one abstract method `method1()`. The B is a functional interface too because it has only one abstract method—the same `method1()` inherited from the A interface. The C is a functional interface because it has only one abstract method, `method1()`, which overrides the abstract `method1()` of the parent interface A. Interface D cannot be a functional interface because it has two abstract methods—`method1()` from the parent interface A and `method5()`.

To help avoid runtime errors, the @FunctionalInterface annotation was introduced in Java 8. It tells the compiler about the intent so the compiler can check and see if there is truly only one abstract method in the annotated interface. This annotation also warns a programmer, who reads the code, that this interface has only one abstract method intentionally. Otherwise, a programmer may waste time adding another abstract method to the interface only to discover at runtime that it cannot be done.

For the same reason, the Runnable and Callable interfaces, which have existed in Java since its early versions, were annotated in Java 8 as @FunctionalInterface. This distinction is made explicit and serves as a reminder to users that these interfaces can be used for creating a function:

```
@FunctionalInterface
interface Runnable {
    void run();
}

@FunctionalInterface
interface Callable<V> {
    V call() throws Exception;
}
```

As with any other interface, the functional interface can be implemented using the anonymous class:

```
Runnable runnable = new Runnable() {
    @Override
    public void run() {
        System.out.println("Hello!");
    }
};
```

An object created this way can later be used as follows:

```
runnable.run();    //prints: Hello!
```

If we look closely at the preceding code, we notice that there is unnecessary overhead. First, there is no need to repeat the interface name, because we declared it already as the type for the object reference. And, second, in the case of a functional interface that has only one abstract method, there is no need to specify the method name that has to be implemented. The compiler and Java runtime can figure it out. All we need is to provide the new functionality. The lambda expressions were introduced especially for this purpose.

What is a lambda expression?

The term lambda comes from lambda calculus—a universal model of computation that can be used to simulate any Turing machine. It was introduced by mathematician Alonzo Church in the 1930s. A **lambda expression** is a function, implemented in Java as an anonymous method. It also allows for the omitting of modifiers, return types, and parameter types. That makes for a very compact notation.

The syntax of lambda expression includes the list of parameters, an arrow token (->), and a body. The list of parameters can be empty, such as (), without parentheses (if there is only one parameter), or a comma-separated list of parameters surrounded by parentheses. The body can be a single expression or a statement block inside the braces ({ }). Let's look at a few examples:

- `() -> 42;` always returns `42`.
- `x -> x*42 + 42;` multiplies the x value by `42`, then adds `42` to the result and returns it.
- `(x, y) -> x * y;` multiplies the passed-in parameters and returns the result.
- `s -> "abc".equals(s);` compares the value of variable s and literal `"abc"`; it returns a `boolean` result value.
- `s -> System.out.println("x=" + s);` prints the s value with the prefix `"x="`.
- `(i, s) -> { i++; System.out.println(s + "=" + i); };` increments the input integer and prints the new value with the prefix `s + "="`, s being the value of the second parameter.

Without functional programming, the only way to pass some functionality as a parameter in Java would be by writing a class that implements an interface, creating its object, and then passing it as a parameter. But even the least-involved style using an anonymous class requires writing too much of the boilerplate code. Using functional interfaces and lambda expressions makes the code shorter, clearer, and more expressive.

For example, lambda expressions allow us to reimplement our preceding example with the `Runnable` interface as follows:

```
Runnable runnable = () -> System.out.println("Hello!");
```

As you can see, creating a functional interface is easy, especially with lambda expressions. But before doing that, consider using one of the 43 functional interfaces provided in the package `java.util.function`. This will not only allow you to writing less code, but will also help other programmers who are familiar with the standard interfaces to understand your code better.

Local-variable syntax for lambda parameters

Until the release of Java 11, there were two ways to declare parameter types—explicitly and implicitly. Here is an explicit version:

```
BiFunction<Double, Integer, Double> f = (Double x, Integer y) -> x / y;
System.out.println(f.apply(3., 2)); //prints: 1.5
```

The following is an implicit parameter type definition:

```
BiFunction<Double, Integer, Double> f = (x, y) -> x / y;
System.out.println(f.apply(3., 2));          //prints: 1.5
```

In the preceding code, the compiler infers the type of the parameters from the interface definition.

In Java 11, another method of parameter type declaration was introduced using the `var` type holder, which is similar to the local variable type holder `var` introduced in Java 10 (see `Chapter 1`, *Getting Started with Java 12*).

The following parameter declaration is syntactically exactly the same as the implicit one before Java 11:

```
BiFunction<Double, Integer, Double> f = (var x, var y) -> x / y;
System.out.println(f.apply(3., 2));          //prints: 1.5
```

The new local variable style syntax allows us to add annotations without defining the parameter type explicitly. Let's add the following dependency to the `pom.xml` file:

```
<dependency>
    <groupId>org.jetbrains</groupId>
    <artifactId>annotations</artifactId>
    <version>16.0.2</version>
</dependency>
```

It allows us to define passed-in variables as non-null:

```
import javax.validation.constraints.NotNull;
import java.util.function.BiFunction;
import java.util.function.Consumer;

BiFunction<Double, Integer, Double> f =
(@NotNull var x, @NotNull var y) -> x / y;
System.out.println(f.apply(3., 2));      //prints: 1.5
```

An annotation communicates to the compiler the programmer's intent, so it can warn the programmer during compilation or execution if the declared intent is violated. For example, we have tried to run the following code:

```
BiFunction<Double, Integer, Double> f = (x, y) -> x / y;
System.out.println(f.apply(null, 2));
```

It failed with `NullPointerException` at runtime. Then we have added the annotation as follows:

```
BiFunction<Double, Integer, Double> f =
        (@NotNull var x, @NotNull var y) -> x / y;
System.out.println(f.apply(null, 2));
```

The result of running the preceding code looks like this:

```
Exception in thread "main" java.lang.IllegalArgumentException:
Argument for @NotNull parameter 'x' of
com/packt/learnjava/ch13_functional/LambdaExpressions
.lambda$localVariableSyntax$1 must not be null
at com.packt.learnjava.ch13_functional.LambdaExpressions
.$$$reportNull$$$0(LambdaExpressions.java)
at com.packt.learnjava.ch13_functional.LambdaExpressions
.lambda$localVariableSyntax$1(LambdaExpressions.java)
at com.packt.learnjava.ch13_functional.LambdaExpressions
.localVariableSyntax(LambdaExpressions.java:59)
at com.packt.learnjava.ch13_functional.LambdaExpressions
.main(LambdaExpressions.java:12)
```

The lambda expression was not even executed.

The advantage of local-variable syntax in the case of lambda parameters becomes clear if we need to use annotations when the parameters are the objects of a class with a really long name. Before Java 11, the code might have looked like the following:

```
BiFunction<SomeReallyLongClassName,
AnotherReallyLongClassName, Double> f =
        (@NotNull SomeReallyLongClassName x,
        @NotNull AnotherReallyLongClassName y) -> x.doSomething(y);
```

We had to declare the type of the variable explicitly because we wanted to add annotations, and the following implicit version would not even compile:

```
BiFunction<SomeReallyLongClassName,
AnotherReallyLongClassName, Double> f =
            (@NotNull x, @NotNull y) -> x.doSomething(y);
```

With Java 11, the new syntax allows us to use the implicit parameter type inference using the type holder `var`:

```
BiFunction<SomeReallyLongClassName,
AnotherReallyLongClassName, Double> f =
            (@NotNull var x, @NotNull var y) -> x.doSomething(y);
```

That is the advantage of and the motivation behind introducing a local-variable syntax for the lambda parameter's declaration. Otherwise, consider staying away from using `var`. If the type of variable is short, using its actual type makes the code easier to understand.

Standard functional interfaces

Most of the interfaces provided in the `java.util.function` package are specializations of the following four interfaces: `Consumer<T>`, `Predicate<T>`, `Supplier<T>`, and `Function<T,R>`. Let's review them and then look at a short overview of the other 39 standard functional interfaces.

Consumer<T>

By looking at the `Consumer<T>` interface definition, you can <indexentry content="standard functional interfaces:Consumer">guess already that this interface has an abstract method that accepts a parameter of type `T` and does not return anything. Well, when only one type is listed, it may define the type of the return value, as in the case of the `Supplier<T>` interface. But the interface name serves as a clue: the **consumer** name indicates that the method of this interface just takes the value and returns nothing, while the **supplier** returns the value. This clue is not precise but helps to jog the memory.

The best source of information about any functional interface is the `java.util.function` package API documentation (https://docs.oracle.com/en/java/javase/12/docs/api/java.base/java/util/function/package-summary.html). If we read it, we learn that the `Consumer<T>` interface has one abstract and one default method:

- `void accept(T t)`: Applies the operation to the given argument
- `default Consumer<T> andThen(Consumer<T> after)`: Returns a composed `Consumer` function that performs, in sequence, the current operation followed by the `after` operation

It means that, for example, we can implement and then execute it as follows:

```
Consumer<String> printResult = s -> System.out.println("Result: " + s);
printResult.accept("10.0");    //prints: Result: 10.0
```

We can also have a factory method that creates the function, for example:

```
Consumer<String> printWithPrefixAndPostfix(String pref, String postf){
    return s -> System.out.println(pref + s + postf);
```

Now we can use it as follows:

```
printWithPrefixAndPostfix("Result: ", " Great!").accept("10.0");
                                        //prints: Result: 10.0 Great!
```

To demonstrate the `andThen()` method, let's create the class `Person`:

```
public class Person {
    private int age;
    private String firstName, lastName, record;
    public Person(int age, String firstName, String lastName) {
        this.age = age;
        this.lastName = lastName;
        this.firstName = firstName;
    }
```

```
        public int getAge() { return age; }
        public String getFirstName() { return firstName; }
        public String getLastName() { return lastName; }
        public String getRecord() { return record; }
        public void setRecord(String fullId) { this.record = record; }
    }
```

You may have noticed that `record` is the only property that has a setting. We will use it to set a personal record in a consumer function:

```
String externalData = "external data";
Consumer<Person> setRecord =
        p -> p.setFullId(p.getFirstName() + " " +
                p.getLastName() + ", " + p.getAge() + ", " + externalData);
```

The `setRecord` function takes the values of the `Person` object properties and some data from an external source and sets the resulting value as the `record` property value. Obviously, it could be done in several other ways, but we do it for demo purposes. Let's also create a function that prints the `record` property:

```
Consumer<Person> printRecord = p -> System.out.println(p.getRecord());
```

The composition of these two functions can be created and executed as follows:

```
Consumer<Person> setRecordThenPrint = setRecord.andThen(printPersonId);
setRecordThenPrint.accept(new Person(42, "Nick", "Samoylov"));
                        //prints: Nick Samoylov, age 42, external data
```

This way, it is possible to create a whole processing pipe of the operations that transform the properties of an object that is passed through the pipe.

Predicate<T>

This functional interface, `Predicate<T>`, has one abstract method, five defaults, and a static method that allows predicates chaining:

- `boolean test(T t)`: Evaluates the provided parameter to see if it meets the criteria or not
- `default Predicate<T> negate()`: Returns the negation of the current predicate
- `static <T> Predicate<T> not(Predicate<T> target)`: Returns the negation of the provided predicate

- default Predicate<T> or(Predicate<T> other): Constructs a logical OR from this predicate and the provided one
- default Predicate<T> and(Predicate<T> other): Constructs a logical AND from this predicate and the provided one
- static <T> Predicate<T> isEqual(Object targetRef): Constructs a predicate that evaluates whether or not two arguments are equal according to Objects.equals(Object, Object)

The basic use of this interface is pretty straightforward:

```
Predicate<Integer> isLessThan10 = i -> i < 10;
System.out.println(isLessThan10.test(7));        //prints: true
System.out.println(isLessThan10.test(12));       //prints: false
```

We can also combine it with the previously created printWithPrefixAndPostfix(String pref, String postf) function:

```
int val = 7;
Consumer<String> printIsSmallerThan10 = printWithPrefixAndPostfix("Is "
                        + val + " smaller than 10? ", " Great!");
printIsSmallerThan10.accept(String.valueOf(isLessThan10.test(val)));
                    //prints: Is 7 smaller than 10? true Great!
```

The other methods (also called **operations**) can be used for creating operational chains (also called **pipelines**) and can be seen in the following examples:

```
Predicate<Integer> isEqualOrGreaterThan10 = isLessThan10.negate();
System.out.println(isEqualOrGreaterThan10.test(7));    //prints: false
System.out.println(isEqualOrGreaterThan10.test(12));   //prints: true

isEqualOrGreaterThan10 = Predicate.not(isLessThan10);
System.out.println(isEqualOrGreaterThan10.test(7));    //prints: false
System.out.println(isEqualOrGreaterThan10.test(12));   //prints: true

Predicate<Integer> isGreaterThan10 = i -> i > 10;
Predicate<Integer> is_lessThan10_OR_greaterThan10 =
                            isLessThan10.or(isGreaterThan10);
System.out.println(is_lessThan10_OR_greaterThan10.test(20));   // true
System.out.println(is_lessThan10_OR_greaterThan10.test(10));   // false

Predicate<Integer> isGreaterThan5 = i -> i > 5;
Predicate<Integer> is_lessThan10_AND_greaterThan5 =
                            isLessThan10.and(isGreaterThan5);
System.out.println(is_lessThan10_AND_greaterThan5.test(3));   // false
System.out.println(is_lessThan10_AND_greaterThan5.test(7));   // true
```

```
Person nick = new Person(42, "Nick", "Samoylov");
Predicate<Person> isItNick = Predicate.isEqual(nick);
Person john = new Person(42, "John", "Smith");
Person person = new Person(42, "Nick", "Samoylov");
System.out.println(isItNick.test(john));          //prints: false
System.out.println(isItNick.test(person));         //prints: true
```

The predicate objects can be chained into more complex logical statements and include all
necessary external data, as was demonstrated before.

Supplier<T>

This functional interface, Supplier<T>, has only one abstract method T get(), which
returns a value. The basic usage can be seen as follows:

```
Supplier<Integer> supply42 = () -> 42;
System.out.println(supply42.get());   //prints: 42
```

It can be chained with the functions discussed in the preceding sections:

```
int input = 7;
int limit = 10;
Supplier<Integer> supply7 = () -> input;
Predicate<Integer> isLessThan10 = i -> i < limit;
Consumer<String> printResult = printWithPrefixAndPostfix("Is " + input +
                              " smaller than " + limit + "? ", " Great!");
printResult.accept(String.valueOf(isLessThan10.test(supply7.get())));
                      //prints: Is 7 smaller than 10? true Great!
```

The Supplier<T> function is typically used as an entry point of data going into a
processing pipeline.

Function<T, R>

The notation of this and other functional interfaces that return values, includes the listing of
the return type as the last in the list of generics (R in this case) and the type of the input data
in front of it (an input parameter of type T in this case). So, the notation Function<T, R>
means that the only abstract method of this interface accepts an argument of type T and
produces a result of type R. Let's look at the online documentation (https://docs.oracle.
com/en/java/javase/12/docs/api/java.base/java/util/function/Function.html).

The Function<T, R> interface has one abstract method, R apply(T), and two methods for operations chaining:

- default <V> Function<T,V> andThen(Function<R, V> after): Returns a composed function that first applies the current function to its input, and then applies the after function to the result.
- default <V> Function<V,R> compose(Function<V, T> before): Returns a composed function that first applies the before function to its input, and then applies the current function to the result.

There is also an identity() method:

- static <T> Function<T,T> identity(): Returns a function that always returns its input argument

Let's review all these methods and how they can be used. Here is an example of the basic usage of the Function<T,R> interface:

```
Function<Integer, Double> multiplyByTen = i -> i * 10.0;
System.out.println(multiplyByTen.apply(1));     //prints: 10.0
```

We can also chain it with all the functions we have discussed in the preceding sections:

```
Supplier<Integer> supply7 = () -> 7;
Function<Integer, Double> multiplyByFive = i -> i * 5.0;
Consumer<String> printResult =
                    printWithPrefixAndPostfix("Result: ", " Great!");
printResult.accept(multiplyByFive.
        apply(supply7.get()).toString()); //prints: Result: 35.0 Great!
```

The andThen() method allows for constructing a complex function from the simpler ones. Notice the divideByTwo.amdThen() line in the following code:

```
Function<Double, Long> divideByTwo =
                            d -> Double.valueOf(d / 2.).longValue();
Function<Long, String> incrementAndCreateString =
                                    l -> String.valueOf(l + 1);
Function<Double, String> divideByTwoIncrementAndCreateString =
                    divideByTwo.andThen(incrementAndCreateString);
printResult.accept(divideByTwoIncrementAndCreateString.apply(4.));
                                    //prints: Result: 3 Great!
```

It describes the sequence of the operations applied to the input value. Notice how the return type of the divideByTwo() function (Long) matches the input type of the incrementAndCreateString() function.

The `compose()` method accomplishes the same result, but in reverse order:

```
Function<Double, String> divideByTwoIncrementAndCreateString =
                    incrementAndCreateString.compose(divideByTwo);
printResult.accept(divideByTwoIncrementAndCreateString.apply(4.));
                                        //prints: Result: 3 Great!
```

Now the sequence of composition of the complex function does not match the sequence of the execution. It may be very convenient in the case where the function `divideByTwo()` is not created yet and you would like to create it in-line. Then the following construct will not compile:

```
Function<Double, String> divideByTwoIncrementAndCreateString =
        (d -> Double.valueOf(d / 2.).longValue())
                            .andThen(incrementAndCreateString);
```

The following line will compile just fine:

```
Function<Double, String> divideByTwoIncrementAndCreateString =
            incrementAndCreateString
                    .compose(d -> Double.valueOf(d / 2.).longValue());
```

It allows for more flexibility while constructing a functional pipeline, so one can build it in a fluent style without breaking the continuous line when creating the next operations.

The `identity()` method is useful when you need to pass in a function that matches the required function signature but does nothing. But it can substitute only the function that returns the same type as the input type. For example:

```
Function<Double, Double> multiplyByTwo = d -> d * 2.0;
System.out.println(multiplyByTwo.apply(2.));   //prints: 4.0

multiplyByTwo = Function.identity();
System.out.println(multiplyByTwo.apply(2.));   //prints: 2.0
```

To demonstrate its usability, let's assume we have the following processing pipeline:

```
Function<Double, Double> multiplyByTwo = d -> d * 2.0;
System.out.println(multiplyByTwo.apply(2.));   //prints: 4.0

Function<Double, Long> subtract7 = d -> Math.round(d - 7);
System.out.println(subtract7.apply(11.0));   //prints: 4

long r = multiplyByTwo.andThen(subtract7).apply(2.);
System.out.println(r);                       //prints: -3
```

Then, we decide that, under certain circumstances, the `multiplyByTwo()` function should do nothing. We could add to it a conditional close that turns it on/off. But if we want to keep the function intact or if this function is passed to us from third-party code, we can just do the following:

```
Function<Double, Double> multiplyByTwo = d -> d * 2.0;
System.out.println(multiplyByTwo.apply(2.));   //prints: 4.0

Function<Double, Long> subtract7 = d -> Math.round(d - 7);
System.out.println(subtract7.apply(11.0));     //prints: 4

multiplyByTwo = Function.identity();

r = multiplyByTwo.andThen(subtract7).apply(2.);
System.out.println(r);                         //prints: -5
```

As you can see, now the `multiplyByTwo()` function does nothing, and the final result is different.

Other standard functional interfaces

The other 39 functional interfaces in the `java.util.function` package are variations of the four interfaces we have just reviewed. These variations are created in order to achieve one or any combination of the following:

- Better performance by avoiding autoboxing and unboxing via the explicit usage of `int`, `double`, or `long` primitives
- Allowing two input parameters and/or a shorter notation

Here are just a few examples:

- `IntFunction<R>` with the method `R apply(int)` provides a shorter notation (without generics for the input parameter type) and avoids autoboxing by requiring the primitive `int` as a parameter.
- `BiFunction<T,U,R>` with the method `R apply(T,U)` allows two input parameters; `BinaryOperator<T>` with the method `T apply(T,T)` allows two input parameters of type `T` and returns a value of the same type, `T`.
- `IntBinaryOperator` with the method `int applAsInt(int,int)` accepts two parameters of type `int` and returns value of type `int`, too.

If you are going to use functional interfaces, we encourage you to study the API of the interfaces of the `java.util.functional` **package** (`https://docs.oracle.com/en/java/javase/12/docs/api/java.base/java/util/function/package-summary.html`).

Lambda expression limitations

There are two aspects of a lambda expression that we would like to point out and clarify:

- If a lambda expression uses a local variable created outside it, this local variable has to be final or effectively final (not reassigned in the same context).
- The `this` keyword in a lambda expression refers to the enclosing context, not the lambda expression itself.

As in an anonymous class, the variable created outside and used inside a lambda expression becomes effectively final and cannot be modified. The following is an example of an error caused by the attempt to change the value of an initialized variable:

```
int x = 7;
//x = 3; //compilation error
Function<Integer, Integer> multiply = i -> i * x;
```

The reason for this restriction is that a function can be passed around and executed in different contexts (different threads, for example), and an attempt to synchronize these contexts would defeat the original idea of the stateless function and the evaluation of the expression, depending only on the input parameters, not on the context variables. That is why all the local variables used in the lambda expression have to be effectively final, meaning that they can either be declared final explicitly or become final by the virtue of not changing the value.

There is one possible workaround for this limitation though. If the local variable is of reference type (but not `String` or primitive wrapping type), it is possible to change its state, even if this local variable is used in the lambda expression:

```
List<Integer> list = new ArrayList();
list.add(7);
int x = list.get(0);
System.out.println(x);   // prints: 7
list.set(0, 3);
x = list.get(0);
System.out.println(x);   // prints: 3
Function<Integer, Integer> multiply = i -> i * list.get(0);
```

This workaround should be used with care because of the danger of unexpected side effects should this lambda be executed in a different context.

The `this` keyword inside an anonymous class refers to the instance of the anonymous class. By contrast, inside the lambda expression, the `this` keyword refers to the instance of the class that surrounds the expression, also called an **enclosing instance**, **enclosing context**, or **enclosing scope**.

Let's create a `ThisDemo` class that illustrates the difference:

```
class ThisDemo {
    private String field = "ThisDemo.field";

    public void useAnonymousClass() {
        Consumer<String> consumer = new Consumer<>() {
            private String field = "Consumer.field";
            public void accept(String s) {
                System.out.println(this.field);
            }
        };
        consumer.accept(this.field);
    }

    public void useLambdaExpression() {
        Consumer<String> consumer = consumer = s -> {
            System.out.println(this.field);
        };
        consumer.accept(this.field);
    }
}
```

If we execute the preceding methods, the output will be as shown in the following code comments:

```
ThisDemo d = new ThisDemo();
d.useAnonymousClass();        //prints: Consumer.field
d.useLambdaExpression();      //prints: ThisDemo.field
```

As you can see, the keyword `this` inside the anonymous class refers to the anonymous class instance, while `this` in a lambda expression refers to the enclosing class instance. A lambda expression just does not have, and cannot have, a field. A lambda expression is not a class instance and cannot be referred by `this`. According to Java's specifications, such an approach *allows more flexibility for implementations* by treating `this` the same as the surrounding context.

Method references

So far, all our functions were short one-liners. Here is another example:

```
Supplier<Integer> input = () -> 3;
Predicate<Integer> checkValue = d -> d < 5;
Function<Integer, Double> calculate = i -> i * 5.0;
Consumer<Double> printResult = d -> System.out.println("Result: " + d);

if(checkValue.test(input.get())){
    printResult.accept(calculate.apply(input.get()));
} else {
    System.out.println("Input " + input.get() + " is too small.");
}
```

If the function consists of two or more lines, we could implement them as follows:

```
Supplier<Integer> input = () -> {
    // as many line of code here as necessary
    return 3;
};
Predicate<Integer> checkValue = d -> {
    // as many line of code here as necessary
    return d < 5;
};
Function<Integer, Double> calculate = i -> {
    // as many lines of code here as necessary
    return i * 5.0;
};
Consumer<Double> printResult = d -> {
    // as many lines of code here as necessary
    System.out.println("Result: " + d);
};
if(checkValue.test(input.get())){
    printResult.accept(calculate.apply(input.get()));
} else {
    System.out.println("Input " + input.get() + " is too small.");
}
```

When the size of a function implementation grows beyond several lines of code, such a code layout may not be easy to read. It may obscure the overall code structure. To avoid the issue, it is possible to move the function implementation into a method and then refer to this method in the lambda expression. For example, let's add one static and one instance method to the class where the lambda expression is used:

```
private int generateInput(){
    // Maybe many lines of code here
```

```
        return 3;
    }
    private static boolean checkValue(double d){
        // Maybe many lines of code here
        return d < 5;
    }
```

Also, to demonstrate the variety of possibilities, let's create another class with one static method and one instance method:

```
class Helper {
    public double calculate(int i){
        // Maybe many lines of code here
        return i* 5;
    }
    public static void printResult(double d){
        // Maybe many lines of code here
        System.out.println("Result: " + d);
    }
}
```

Now we can rewrite our last example as follows:

```
Supplier<Integer> input = () -> generateInput();
Predicate<Integer> checkValue = d -> checkValue(d);
Function<Integer, Double> calculate = i -> new Helper().calculate(i);
Consumer<Double> printResult = d -> Helper.printResult(d);

if(checkValue.test(input.get())){
    printResult.accept(calculate.apply(input.get()));
} else {
    System.out.println("Input " + input.get() + " is too small.");
}
```

As you can see, even if each function consists of many lines of code, such a structure keeps the code easy to read. Yet, when a one-line lambda expression consists of a reference to an existing method, it is possible to further simplify the notation by using method reference without listing the parameters.

The syntax of the method reference is `Location::methodName`, where `Location` indicates in which object or class the `methodName` method belongs, and the two colons (`::`) serve as a separator between the location and the method name. Using method reference notation, the preceding example can be rewritten as follows:

```
Supplier<Integer> input = this::generateInput;
Predicate<Integer> checkValue = MethodReferenceDemo::checkValue;
Function<Integer, Double> calculate = new Helper()::calculate;
```

```
Consumer<Double> printResult = Helper::printResult;

if(checkValue.test(input.get())){
    printResult.accept(calculate.apply(input.get()));
} else {
    System.out.println("Input " + input.get() + " is too small.");
}
```

You have probably noticed that we have intentionally used different locations, two instance methods, and two static methods in order to demonstrate the variety of possibilities. If it feels like too much to remember, the good news is that a modern IDE (IntelliJ IDEA is one example) can do it for you and convert the code you are writing to the most compact form. You have just to accept with the IDE's suggestion.

Summary

This chapter introduced the reader to functional programming by explaining and demonstrating the concept of functional interface and lambda expressions. The overview of standard functional interfaces that comes with JDK helps the reader to avoid writing custom code, while the method reference notation allows the reader to write well-structured code that is easy to understand and maintain.

In the next chapter, we will talk about data streams processing. We will define what data streams are, and look at how to process their data and how to chain stream operations in a pipeline. Specifically, we will discuss the stream's initialization and operations (methods), how to connect them in a fluent style, and how to create parallel streams.

Quiz

1. What is a functional interface? Select all that apply:

    a. A collection of functions
    b. An interface that has only one method
    c. Any interface that has only one abstract method
    d. Any library written in Java

2. What is a lambda expression? Select all that apply:

    a. A function, implemented as an anonymous method without modifiers, return types, and parameter types

    b. A functional interface implementation

    c. Any implementation in a lambda calculus style

    d. The notation that includes the list of parameters, an arrow token `->`, and a body that consists of a single statement or a block of statements

3. How many input parameters does the implementation of the `Consumer<T>` interface have?

4. What is the type of the return value in the implementation of the `Consumer<T>` interface?

5. How many input parameters does the implementation of the `Predicate<T>` interface have?

6. What is the type of the return value in the implementation of the `Predicate<T>` interface?

7. How many input parameters does the implementation of the `Supplier<T>` interface have?

8. What is the type of the return value in the implementation of the `Supplier<T>` interface?

9. How many input parameters does the implementation of the `Function<T,R>` interface have?

10. What is the type of the return value in the implementation of the `Function<T,R>` interface?

11. In a lambda expression, what does the keyword `this` refer to?

12. What is method reference syntax?

14
Java Standard Streams

In this chapter, we will talk about processing data streams, which are different from the I/O streams we reviewed in `Chapter 5`, *Strings, Input/Output, and Files*. We will define what data streams are, how to process their elements using methods (operations) of the `java.util.stream.Stream` object, and how to chain (connect) stream operations in a pipeline. We will also discuss stream initialization and how to process streams in parallel.

The following topics will be covered in this chapter, as follows:

- Streams as a source of data and operations
- Stream initialization
- Operations (methods)
- Numeric stream interfaces
- Parallel streams

Streams as a source of data and operations

Lambda expressions, described and demonstrated in the previous chapter, together with functional interfaces added a powerful functional programming capability to Java. They allow passing behavior (functions) as parameters to libraries optimized for the performance of the data processing. This way, an application programmer can concentrate on the business aspects of the developed system, leaving the performance aspects to the specialists – the authors of the library. One example of such a library is `java.util.stream`, which is going to be the focus of this chapter.

In Chapter 5, *Strings, Input/Output, and Files,* we talked about I/O streams as the source of data, but beyond that, they are not of much help for the further processing of the data. And they are byte- or character-based, not object-based. You can create a stream of objects only after the objects have been programmatically created and serialized first. The I/O streams are just connections to external resources, mostly files and not much else. However, sometimes it is possible to make a transition from an I/O stream to java.util.stream.Stream. For example, the BufferedReader class has the lines() method that converts the underlying character-based stream into a Stream<String> object.

The streams of the java.util.stream package, on the other hand, are oriented on processing collections of objects. In Chapter 6, *Data Structures, Generics, and Popular Utilities,* we described two methods of the Collection interface that allow reading collection elements as elements of a stream: default Stream<E> stream() and default Stream<E> parallelStream(). We also have mentioned the stream() method of java.util.Arrays. It has the following eight overloaded versions that convert an array or a part of it into a stream of the corresponding data types:

- static DoubleStream stream(double[] array)
- static DoubleStream stream(double[] array, int startInclusive, int endExclusive)
- static IntStream stream(int[] array)
- static IntStream stream(int[] array, int startInclusive, int endExclusive)
- static LongStream stream(long[] array)
- static LongStream stream(long[] array, int startInclusive, int endExclusive)
- static <T> Stream<T> stream(T[] array)
- static <T> Stream<T> stream(T[] array, int startInclusive, int endExclusive)

Let's now look at the streams of the package java.util.stream closer. The best way to understand what a stream is to compare it with a collection. The latter is a data structure stored in memory. Every collection element is computed before being added to the collection. By contrast, an element emitted by a stream exists somewhere else, in the source, and is computed on demand. So, a collection can be a source for a stream.

A `Stream` object is an implementation of `Stream` interface, `IntStream`, `LongStream`, or `DoubleStream`; the last three are called **numeric streams**. The methods of the `Stream` interface are also available in numeric streams. Some of the numeric streams have a few extra methods, such as `average()` and `sum()`, specific numeric values. In this chapter, we are going to speak mostly about the `Stream` interface and its methods, but everything we will cover is equally applicable to numeric streams too.

A stream *produces* (or *emits*) stream elements as soon as the previously emitted element has been processed. It allows declarative presentation of methods (operations) that can be applied to the emitted elements, also in parallel. Today, when the machine learning requirements of large dataset processing are becoming ubiquitous, this feature reinforces the position of Java among the few modern programming languages of choice.

Stream initialization

There are many ways to create and initialize a stream – an object of type `Stream` or any of the numeric interfaces. We grouped them by classes and interfaces that have `Stream`-creating methods. We did it for the reader's convenience, so it would be easier for the reader to remember and find them if need be.

Stream interface

This group of `Stream` factories is composed of static methods that belong to the `Stream` interface.

empty()

The `Stream<T> empty()` method creates an empty stream that does not emit any element:

```
Stream.empty().forEach(System.out::println);    //prints nothing
```

The `Stream` method `forEach()` acts similarly to the `Collection` method `forEach()` and applies the passed-in function to each of the stream elements:

```
new ArrayList().forEach(System.out::println);   //prints nothing
```

The result is the same as creating a stream from an empty collection:

```
new ArrayList().stream().forEach(System.out::println);   //prints nothing
```

Without any element emitted, nothing happens. We will discuss the `Stream` method `forEach()` in the *Terminal operations* section.

of(T... values)

The `of(T... values)` method accepts varargs and can also create an empty stream:

```
Stream.of().forEach(System.out::print);          //prints nothing
```

But it is most often used for initializing a non-empty stream:

```
Stream.of(1).forEach(System.out::print);          //prints: 1
Stream.of(1,2).forEach(System.out::print);        //prints: 12
Stream.of("1 ","2").forEach(System.out::print);   //prints: 1 2
```

Notice the method reference used for the invocation of the `println()` and `print()` methods.

Another way to use the `of(T... values)` method is as follows:

```
String[] strings = {"1 ", "2"};
Stream.of(strings).forEach(System.out::print);    //prints: 1 2
```

If there is no type specified for the `Stream` object, the compiler does not complain if the array contains a mix of types:

```
Stream.of("1 ", 2).forEach(System.out::print);    //prints: 1 2
```

Adding generics that declare the expected element type causes an exception when at least one of the listed elements has a different type:

```
//Stream<String> stringStream = Stream.of("1 ", 2);   //compile error
```

Generics help a programmer to avoid many mistakes, so they should be added wherever possible.

The `of(T... values)` method can also be used for the concatenation of multiple streams. Let's assume, for example, that we have the following four streams that we would like to concatenate into one:

```
Stream<Integer> stream1 = Stream.of(1, 2);
Stream<Integer> stream2 = Stream.of(2, 3);
Stream<Integer> stream3 = Stream.of(3, 4);
Stream<Integer> stream4 = Stream.of(4, 5);
```

We would like to concatenate them into a new stream that emits values $1, 2, 2, 3, 3, 4, 4, 5$. First, we try the following code:

```
Stream.of(stream1, stream2, stream3, stream4)
     .forEach(System.out::print);
               //prints: java.util.stream.ReferencePipeline$Head@58ceff1j
```

It does not do what we hoped for. It treats each stream as an object of the internal class `java.util.stream.ReferencePipeline` that is used in the `Stream` interface implementation. So, we need to add the `flatMap()` operation to convert each stream element into a stream (we describe it in the *Intermediate operations* section):

```
Stream.of(stream1, stream2, stream3, stream4)
     .flatMap(e -> e).forEach(System.out::print);    //prints: 12233445
```

The function we passed into `flatMap()` as a parameter (e -> e) looks like it's doing nothing, but that is because each element of the stream is a stream already, so there is no need to transform it. By returning an element as the result of the `flatMap()` operation, we tell the pipeline to treat the return value as a `Stream` object.

ofNullable(T t)

The `ofNullable(T t)` method returns a `Stream<T>` emitting a single element if the passed-in parameter t is not `null`; otherwise, it returns an empty `Stream`. To demonstrate the usage of the `ofNullable(T t)` method, we have created the following method:

```
void printList1(List<String> list){
     list.stream().forEach(System.out::print);
}
```

We have executed this method twice - with the parameter list equal to `null` and to a `List` object. Here are the results:

```
//printList1(null);                          //NullPointerException
List<String> list = List.of("1 ", "2");
printList1(list);                            //prints: 1 2
```

Notice how the first call to the `printList1()` method generates `NullPointerException`. To avoid the exception, we could implement the method as follows:

```
void printList1(List<String> list){
     (list == null ? Stream.empty() : list.stream())
                    .forEach(System.out::print);
}
```

The same result can be achieved with the ofNullable(T t) method:

```
void printList2(List<String> list){
    Stream.ofNullable(list).flatMap(l -> l.stream())
                          .forEach(System.out::print);
}
```

Notice how we have added flatMap() because, otherwise, the Stream element that flows into forEach() would be a List object. We will talk more about the flatMap() method in the *Intermediate operations* section. The function passed into the flatMap() operation in the preceding code can be expressed as a method reference too:

```
void printList4(List<String> list){
    Stream.ofNullable(list).flatMap(Collection::stream)
                          .forEach(System.out::print);
}
```

iterate(Object, UnaryOperator)

Two static methods of the Stream interface allow generating a stream of values using an iterative process similar to the traditional for loop, as follows:

- Stream<T> iterate(T seed, UnaryOperator<T> func): Creates an infinite sequential stream based on the iterative application of the second parameter, the function func, to the first parameter seed, producing a stream of values seed, f(seed), f(f(seed)), an so on
- Stream<T> iterate(T seed, Predicate<T> hasNext, UnaryOperator<T> next): Creates a finite sequential stream based on the iterative application of the third parameter, the next function, to the first parameter seed, producing a stream of values seed, f(seed), f(f(seed)), and so on, as long as the third parameter, the function hasNext, returns true

The following code demonstrates the usage of these methods, as follows:

```
Stream.iterate(1, i -> ++i).limit(9)
    .forEach(System.out::print); //prints: 123456789

Stream.iterate(1, i -> i < 10, i -> ++i)
    .forEach(System.out::print);       //prints: 123456789
```

Notice that we were forced to add an intermediate operator `limit(int n)` to the first pipeline to avoid generating an infinite number of generated values. We will talk more about this method in the *Intermediate operations* section.

concat (Stream a, Stream b)

The `Stream<T> concat(Stream<> a, Stream<T> b)` static method of the `Stream` interface creates a stream of values based on two streams, a and b, passed in as parameters. The newly created stream consists of all the elements of the first parameter, a, followed by all the elements of the second parameter, b. The following code demonstrates this method:

```
Stream<Integer> stream1 = List.of(1, 2).stream();
Stream<Integer> stream2 = List.of(2, 3).stream();
Stream.concat(stream1, stream2)
  .forEach(System.out::print); //prints: 1223
```

Notice that element 2 is present in both original streams and consequently is emitted twice by the resulting stream.

generate (Supplier)

The static method `Stream<T> generate(Supplier<T> supplier)` of the `Stream` interface creates an infinite stream where each element is generated by the provided `Supplier<T>` function. The following are two examples:

```
Stream.generate(() -> 1).limit(5)
  .forEach(System.out::print);        //prints: 11111

Stream.generate(() -> new Random().nextDouble()).limit(5)
      .forEach(System.out::println);       //prints: 0.38575117472619247
                                           //        0.5055765386778835
                                           //        0.6528038976983277
                                           //        0.4422354489467244
                                           //        0.06770955839148762
```

If you run this code, you will probably get different results because of the random (pseudo-random) nature of the generated values.

Since the created stream is infinite, we have added a `limit(int n)` operation that allows only the specified number of stream elements to flow through. We will talk more about this method in the *Intermediate operations* section.

Stream.Builder interface

The `Stream.Builder<T> builder()` static method returns an internal (located inside the `Stream` interface) interface `Builder` that can be used to construct a `Stream` object. The interface `Builder` extends the `Consumer` interface and has the following methods:

- `default Stream.Builder<T> add(T t)`: Calls the `accept(T)` method and returns the (`Builder` object), thus allowing chaining `add(T t)` methods in a fluent dot- connected style
- `void accept(T t)`: Adds an element to the stream (this method comes from the `Consumer` interface)
- `Stream<T> build()`: Transitions this builder from the constructing state to the `built` state; after this method is called, no new elements can be added to this stream

The usage of the `add(T t)` method is straightforward:

```
Stream.<String>builder().add("cat").add(" dog").add(" bear")
       .build().forEach(System.out::print);        //prints: cat dog bear
```

Please notice how we have added the generics `<String>` in front of the `builder()` method. This way, we tell the builder that the stream we are creating will have `String` type elements. Otherwise, it will add the elements as `Object` types and will not make sure that the added elements are of the `String` type.

The `accept(T t)` method is used when the builder is passed as a parameter of the `Consumer<T>` type or when you do not need to chain the methods that add the elements. For example, the following is a code example:

```
Stream.Builder<String> builder = Stream.builder();
List.of("1", "2", "3").stream().forEach(builder);
builder.build().forEach(System.out::print);        //prints: 123
```

The `forEach(Consumer<T> consumer)` method accepts a `Consumer` function that has the `accept(T t)` method. Every time an element is emitted by the stream, the `forEach()` method receives it and passes it to the `accept(T t)` method of the `Builder` object. Then, when the `build()` method is called in the next line, the `Stream` object is created and starts emitting the elements added earlier by the `accept(T t)` method. The emitted elements are passed to the `forEach()` method, which then prints them one by one.

And here is an example of an explicit usage of the `accept(T t)` method:

```
List<String> values = List.of("cat", " dog", " bear");
Stream.Builder<String> builder = Stream.builder();
for(String s: values){
    if(s.contains("a")){
        builder.accept(s);
    }
}
builder.build().forEach(System.out::print);          //prints: cat bear
```

This time, we decided not to add all the list elements to the stream, but only those that contain the character a. As was expected, the created stream contains only the `cat` and `bear` elements. Also, notice how we use `<String>` generics to make sure that all the stream elements are of the `String` type.

Other classes and interfaces

In Java 8, two default methods were added to the `java.util.Collection` interface, as follows:

- `Stream<E> stream()`: Returns a stream of the elements of this collection
- `Stream<E> parallelStream()`: Returns (possibly) a parallel stream of the elements of this collection; *possibly*, because the JVM attempts to split the stream into several chunks and process them in parallel (if there are several CPUs) or virtually parallel (using CPU time-sharing); but it is not always possible and depends in part on the nature of the requested processing

It means that all the collection interfaces that extend this interface, including `Set` and `List`, have these methods. For example:

```
List.of("1", "2", "3").stream().forEach(builder);
List.of("1", "2", "3").parallelStream().forEach(builder);
```

We will talk about parallel streams in the *Parallel processing* section.

We have described eight static overloaded `stream()` methods of the `java.util.Arrays` class at the beginning of the *Streams as a source of data and operations* section. Here is an example of another way of creating a stream, using the subset of an array:

```
int[] arr = {1, 2, 3, 4, 5};
Arrays.stream(arr, 2, 4).forEach(System.out::print); //prints: 34
```

The `java.util.Random` class allows creating numeric streams of pseudo-random values, as follows:

- `DoubleStream doubles()`: Creates an unlimited stream of `double` values between 0 (inclusive) and 1 (exclusive)
- `IntStream ints()` and `LongStream longs()`: Create an unlimited stream of corresponding type values
- `DoubleStream doubles(long streamSize)`: Creates a stream (of the specified size) of `double` values between 0 (inclusive) and 1 (exclusive)
- `IntStream ints(long streamSize)` and `LongStream longs(long streamSize)`: Creates a stream of the specified size of the corresponding type values
- `IntStream ints(int randomNumberOrigin, int randomNumberBound)`: Creates an unlimited stream of `int` values between `randomNumberOrigin` (inclusive) and `randomNumberBound` (exclusive)
- `LongStream longs(long randomNumberOrigin, long randomNumberBound)`: Creates an unlimited stream of `long` values between `randomNumberOrigin` (inclusive) and `randomNumberBound` (exclusive)
- `DoubleStream doubles(long streamSize, double randomNumberOrigin, double randomNumberBound)`: Creates a stream of the specified size of `double` values between `randomNumberOrigin` (inclusive) and `randomNumberBound` (exclusive)

Here is an example of one of the preceding methods:

```
new Random().ints(5, 8).limit(5)
          .forEach(System.out::print);    //prints: 56757
```

The `java.nio.file.Files` class has six static methods creating streams of lines and paths, as follows:

- `Stream<String> lines(Path path)`: Creates a stream of lines from the file specified by the provided path
- `Stream<String> lines(Path path, Charset cs)`: Creates a stream of lines from the file specified by the provided path; bytes from the file are decoded into characters using the provided charset
- `Stream<Path> list(Path dir)`: Creates a stream of files and directories in the specified directory

- `Stream<Path> walk(Path start, FileVisitOption... options)`:
  Creates a stream of files and directories of the file tree that starts with `Path start`
- `Stream<Path> walk(Path start, int maxDepth, FileVisitOption... options)`: Creates a stream of files and directories of the file tree that starts with `Path start` down to the specified depth `maxDepth`
- `Stream<Path> find(Path start, int maxDepth, BiPredicate<Path, BasicFileAttributes> matcher, FileVisitOption... options)`: Creates a stream of files and directories (that match the provided predicate) of the file tree that starts with `Path start` down to the specified depth, specified by `maxDepth` value

Other classes and methods that create streams include the following:

- The `java.util.BitSet` class has the `IntStream stream()` method, which creates a stream of indices for which this `BitSet` contains a bit in the set state.
- The `java.io.BufferedReader` class has the `Stream<String> lines()` method, which creates a stream of lines from this `BufferedReader` object, typically from a file.
- The `java.util.jar.JarFile` class has the `Stream<JarEntry> stream()` method that creates a stream of the ZIP-file entries.
- The `java.util.regex.Pattern` class has the `Stream<String> splitAsStream(CharSequence input)` method, which creates a stream from the provided sequence around matches of this pattern.
- A `java.lang.CharSequence` interface has two methods, as follows:
  - `default IntStream chars()`: creates a stream of `int` zero-extending the char values
  - `default IntStream codePoints()`: creates a stream of code point values from this sequence

There is also a `java.util.stream.StreamSupport` class that contains static low-level utility methods for library developers. But we won't be reviewing it, as this is outside the scope of this book.

Operations (methods)

Many methods of the `Stream` interface, those that have a functional interface type as a parameter, are called **operations** because they are not implemented as traditional methods. Their functionality is passed into the method as a function. The operations are just shells that call a method of the functional interface assigned as the type of the parameter method.

For example, let's look at the `Stream<T> filter (Predicate<T> predicate)` method. Its implementation is based on the call to the method boolean `test (T t)` of the `Predicate<T>` function. So, instead of saying, *We use the* `filter()` *method of the* `Stream` *object to select some of the stream elements and skip others,* programmers prefer to say, *We apply an operation filter that allows some of the stream elements to get through and skip others.* It describes the nature of the action (operation), not the particular algorithm, which is unknown until the method receives a particular function. There are two groups of operations in the `Stream` interface, as follows:

- **Intermediate operations**: Instance methods that return a `Stream` object
- **Terminal operations:** Instance methods that return some type other than `Stream`

Stream processing is organized typically as a pipeline using a fluent (dot-connected) style. A `Stream` creating method or another stream source starts such a pipeline. A terminal operation produces the final result or a side-effect and eponymously ends the pipeline. An intermediate operation can be placed between the originating `Stream` object and the terminal operation.

An intermediate operation processes stream elements (or not, in some cases) and returns the modified (or not) `Stream` object, so the next intermediate or terminal operation can be applied. Examples of intermediate operations are the following:

- `Stream<T> filter(Predicate<T> predicate)`: Selects only elements matching a criterion
- `Stream<R> map(Function<T,R> mapper)`: Transforms elements according to the passed-in function; please notice that the type of the returned `Stream` object may be quite different from the input type
- `Stream<T> distinct()`: Removes duplicates
- `Stream<T> limit(long maxSize)`: Limits a stream to the specified number of elements
- `Stream<T> sorted()`: Arranges the stream elements in a certain order

We will discuss some other intermediate operations in the *Intermediate operations* section.

The processing of the stream elements actually begins only when a terminal operation starts executing. Then all the intermediate operations (if present) start processing in sequence. As soon as the terminal operation has finished execution, the stream closes and cannot be reopened.

Examples of terminal operations are `forEach()`, `findFirst()`, `reduce()`, `collect()`, `sum()`, `max()`, and other methods of the `Stream` interface that do not return the `Stream` object. We will discuss them in the *Terminal operations* section.

All the `Stream` operations support parallel processing, which is especially helpful in the case of a large amount of data processed on a multi-core computer. We will discuss it in the *Parallel streams* section.

Intermediate operations

As we have mentioned already, an intermediate operation returns a `Stream` object that emits the same or modified values and may even be of a different type than the stream source.

The intermediate operations can be grouped by their functionality in four categories of operations that perform **filtering**, **mapping**, **sorting**, or **peeking**.

Filtering

This group includes operations that remove duplicates, skip some of the elements, limit the number of processed elements, and select for the further processing only those that pass certain criteria, as follows:

- `Stream<T> distinct()`: Compares stream elements using `method Object.equals(Object)` and skips duplicates
- `Stream<T> skip(long n)`: Ignores the provided number of stream elements that are emitted first
- `Stream<T> limit(long maxSize)`: Allows only the provided number of stream elements to be processed
- `Stream<T> filter(Predicate<T> predicate)`: Allows only those elements to be processed that result in `true` when processed by the provided `Predicate` function

- `default Stream<T> dropWhile(Predicate<T> predicate)`: Skips those first elements of the stream that result in `true` when processed by the provided `Predicate` function
- `default Stream<T> takeWhile(Predicate<T> predicate)`: Allows only those first elements of the stream to be processed that result in `true` when processed by the provided `Predicate` function

The following is the code that demonstrates how the operations just described work:

```
Stream.of("3", "2", "3", "4", "2").distinct()
                       .forEach(System.out::print);      //prints: 324

List<String> list = List.of("1", "2", "3", "4", "5");
list.stream().skip(3).forEach(System.out::print);       //prints: 45

list.stream().limit(3).forEach(System.out::print);      //prints: 123

list.stream().filter(s -> Objects.equals(s, "2"))
             .forEach(System.out::print);                //prints: 2

list.stream().dropWhile(s -> Integer.valueOf(s) < 3)
             .forEach(System.out::print);                //prints: 345

list.stream().takeWhile(s -> Integer.valueOf(s) < 3)
             .forEach(System.out::print);                //prints: 12
```

Notice that we were able to reuse the source `List<String>` object, but could not reuse the `Stream` object. Once a `Stream` object is closed, it cannot be reopened.

Mapping

This group includes arguably the most important intermediate operations. They are the only intermediate operations that modify the elements of the stream. They **map** (transform) the original stream element value to a new one, as follows:

- `Stream<R> map(Function<T, R> mapper)`: Applies the provided function to each element of type `T` of the stream and produces a new element value of type `R`
- `IntStream mapToInt(ToIntFunction<T> mapper)`: Applies the provided function to each element of type `T` of the stream and produces a new element value of type `int`
- `LongStream mapToLong(ToLongFunction<T> mapper)`: Applies the provided function to each element of type `T` of the stream and produces a new element value of type `long`

- `DoubleStream mapToDouble(ToDoubleFunction<T> mapper)`: Applies the provided function to each element of type `T` of the stream and produces a new element value of type `double`

- `Stream<R> flatMap(Function<T, Stream<R>> mapper)`: Applies the provided function to each element of type `T` of the stream and produces a `Stream<R>` object that emits elements of type `R`

- `IntStream flatMapToInt(Function<T, IntStream> mapper)`: Applies the provided function to each element of type `T` of the stream and produces a `IntStream` object that emits elements of type `int`

- `LongStream flatMapToLong(Function<T, LongStream> mapper)`: Applies the provided function to each element of type `T` of the stream and produces a `LongStream` object that emits elements of type `long`

- `DoubleStream flatMapToDouble(Function<T, DoubleStream> mapper)`: Applies the provided function to each element of type `T` of the stream and produces a `DoubleStream` object that emits elements of type `double`

The following are examples of the usage of these operations, as follows:

```
List<String> list = List.of("1", "2", "3", "4", "5");
list.stream().map(s -> s + s)
            .forEach(System.out::print);     //prints: 1122334455

list.stream().mapToInt(Integer::valueOf)
            .forEach(System.out::print);     //prints: 12345

list.stream().mapToLong(Long::valueOf)
            .forEach(System.out::print);     //prints: 12345

list.stream().mapToDouble(Double::valueOf)
            .mapToObj(Double::toString)
            .map(s -> s + " ")
            .forEach(System.out::print);   //prints: 1.0 2.0 3.0 4.0 5.0

list.stream().mapToInt(Integer::valueOf)
            .flatMap(n -> IntStream.iterate(1, i -> i < n, i -> ++i))
            .forEach(System.out::print);          //prints: 1121231234

list.stream().map(Integer::valueOf)
            .flatMapToInt(n ->
                IntStream.iterate(1, i -> i < n, i -> ++i))
            .forEach(System.out::print);          //prints: 1121231234

list.stream().map(Integer::valueOf)
            .flatMapToLong(n ->
```

```
            LongStream.iterate(1, i -> i < n, i -> ++i))
        .forEach(System.out::print);            //prints: 1121231234

list.stream().map(Integer::valueOf)
            .flatMapToDouble(n ->
                DoubleStream.iterate(1, i -> i < n, i -> ++i))
            .mapToObj(Double::toString)
            .map(s -> s + " ")
            .forEach(System.out::print);
                    //prints: 1.0 1.0 2.0 1.0 2.0 3.0 1.0 2.0 3.0 4.0
```

In the last example, converting the stream to `DoubleStream`, we transformed each numeric value to a `String` object and added whitespace, so the result can be printed with whitespace between the numbers. These examples are very simple: just conversion with minimal processing. But in real life, each `map()` or `flatMap()` operation typically accepts a more complex function that does something more useful.

Sorting

The following two intermediate operations sort the stream elements, as follows:

- `Stream<T> sorted()`: Sorts stream elements in natural order (according to their `Comparable` interface implementation)
- `Stream<T> sorted(Comparator<T> comparator)`: Sorts stream elements in order according to the provided `Comparator<T>` object

Naturally, these operations cannot be finished until all the elements are emitted, so such processing creates a lot of overhead, slows down performance, and has to be used for small streams.

Here is demo code:

```
List<String> list = List.of("2", "1", "5", "4", "3");
list.stream().sorted().forEach(System.out::print);  //prints: 12345
list.stream().sorted(Comparator.reverseOrder())
            .forEach(System.out::print);            //prints: 54321
```

Peeking

An intermediate `Stream<T> peek(Consumer<T> action)` operation applies the provided `Consumer<T>` function to each stream element but does not change the stream values (`Consumer<T>` returns `void`). This operation is used for debugging. The following code shows how it works:

```
List<String> list = List.of("1", "2", "3", "4", "5");
list.stream()
  .peek(s -> System.out.print("3".equals(s) ? 3 : 0))
  .forEach(System.out::print); //prints: 0102330405
```

Terminal operations

Terminal operations are the most important operations in a stream pipeline. It is possible to accomplish everything in them without using any other operations.

We have used already the `forEach(Consumer<T>)` terminal operation to print each element. It does not return a value, thus it is used for its side-effects. But the `Stream` interface has many more powerful terminal operations that do return values.

Chief among them is the `collect()` operation, which has two forms as follows:

- `R collect(Collector<T, A, R> collector)`
- `R collect(Supplier<R> supplier, BiConsumer<R, T> accumulator, BiConsumer<R, R> combiner)`

It allows composing practically any process that can be applied to a stream. The classic example is as follows:

```
List<String> list = Stream.of("1", "2", "3", "4", "5")
                          .collect(ArrayList::new,
                                   ArrayList::add,
                                   ArrayList::addAll);
System.out.println(list);  //prints: [1, 2, 3, 4, 5]
```

This example is used in such a way as to be suitable for parallel processing. The first parameter of the `collect()` operation is a function that produces a value based on the stream element. The second parameter is the function that accumulates the result. And the third parameter is the function that combines the accumulated results from all the threads that processed the stream.

But having only one such generic terminal operation would force programmers to write the same functions repeatedly. That is why the API authors added the `Collectors` class, which generates many specialized `Collector` objects without the need to create three functions for every `collect()` operation.

In addition to that, the API authors added to the `Stream` interface various even more specialized terminal operations that are much simpler and easier to use. In this section, we will review all the terminal operations of the `Stream` interface and, in the `Collect` subsection, look at the plethora of `Collector` objects produced by the `Collectors` class. We start with the most simple terminal operation that allows processing each element of this stream one at a time.

In our examples, we are going to use the following class: `Person`:

```java
public class Person {
    private int age;
    private String name;
    public Person(int age, String name) {
        this.age = age;
        this.name = name;
    }
    public int getAge() {return this.age; }
    public String getName() { return this.name; }
    @Override
    public String toString() {
        return "Person{" + "name='" + this.name + "'" +
                                   ", age=" + age + "}";
    }
}
```

Processing each element

There are two terminal operations in this group, as follows:

- `void forEach(Consumer<T> action)`: Applies the provided action for each element of this stream
- `void forEachOrdered(Consumer<T> action)`: Applies the provided action for each element of this stream in the order defined by the source, regardless of whether the stream is sequential or parallel

If the order in which you need the elements to be processed is important and it has to be the order in which values are arranged at the source, use the second method, especially if you can foresee that it is possible your code is going to be executed on a computer with several CPUs. Otherwise, use the first one, as we did in all our examples.

Let's see an example of using the `forEach()` operation for reading comma-separated values (age and name) from a file and creating `Person` objects. We have placed the following `persons.csv` file (*csv* stands for *comma-separated values*) file in the `resources` folder:

```
23 , Ji m
   2 5 , Bob
  15 , Jill
17 , Bi ll
```

We have added spaces inside and outside the values in order to take this opportunity to show you some simple but very useful tips for working with real-life data.

First, we will just read the file and display its content line by line, but only those lines that contain the letter J:

```
Path path = Paths.get("src/main/resources/persons.csv");
try (Stream<String> lines = Files.newBufferedReader(path).lines()) {
    lines.filter(s -> s.contains("J"))
         .forEach(System.out::println);   //prints: 23 , Ji m
                                          //              15 , Jill
} catch (IOException ex) {
    ex.printStackTrace();
}
```

That is a typical way of using the `forEach()` operation: to process each element independently. This code also provides an example of a try-with-resources construct that closes the `BufferedReader` object automatically.

And the following is how an inexperienced programmer might write the code that reads the stream elements from the `Stream<String> lines` object and creates a list of `Person` objects:

```
List<Person> persons = new ArrayList<>();
lines.filter(s -> s.contains("J")).forEach(s -> {
    String[] arr = s.split(",");
    int age = Integer.valueOf(StringUtils.remove(arr[0], ' '));
    persons.add(new Person(age, StringUtils.remove(arr[1], ' ')));
});
```

You can see how the `split()` method is used to break each line by a comma that separates the values and how the `org.apache.commons.lang3.StringUtils.remove()` method removes spaces from each value. Although this code works well in small examples on a single-core computer, it might create unexpected results with a long stream and parallel processing.

That is the reason that lambda expressions require all the variables to be final or effectively final because the same function can be executed in a different context.

The following is a correct implementation of the preceding code:

```
List<Person> persons = lines.filter(s -> s.contains("J"))
        .map(s -> s.split(","))
        .map(arr -> {
            int age = Integer.valueOf(StringUtils.remove(arr[0], ' '));
            return new Person(age, StringUtils.remove(arr[1], ' '));
        }).collect(Collectors.toList());
```

To improve readability, we can create a method that does the job of mapping:

```
private Person createPerson(String[] arr){
    int age = Integer.valueOf(StringUtils.remove(arr[0], ' '));
    return new Person(age, StringUtils.remove(arr[1], ' '));
}
```

Now we can use it as follows:

```
List<Person> persons = lines.filter(s -> s.contains("J"))
                    .map(s -> s.split(","))
                    .map(this::createPerson)
                    .collect(Collectors.toList());
```

As you can see, we have used the `collect()` operator and the `Collector` function created by the `Collectors.toList()` method. We will see more functions created by the `Collectors` class in the *Collect* subsection.

Counting all elements

The `long count()` terminal operation of the `Stream` interface looks straightforward and benign. It returns the number of elements in this stream. Those who are used to working with collections and arrays may use the `count()` operation without thinking twice. The following code snippet demonstrates a caveat:

```
long count = Stream.of("1", "2", "3", "4", "5")
                   .peek(System.out::print)
                   .count();
System.out.print(count);              //prints: 5
```

If we run the preceding code, the result will look as follows:

As you see, the code that implements the `count()` method was able to determine the stream size without executing all the pipeline. The `peek()` operation did not print anything, which proves that elements were not emitted. So, if you expected to see the values of the stream printed, you might be puzzled and expect that the code has some kind of defect.

Another caveat is that it is not always possible to determine the stream size at the source. Besides, the stream may be infinite. So, you have to use `count()` with care.

Another possible way to determine the stream size is by using the `collect()` operation:

```
long count = Stream.of("1", "2", "3", "4", "5")
                   .peek(System.out::print)          //prints: 12345
                   .collect(Collectors.counting());
System.out.println(count);                           //prints: 5
```

The following screenshot that shows what happens after the preceding code example has been run:

As you can see, the `collect()` operation does not calculate the stream size at the source. That is because the `collect()` operation is not as specialized as the `count()` operation. It just applies the passed-in collector to the stream. And the collector just counts the elements provided to it by the `collect()` operation.

Match all, any, none

There are three seemingly very similar terminal operations that allow us to assess whether all, any, or none of the stream elements have a certain value, as follows:

- `boolean allMatch(Predicate<T> predicate)`: Returns `true` when each of the stream elements returns `true` when used as a parameter of the provided `Predicate<T>` function
- `boolean anyMatch(Predicate<T> predicate)`: Returns `true` when one of the stream elements returns `true` when used as a parameter of the provided `Predicate<T>` function
- `boolean noneMatch(Predicate<T> predicate)`: Returns `true` when none of this stream elements returns `true` when used as a parameter of the provided `Predicate<T>` function

The following are examples of their usage:

```
List<String> list = List.of("1", "2", "3", "4", "5");
boolean found = list.stream()
                    .peek(System.out::print)            //prints: 123
                    .anyMatch(e -> "3".equals(e));
System.out.println(found);                              //prints: true

boolean noneMatches = list.stream()
                        .peek(System.out::print)        //prints: 123
                        .noneMatch(e -> "3".equals(e));
System.out.println(noneMatches);                        //prints: false

boolean allMatch = list.stream()
                      .peek(System.out::print)          //prints: 1
                      .allMatch(e -> "3".equals(e));
System.out.println(allMatch);                           //prints: false
```

Please notice that all these operations are optimized so as not to process all the stream elements if the result can be determined early.

Find any or first

The following terminal operations allow finding any element or the first element of the stream correspondingly, as follows:

- `Optional<T> findAny()`: Returns an `Optional` with the value of any element of the stream, or an empty `Optional` if the stream is empty
- `Optional<T> findFirst()`: Returns an `Optional` with the value of the first element of the stream, or an empty `Optional` if the stream is empty

The following examples illustrate these operations:

```
List<String> list = List.of("1", "2", "3", "4", "5");
Optional<String> result = list.stream().findAny();
System.out.println(result.isPresent());    //prints: true
System.out.println(result.get());          //prints: 1

result = list.stream()
            .filter(e -> "42".equals(e))
            .findAny();
System.out.println(result.isPresent());    //prints: false
//System.out.println(result.get());        //NoSuchElementException

result = list.stream().findFirst();
System.out.println(result.isPresent());    //prints: true
System.out.println(result.get());          //prints: 1
```

In the first and third of the preceding examples, the `findAny()` and `findFirst()` operations produce the same result: they both find the first element of the stream. But in parallel processing, the result may be different.

When the stream is broken into several parts for parallel processing, the `findFirst()` operation always returns the first element of the stream, while the `findAny()` operation returns the first element only in one of the processing threads.

Now let's talk about `class java.util.Optional` in more detail.

Optional class

The object of `java.util.Optional` is used to avoid returning `null` (as it may cause `NullPointerException`). Instead, an `Optional` object provides methods that allow checking the presence of a value and substituting it with a predefined value if the return value is `null`. For example:

```java
List<String> list = List.of("1", "2", "3", "4", "5");
String result = list.stream()
                    .filter(e -> "42".equals(e))
                    .findAny()
                    .or(() -> Optional.of("Not found"))
                    .get();
System.out.println(result);                          //prints: Not found

result = list.stream()
             .filter(e -> "42".equals(e))
             .findAny()
             .orElse("Not found");
System.out.println(result);                          //prints: Not found

Supplier<String> trySomethingElse = () -> {
    //Code that tries something else
    return "43";
};
result = list.stream()
             .filter(e -> "42".equals(e))
             .findAny()
             .orElseGet(trySomethingElse);
System.out.println(result);                          //prints: 43

list.stream()
    .filter(e -> "42".equals(e))
    .findAny()
    .ifPresentOrElse(System.out::println,
        () -> System.out.println("Not found")); //prints: Not found
```

As you can see, if the `Optional` object is empty, then the following applies, as follows:

- The `or()` method of the `Optional` class allows returning an alternative `Optional` object.
- The `orElse()` method allows returning an alternative value.

- The orElseGet() method allows providing the Supplier function, which returns an alternative value.
- The ifPresentOrElse() method allows providing two functions: one that consumes the value from the Optional object, and another one that does something else in the case the Optional object is empty.

Min and max

The following terminal operations return the minimum or maximum value of stream elements, if present, as follows:

- Optional<T> min(Comparator<T> comparator): Returns the minimum element of this stream using the provided Comparator object
- Optional<T> max(Comparator<T> comparator): Returns the maximum element of this stream using the provided Comparator object

The following code demonstrates this:

```
List<String> list = List.of("a", "b", "c", "c", "a");
String min = list.stream()
                 .min(Comparator.naturalOrder())
                 .orElse("0");
System.out.println(min);      //prints: a

String max = list.stream()
  .max(Comparator.naturalOrder())
                 .orElse("0");
System.out.println(max);      //prints: c
```

As you can see, in the case of non-numerical values, the minimum element is the one that is the first when ordered from the left to the right according to the provided comparator. And the maximum, accordingly, is the last element. In the case of numeric values, the minimum and maximum are just that: the smallest and the biggest numbers among the stream elements:

```
int mn = Stream.of(42, 77, 33)
                .min(Comparator.naturalOrder())
                .orElse(0);
System.out.println(mn);     //prints: 33

int mx = Stream.of(42, 77, 33)
                .max(Comparator.naturalOrder())
                .orElse(0);
System.out.println(mx);     //prints: 77
```

Let's look at another example, using the `Person` class. The task is to find the oldest person in the following list:

```
List<Person> persons = List.of(new Person(23, "Bob"),
 new Person(33, "Jim"),
 new Person(28, "Jill"),
 new Person(27, "Bill"));
```

In order to do that, we can create the following `Compartor<Person>` that compares `Person` objects only by age:

```
Comparator<Person> perComp = (p1, p2) -> p1.getAge() - p2.getAge();
```

Then, using this comparator, we can find the oldest person:

```
Person theOldest = persons.stream()
                          .max(perComp)
                          .orElse(null);
System.out.println(theOldest);    //prints: Person{name='Jim', age=33}
```

To array

The following two terminal operations generate an array that contains stream elements, as follows:

- `Object[] toArray()`: Creates an array of objects; each object is an element of the stream
- `A[] toArray(IntFunction<A[]> generator)`: Creates an array of stream elements using the provided function

Let's look at some examples:

```
List<String> list = List.of("a", "b", "c");
Object[] obj = list.stream().toArray();
Arrays.stream(obj).forEach(System.out::print);    //prints: abc

String[] str = list.stream().toArray(String[]::new);
Arrays.stream(str).forEach(System.out::print);    //prints: abc
```

The first example is straightforward. It converts elements to an array of the same type. As for the second example, the representation of `IntFunction` as `String[]::new` is probably not obvious, so let's walk through it. The `String[]::new` is a method reference that represents the lambda expression `i -> new String[i]` because the `toArray()` operation receives from the stream not the elements, but their count:

```
String[] str = list.stream().toArray(i -> new String[i]);
```

We can prove it by printing an `i` value:

```
String[] str = list.stream()
                .toArray(i -> {
                        System.out.println(i);     //prints: 3
                        return  new String[i];
                });
```

The `i -> new String[i]` expression is an `IntFunction<String[]>` that, according to its documentation, accepts an `int` parameter and returns the result of the specified type. It can be defined using an anonymous class as follows:

```
IntFunction<String[]> intFunction = new IntFunction<String[]>() {
        @Override
        public String[] apply(int i) {
                return new String[i];
        }
};
```

The `java.util.Collection` interface has a very similar method that converts the collection to an array:

```
List<String> list = List.of("a", "b", "c");
String[] str = list.toArray(new String[lits.size()]);
Arrays.stream(str).forEach(System.out::print);     //prints: abc
```

The only difference is that the `toArray()` of the `Stream` interface accepts a function, while the `toArray()` of the `Collection` interface takes an array.

Reduce

This terminal operation is called `reduce` because it processes all the stream elements and produces one value, thus reducing all the stream elements to one value. But this is not the only operation that does it. The `collect` operation reduces all the values of the stream element to one result as well. And, in a way, all terminal operations are reductive. They produce one value after processing many elements.

So, you may look at `reduce` and `collect` as synonyms that help to add structure and classification to many operations available in the `Stream` interface. Also, operations in the `reduce` group can be viewed as specialized versions of the `collect` operation because `collect()` can be tailored to provide the same functionality as the `reduce()` operation.

That said, let's look at a group of `reduce` operations, as follows:

- `Optional<T> reduce(BinaryOperator<T> accumulator)`: Reduces the elements of the stream using the provided associative function that aggregates the elements; returns an `Optional` with the reduced value if available
- `T reduce(T identity, BinaryOperator<T> accumulator)`: Provides the same functionality as the previous `reduce()` version but with the identity parameter used as the initial value for an accumulator or a default value if a stream is empty
- `U reduce(U identity, BiFunction<U,T,U> accumulator, BinaryOperator<U> combiner)`: Provides the same functionality as the previous `reduce()` versions but, in addition, uses the `combiner` function to aggregate the results when this operation is applied to a parallel stream; if the stream is not parallel, the `combiner` function is not used

To demonstrate the `reduce()` operation, we are going to use the same `Person` class we have used before and the same list of `Person` objects as the source for our stream examples:

```
List<Person> persons = List.of(new Person(23, "Bob"),
                               new Person(33, "Jim"),
                               new Person(28, "Jill"),
                               new Person(27, "Bill"));
```

Let's find the oldest person in this list using the `reduce()` operation:

```
Person theOldest = list.stream()
                .reduce((p1, p2) -> p1.getAge() > p2.getAge() ? p1 : p2)
                .orElse(null);
System.out.println(theOldest);     //prints: Person{name='Jim', age=33}
```

The implementation is somewhat surprising, isn't it? The `reduce()` operation takes an accumulator, but it seems it did not accumulate anything. Instead, it compares all stream elements. Well, the accumulator saves the result of the comparison and provides it as the first parameter for the next comparison (with the next element). You can say that the accumulator, in this case, accumulates the results of all previous comparisons.

Let's now accumulate something explicitly. Let's assemble all the names from a list of persons in one comma-separated list:

```
String allNames = list.stream()
                   .map(p -> p.getName())
                   .reduce((n1, n2) -> n1 + ", " + n2)
                   .orElse(null);
System.out.println(allNames);           //prints: Bob, Jim, Jill, Bill
```

The notion of accumulation, in this case, makes a bit more sense, doesn't it?

Now let's use the `identity` value to provide some initial value:

```
String all = list.stream()
                .map(p -> p.getName())
                .reduce("All names: ", (n1, n2) -> n1 + ", " + n2);
System.out.println(all);    //prints: All names: , Bob, Jim, Jill, Bill
```

Notice that this version of the `reduce()` operation returns `value`, not the `Optional` object. That is because, by providing the initial value, we guarantee that at least this value will be present in the result if the stream turns out to be empty. But the resulting string does not look as pretty as we hoped. Apparently, the provided initial value is treated as any other stream element, and a comma is added after it by the accumulator we have created. To make the result look pretty again, we could use the first version of the `reduce()` operation again and add the initial value this way:

```
String all = "All names: " + list.stream()
                     .map(p -> p.getName())
                     .reduce((n1, n2) -> n1 + ", " + n2)
                     .orElse(null);
System.out.println(all);      //prints: All names: Bob, Jim, Jill, Bill
```

Or we can use a space as a separator instead of a comma:

```
String all = list.stream()
                 .map(p -> p.getName())
                 .reduce("All names:", (n1, n2) -> n1 + " " + n2);
System.out.println(all);      //prints: All names: Bob Jim Jill Bill
```

Now the result looks better. While demonstrating the `collect()` operation in the next subsection, we will show a better way to create a comma-separated list of values with a prefix.

Meanwhile, let's continue to review the `reduce()` operation and look at its third form: the one with three parameters: `identity`, `accumulator`, and `combiner`. Adding the combiner to the `reduce()` operation does not change the result:

```
String all = list.stream()
                 .map(p -> p.getName())
                 .reduce("All names:", (n1, n2) -> n1 + " " + n2,
                                       (n1, n2) -> n1 + " " + n2 );
System.out.println(all);      //prints: All names: Bob Jim Jill Bill
```

That is because the stream is not parallel and the combiner is used only with a parallel stream. If we make the stream parallel, the result changes:

```
String all = list.parallelStream()
                 .map(p -> p.getName())
                 .reduce("All names:", (n1, n2) -> n1 + " " + n2,
                                       (n1, n2) -> n1 + " " + n2 );
System.out.println(all);
   //prints: All names: Bob All names: Jim All names: Jill All names: Bill
```

Apparently, for a parallel stream, the sequence of elements is broken into subsequences, each processed independently, and their results aggregated by the combiner. While doing that, the combiner adds the initial value (identity) to each of the results. Even if we remove the combiner, the result of the parallel stream processing remains the same, because a default combiner behavior is provided:

```
String all = list.parallelStream()
                   .map(p -> p.getName())
                   .reduce("All names:", (n1, n2) -> n1 + " " + n2);
System.out.println(all);
    //prints: All names: Bob All names: Jim All names: Jill All names: Bill
```

In the previous two forms of the `reduce()` operations, the identity value was used by the accumulator. In the third form, the identity value is used by the combiner (notice, the U type is the combiner type). To get rid of the repetitive identity value in the result, we have decided to remove it (and the trailing space) from the second parameter in the combiner:

```
String all = list.parallelStream().map(p->p.getName())
                   .reduce("All names:", (n1, n2) -> n1 + " " + n2,
           (n1, n2) -> n1 + " " + StringUtils.remove(n2, "All names: "));
System.out.println(all);        //prints: All names: Bob Jim Jill Bill
```

The result is as expected.

In our string-based examples so far, the identity has not just been an initial value. It also served as an identifier (a label) in the resulting string. But when the elements of the stream are numeric, the identity looks more like just an initial value. Let's look at the following example:

```
List<Integer> ints = List.of(1, 2, 3);
int sum = ints.stream()
               .reduce((i1, i2) -> i1 + i2)
               .orElse(0);
System.out.println(sum);                         //prints: 6
sum = ints.stream()
           .reduce(Integer::sum)
           .orElse(0);
System.out.println(sum);                         //prints: 6
sum = ints.stream()
           .reduce(10, Integer::sum);
System.out.println(sum);                         //prints: 16
sum = ints.stream()
           .reduce(10, Integer::sum, Integer::sum);
System.out.println(sum);                         //prints: 16
```

The first two of the pipelines are exactly the same, except that the second pipeline uses a method reference. The third and the fourth pipelines have the same functionality too. They both use an initial value of 10. Now the first parameter makes more sense as the initial value than the identity, doesn't it? In the fourth pipeline, we added a combiner, but it is not used because the stream is not parallel. Let's make it parallel and see what happens:

```
List<Integer> ints = List.of(1, 2, 3);
int sum = ints.parallelStream()
              .reduce(10, Integer::sum, Integer::sum);
System.out.println(sum);                          //prints: 36
```

The result is 36 because the initial value of 10 was added three times, with each partial result. Apparently, the stream was broken into three subsequences. But it is not always the case, as the number of the subsequences changes as the stream grows and the number of CPUs on the computer increases. That is why you cannot rely on a certain fixed number of subsequences and it is better not to use a non-zero initial value with parallel streams:

```
List<Integer> ints = List.of(1, 2, 3);
int sum = ints.parallelStream()
              .reduce(0, Integer::sum, Integer::sum);
System.out.println(sum);                          //prints: 6
sum = 10 + ints.parallelStream()
              .reduce(0, Integer::sum, Integer::sum);
System.out.println(sum);                          //prints: 16
```

As you can see, we have set the identity to 0, so every subsequence will get it, but the total is not affected when the result from all the processing threads is assembled by the combinator.

Collect

Some of the usages of the collect() operation are very simple and can be easily mastered by any beginner, while other cases can be complex and not easy to understand even for a seasoned programmer. Together with the operations discussed already, the most popular cases of collect() usage we present in this section are more than enough for all the needs a beginner may have and will cover most needs of a more experienced professional. Together with the operations of numeric streams (see the *Numeric stream interfaces* section), they cover all the needs a mainstream programmer will ever have.

As we have mentioned already, the `collect()` operation is very flexible and allows us to customize stream processing. It has two forms, as follows:

- `R collect(Collector<T, A, R> collector)`: Processes the stream elements of type `T` using the provided `Collector` and producing the result of type `R` via an intermediate accumulation of type `A`
- `R collect(Supplier<R> supplier, BiConsumer<R, T> accumulator, BiConsumer<R, R> combiner)`: Processes the stream elements of type `T` using the provided functions:
  - `Supplier<R> supplier`: Creates a new result container
  - `BiConsumer<R, T> accumulator`: A stateless function that adds an element to the result container
  - `BiConsumer<R, R> combiner`: A stateless function that merges two partial result containers: adds the elements from the second result container into the first result container

Let's look at the second form of the `collect()` operation first. It is very similar to the `reduce()` operation with three parameters we have just demonstrated: `supplier`, `accumulator`, and `combiner`. The biggest difference is that the first parameter in the `collect()` operation is not an identity or the initial value, but the container, an object, that is going to be passed between functions and which maintains the state of the processing.

Let's demonstrate how it works by selecting the oldest person from the list of `Person` objects. For the following example, we are going to use the familiar `Person` class as the container but add to it a constructor without parameters with two setters:

```
public Person(){}
public void setAge(int age) { this.age = age;}
public void setName(String name) { this.name = name; }
```

Adding a constructor without parameters and setters is necessary because the `Person` object as a container should be creatable at any moment without any parameters and should be able to receive and keep the partial results: the `name` and `age` of the person who is the oldest, so far. The `collect()` operation will use this container while processing each element and, after the last element is processed, will contain the name and the age of the oldest person.

We will use again the same list of persons:

```
List<Person> list = List.of(new Person(23, "Bob"),
                            new Person(33, "Jim"),
                            new Person(28, "Jill"),
                            new Person(27, "Bill"));
```

And here is the `collect()` operation that finds the oldest person in the list:

```
BiConsumer<Person, Person> accumulator = (p1, p2) -> {
    if(p1.getAge() < p2.getAge()){
        p1.setAge(p2.getAge());
        p1.setName(p2.getName());
    }
};
BiConsumer<Person, Person> combiner = (p1, p2) -> {
    System.out.println("Combiner is called!");
    if(p1.getAge() < p2.getAge()){
        p1.setAge(p2.getAge());
        p1.setName(p2.getName());
    }
};
Person theOldest = list.stream()
                        .collect(Person::new, accumulator, combiner);
System.out.println(theOldest);      //prints: Person{name='Jim', age=33}
```

We tried to inline the functions in the operation call, but it looked a bit difficult to read, so we decided to create functions first and then use them in the `collect()` operation. The container, a `Person` object, is created only once before the first element is processed. In this sense, it is similar to the initial value of the `reduce()` operation. Then it is passed to the accumulator, which compares it to the first element. The `age` field in the container was initialized to the default value of zero and thus the `age` and `name` of the first element were set in the container as the parameters of the oldest person, so far. When the second stream element (`Person` object) is emitted, its `age` value is compared to the `age` value currently stored in the container, and so on, until all elements of the stream are processed. The result is shown in the previous comments.

When the stream is sequential, the combiner is never called. But when we make it parallel (`list.parallelStream()`), the message **Combiner is called!** is printed three times. Well, as in the case of the `reduce()` operation, the number of partial results may vary, depending on the number of CPUs and the internal logic of the `collect()` operation implementation. So, the **Combiner is called!** message can be printed any number of times.

Now let's look at the first form of the `collect()` operation. It requires an object of the class that implements the `java.util.stream.Collector<T,A,R>` interface, where `T` is the stream type, `A` is the container type, and `R` is the result type. You can use one of the following methods `of()` (from the `Collector` interface) to create the necessary `Collector` object:

```
static Collector<T,R,R> of(Supplier<R> supplier,
                           BiConsumer<R,T> accumulator,
                           BinaryOperator<R> combiner,
                           Collector.Characteristics... characteristics)
```

Or

```
static Collector<T,A,R> of(Supplier<A> supplier,
                           BiConsumer<A,T> accumulator,
                           BinaryOperator<A> combiner,
                           Function<A,R> finisher,
                           Collector.Characteristics... characteristics).
```

The functions you have to pass to the preceding methods are similar to those we have demonstrated already. But we are not going to do this, for two reasons. First, it is more involved and pushes us beyond the scope of this book, and, second, before doing that, you have to look in the `java.util.stream.Collectors` class, which provides many ready-to-use collectors.

As we have mentioned already, together with the operations discussed so far and the numeric streams operations we are going to present in the next section, ready-to-use collectors cover the vast majority of processing needs in mainstream programming, and there is a good chance you will never need to create a custom collector.

Collectors

The `java.util.stream.Collectors` class provides more than 40 methods that create `Collector` objects. We are going to demonstrate only the simplest and most popular ones, as follows:

- `Collector<T,?,List<T>> toList()`: Creates a collector that generates a `List` object from stream elements
- `Collector<T,?,Set<T>> toSet()`: Creates a collector that generates a `Set` object from stream elements
- `Collector<T,?,Map<K,U>> toMap (Function<T,K> keyMapper, Function<T,U> valueMapper)`: Creates a collector that generates a `Map` object from stream elements

- `Collector<T,?,C> toCollection (Supplier<C> collectionFactory)`: Creates a collector that generates a `Collection` object of the type provided by `Supplier<C> collectionFactory`

- `Collector<CharSequence,?,String> joining()`: Creates a collector that generates a `String` object by concatenating stream elements

- `Collector<CharSequence,?,String> joining (CharSequence delimiter)`: Creates a collector that generates a delimiter-separated `String` object from stream elements

- `Collector<CharSequence,?,String> joining (CharSequence delimiter, CharSequence prefix, CharSequence suffix)`: Creates a collector that generates a delimiter-separated `String` object from the stream elements and adds the specified `prefix` and `suffix`

- `Collector<T,?,Integer> summingInt (ToIntFunction<T>)`: Creates a collector that calculates the sum of the results generated by the provided function applied to each element; the same method exists for `long` and `double` types

- `Collector<T,?,IntSummaryStatistics> summarizingInt (ToIntFunction<T>)`: Creates a collector that calculates the sum, min, max, count, and average of the results generated by the provided function applied to each element; the same method exists for `long` and `double` types

- `Collector<T,?,Map<Boolean,List<T>>> partitioningBy (Predicate<? super T> predicate)`: Creates a collector that separates the elements using the provided `Predicate` function

- `Collector<T,?,Map<K,List<T>>> groupingBy(Function<T,U>)`: Creates a collector that groups elements into a `Map` with keys generated by the provided function

The following demo code shows how to use the collectors created by the methods listed earlier. First, we demonstrate usage of the `toList()`, `toSet()`, `toMap()`, and `toCollection()` methods:

```
List<String> ls = Stream.of("a", "b", "c")
                    .collect(Collectors.toList());
System.out.println(ls);                    //prints: [a, b, c]

Set<String> set = Stream.of("a", "a", "c")
                    .collect(Collectors.toSet());
System.out.println(set);                   //prints: [a, c]

List<Person> list = List.of(new Person(23, "Bob"),
                            new Person(33, "Jim"),
```

```
                              new Person(28, "Jill"),
                              new Person(27, "Bill"));
    Map<String, Person> map = list.stream()
                              .collect(Collectors
                              .toMap(p -> p.getName() + "-" +
                                     p.getAge(), p -> p));
    System.out.println(map); //prints: {Bob-23=Person{name='Bob', age:23},
                    //          Bill-27=Person{name='Bill', age:27},
                    //          Jill-28=Person{name='Jill', age:28},
                    //          Jim-33=Person{name='Jim', age:33}}

    Set<Person> personSet = list.stream()
                            .collect(Collectors
                            .toCollection(HashSet::new));
    System.out.println(personSet);  //prints: [Person{name='Bill', age=27},
                    //          Person{name='Jim', age=33},
                    //          Person{name='Bob', age=23},
                    //          Person{name='Jill', age=28}]
```

The `joining()` method allows concatenating `Character` and `String` values in a delimited list with `prefix` and `suffix`:

```
    List<String> list1 = List.of("a", "b", "c", "d");
    String result = list1.stream()
                    .collect(Collectors.joining());
    System.out.println(result);                    //prints: abcd

    result = list1.stream()
                .collect(Collectors.joining(", "));
    System.out.println(result);                    //prints: a, b, c, d

    result = list1.stream()
                .collect(Collectors.joining(", ", "The result: ", ""));
    System.out.println(result);         //prints: The result: a, b, c, d

    result = list1.stream()
            .collect(Collectors.joining(", ", "The result: ", ". The End."));
    System.out.println(result);    //prints: The result: a, b, c, d. The End.
```

Now let's turn to the `summingInt()` and `summarizingInt()` methods. They create collectors that calculate the sum and other statistics of the `int` values produced by the provided function applied to each element:

```
    List<Person> list2 = List.of(new Person(23, "Bob"),
                                 new Person(33, "Jim"),
                                 new Person(28, "Jill"),
                                 new Person(27, "Bill"));
    int sum = list2.stream()
```

```
              .collect(Collectors.summingInt(Person::getAge));
System.out.println(sum);                      //prints: 111

IntSummaryStatistics stats = list2.stream()
          .collect(Collectors.summarizingInt(Person::getAge));
System.out.println(stats); //prints: IntSummaryStatistics{count=4,
                           //sum=111, min=23, average=27.750000, max=33}
System.out.println(stats.getCount());   //prints: 4
System.out.println(stats.getSum());     //prints: 111
System.out.println(stats.getMin());     //prints: 23
System.out.println(stats.getAverage()); //prints: 27.750000
System.out.println(stats.getMax());     //prints: 33
```

There are also the `summingLong()`, `summarizingLong()`, `summingDouble()`, and `summarizingDouble()` methods.

The `partitioningBy()` method creates a collector that groups the elements by the provided criteria and put the groups (lists) in a `Map` object with a `boolean` value as the key:

```
Map<Boolean, List<Person>> map2 = list2.stream()
        .collect(Collectors.partitioningBy(p -> p.getAge() > 27));
System.out.println(map2);
    //{false=[Person{name='Bob', age=23}, Person{name='Bill', age=27},
    //  true=[Person{name='Jim', age=33}, Person{name='Jill', age=28}]}
```

As you can see, using the `p.getAge() > 27` criteria, we were able to put all the persons in two groups: one is below or equals to 27 years of `age` (the key is `false`), and another is above 27 (the key is `true`).

And, finally, the `groupingBy()` method allows grouping elements by a value and puts the groups (lists) in a `Map` object with this value as a key:

```
List<Person> list3 = List.of(new Person(23, "Bob"),
                             new Person(33, "Jim"),
                             new Person(23, "Jill"),
                             new Person(33, "Bill"));
Map<Integer, List<Person>> map3 = list3.stream()
                      .collect(Collectors.groupingBy(Person::getAge));
System.out.println(map3);
    // {33=[Person{name='Jim', age=33}, Person{name='Bill', age=33}],
    //  23=[Person{name='Bob', age=23}, Person{name='Jill', age=23}]}
```

To be able to demonstrate this method, we changed our list of `Person` objects by setting the `age` on each of them either to 23 or to 33. The result is two groups ordered by their `age`.

There are also overloaded `toMap()`, `groupingBy()`, and `partitioningBy()` methods as well as the following, often overloaded, methods that create corresponding `Collector` objects, as follows:

- `counting()`
- `reducing()`
- `filtering()`
- `toConcurrentMap()`
- `collectingAndThen()`
- `maxBy()`, `minBy()`
- `mapping()`, `flatMapping()`
- `averagingInt()`, `averagingLong()`, `averagingDouble()`
- `toUnmodifiableList()`, `toUnmodifiableMap()`, `toUnmodifiableSet()`

If you cannot find the operation you need among those discussed in this book, search the `Collectors` API first, before building your own `Collector` object.

Numeric stream interfaces

As we have mentioned already, all three numeric interfaces, `IntStream`, `LongStream`, and `DoubleStream`, have methods similar to the methods in the `Stream` interface, including the methods of the `Stream.Builder` interface. This means that everything we have discussed so far in this chapter equally applies to any numeric stream interfaces. That is why in this section we will only talk about those methods that are not present in the `Stream` interface, as follows:

- The `range(lower, upper)` and `rangeClosed(lower, upper)` methods in the `IntStream` and `LongStream` interfaces allow creating a stream from the values in the specified range
- The `boxed()` and `mapToObj()` intermediate operations convert a numeric stream to `Stream`

- The `mapToInt()`, `mapToLong()`, and `mapToDouble()` intermediate operations convert a numeric stream of one type to a numeric stream of another type
- The intermediate operations `flatMapToInt()`, `flatMapToLong()`, and `flatMapToDouble()` convert a stream to a numeric stream
- The `sum()` and `average()` terminal operations calculate the sum and average of numeric stream elements

Creating a stream

In addition to the methods of the `Stream` interface that create streams, the `IntStream` and `LongStream` interfaces allow creating a stream from the values in the specified range.

range(), rangeClosed()

The `range(lower, upper)` method generates all values sequentially, starting from the `lower` value and ending with the value just before `upper`:

```
IntStream.range(1, 3).forEach(System.out::print);    //prints: 12
LongStream.range(1, 3).forEach(System.out::print);   //prints: 12
```

The `rangeClosed(lower, upper)` method generates all the values sequentially, starting from the `lower` value and ending with the `upper` value:

```
IntStream.rangeClosed(1, 3).forEach(System.out::print); //prints: 123
LongStream.rangeClosed(1, 3).forEach(System.out::print);  //prints: 123
```

Intermediate operations

In addition to the intermediate operations of the `Stream` interface, the `IntStream`, `LongStream`, and `DoubleStream` interfaces also have number-specific intermediate operations: `boxed()`, `mapToObj()`, `mapToInt()`, `mapToLong()`, `mapToDouble()`, `flatMapToInt()`, `flatMapToLong()`, and `flatMapToDouble()`.

boxed(), mapToObj()

The boxed() intermediate operation converts (boxes) elements of the primitive numeric type to the corresponding wrapper type:

```
//IntStream.range(1, 3).map(Integer::shortValue) //compile error
//                    .forEach(System.out::print);

IntStream.range(1, 3)
          .boxed()
          .map(Integer::shortValue)
          .forEach(System.out::print);          //prints: 12

//LongStream.range(1, 3).map(Long::shortValue)   //compile error
//                    .forEach(System.out::print);

LongStream.range(1, 3)
          .boxed()
          .map(Long::shortValue)
          .forEach(System.out::print);          //prints: 12
//DoubleStream.of(1).map(Double::shortValue)      //compile error
//                  .forEach(System.out::print);

DoubleStream.of(1)
          .boxed()
          .map(Double::shortValue)
          .forEach(System.out::print);          //prints: 1
```

In the preceding code, we have commented out the lines that generate compilation error because the elements generated by the range() method are of primitive types. The boxed() operation converts a primitive value to the corresponding wrapping type, so it can be processed as a reference type. The mapToObj() intermediate operation does a similar transformation, but it is not as specialized as the boxed() operation and allows using an element of primitive type to produce an object of any type:

```
IntStream.range(1, 3)
          .mapToObj(Integer::valueOf)
          .map(Integer::shortValue)
          .forEach(System.out::print);          //prints: 12

IntStream.range(42, 43)
        .mapToObj(i -> new Person(i, "John"))
        .forEach(System.out::print); //prints: Person{name='John', age=42}

LongStream.range(1, 3)
          .mapToObj(Long::valueOf)
          .map(Long::shortValue)
```

```
        .forEach(System.out::print);              //prints: 12

DoubleStream.of(1)
        .mapToObj(Double::valueOf)
        .map(Double::shortValue)
        .forEach(System.out::print);              //prints: 1
```

In the preceding code, we have added the map() operation just to prove that the mapToObj() operation does the job and creates an object of the wrapping type as expected. Also, by adding the pipeline that produces Person objects, we have demonstrated how the mapToObj() operation can be used to create an object of any type.

mapToInt(), mapToLong(), mapToDouble()

The mapToInt(), mapToLong(), mapToDouble() intermediate operations allow converting a numeric stream of one type to a numeric stream of another type. For the sake of example, we convert a list of String values to a numeric stream of different types by mapping each String value to its length:

```
List<String> list = List.of("one", "two", "three");
list.stream()
    .mapToInt(String::length)
    .forEach(System.out::print);              //prints: 335

list.stream()
    .mapToLong(String::length)
    .forEach(System.out::print);              //prints: 335

list.stream()
    .mapToDouble(String::length)
    .forEach(d -> System.out.print(d + " "));  //prints: 3.0 3.0 5.0

list.stream()
    .map(String::length)
    .map(Integer::shortValue)
    .forEach(System.out::print);              //prints: 335
```

The elements of the created numeric streams are of the primitive type:

```
//list.stream().mapToInt(String::length)
//              .map(Integer::shortValue) //compile error
//              .forEach(System.out::print);
```

And, as we are on this topic, if you would like to convert elements to a numeric wrapping type, the intermediate `map()` operation is the way to do it (instead of `mapToInt()`):

```
list.stream().map(String::length)
        .map(Integer::shortValue)
        .forEach(System.out::print);        //prints: 335
```

flatMapToInt(), flatMapToLong(), flatMapToDouble()

The `flatMapToInt()`, `flatMapToLong()`, `flatMapToDouble()` intermediate operations produce a numeric stream of the corresponding type:

```
List<Integer> list = List.of(1, 2, 3);
list.stream()
    .flatMapToInt(i -> IntStream.rangeClosed(1, i))
    .forEach(System.out::print);                //prints: 112123

list.stream()
    .flatMapToLong(i -> LongStream.rangeClosed(1, i))
    .forEach(System.out::print);                //prints: 112123

list.stream()
    .flatMapToDouble(DoubleStream::of)
    .forEach(d -> System.out.print(d + " "));   //prints: 1.0 2.0 3.0
```

As you can see in the preceding code, we have used `int` values in the original stream. But it can be a stream of any type:

```
List.of("one", "two", "three")
    .stream()
    .flatMapToInt(s -> IntStream.rangeClosed(1, s.length()))
    .forEach(System.out::print);            //prints: 12312312345
```

Terminal operations

Numeric-specific terminal operations are pretty straightforward. There are two of them, as follows:

- `sum()`: Calculates the sum of numeric stream elements
- `average()`: Calculates the average of numeric stream elements

sum(), average()

If you need to calculate a sum or an average of the values of numeric stream elements, the only requirement for the stream is that it should not be infinite. Otherwise, the calculation never finishes. The following are examples of these operations usages:

```
int sum = IntStream.empty()
                  .sum();
System.out.println(sum);              //prints: 0

sum = IntStream.range(1, 3)
              .sum();
System.out.println(sum);              //prints: 3

double av = IntStream.empty()
                    .average()
                    .orElse(0);
System.out.println(av);               //prints: 0.0

av = IntStream.range(1, 3)
            .average()
            .orElse(0);
System.out.println(av);               //prints: 1.5

long suml = LongStream.range(1, 3)
                    .sum();
System.out.println(suml);             //prints: 3

double avl = LongStream.range(1, 3)
                      .average()
                      .orElse(0);
System.out.println(avl);              //prints: 1.5

double sumd = DoubleStream.of(1, 2)
                        .sum();
System.out.println(sumd);             //prints: 3.0

double avd = DoubleStream.of(1, 2)
                        .average()
                        .orElse(0);
System.out.println(avd);              //prints: 1.5
```

As you can see, using these operations on an empty stream is not a problem.

Parallel streams

We have seen that changing from a sequential stream to a parallel stream can lead to incorrect results if the code was not written and tested for processing a parallel stream. The following are a few more considerations related to parallel streams.

Stateless and stateful operations

There are **stateless operations**, such as `filter()`, `map()`, and `flatMap()`, which do not keep data around (do not maintain state) while moving from processing from one stream element to the next. And there are stateful operations, such as `distinct()`, `limit()`, `sorted()`, `reduce()`, and `collect()`, that can pass the state from previously processed elements to the processing of the next element.

Stateless operations usually do not pose a problem while switching from a sequential stream to a parallel one. Each element is processed independently and the stream can be broken into any number of sub-streams for independent processing. With stateful operations, the situation is different. To start with, using them for an infinite stream may never finish processing. Also, while discussing the `reduce()` and `collect()` stateful operations, we have demonstrated how switching to a parallel stream can produce a different result if the initial value (or identity) is set without parallel processing in mind.

And there are performance considerations too. Stateful operations often require processing all the stream elements in several passes using buffering. For large streams, it may tax JVM resources and slow down, if not completely shut down, the application.

That is why a programmer should not take switching from a sequential to a parallel streams lightly. If stateful operations are involved, the code has to be designed and tested to be able to perform parallel stream processing without negative effects.

Sequential or parallel processing?

As we indicated in the previous section, parallel processing may or may not produce better performance. You have to test every use case before deciding on using parallel streams. Parallelism can yield better performance, but the code has to be designed and possibly optimized to do it. And each assumption has to be tested in an environment that is as close to the production as possible.

However, there are a few considerations you can take into account while deciding between sequential and parallel processing, as follows:

- Small streams are typically processed faster sequentially (well, what is *small* for your environment should be determined through testing and by measuring the performance)
- If stateful operations cannot be replaced with stateless ones, carefully design your code for the parallel processing or just avoid it
- Consider parallel processing for procedures that require extensive calculations, but think about bringing the partial results together for the final result

Summary

In this chapter, we have talked about data-streams processing, which is different from processing the I/O streams we reviewed in Chapter 5, *Strings, Input/Output, and Files*. We defined what the data streams are, how to process their elements using stream operations, and how to chain (connect) stream operations in a pipeline. We also discussed stream initialization and how to process streams in parallel.

In the next chapter, the reader will be introduced to The **Reactive Manifesto**, its thrust, and examples of its implementations. We will discuss the difference between reactive and responsive systems and what **asynchronous** and **non-blocking** processings are. We will also talk about **Reactive Streams** and **RxJava**.

Quiz

1. What is the difference between I/O streams and java.util.stream.Stream? Select all that apply:
    a. I/O streams are oriented toward data delivery, while Stream is oriented toward data processing
    b. Some I/O streams can be transformed into Stream
    c. I/O streams can read from a file, while Stream cannot
    d. I/O streams can write to a file, while Stream cannot

2. What do the `empty()` and `of(T... values)` `Stream` methods have in common?

3. What type are the elements emitted by the `Stream.ofNullable(Set.of(1,2,3)` stream?

4. What does the following code print?

```
Stream.iterate(1, i -> i + 2)
      .limit(3)
      .forEach(System.out::print);
```

5. What does the following code print?

```
Stream.concat(Set.of(42).stream(),
              List.of(42).stream()).limit(1)
                                   .forEach(System.out::print);
```

6. What does the following code print?

```
Stream.generate(() -> 42 / 2)
      .limit(2)
      .forEach(System.out::print);
```

7. Is `Stream.Builder` a functional interface?

8. How many elements does the following stream emit?

```
new Random().doubles(42).filter(d -> d >= 1)
```

9. What does the following code print?

```
Stream.of(1,2,3,4)
      .skip(2)
      .takeWhile(i -> i < 4)
      .forEach(System.out::print);
```

10. What is the value of d in the following code?

```
double d = Stream.of(1, 2)
                 .mapToDouble(Double::valueOf)
                 .map(e -> e / 2)
                 .sum();
```

11. What is the value of the s string in the following code?

```
String s = Stream.of("a","X","42").sorted()
 .collect(Collectors.joining(","));
```

12. What is the result of the following code?

```
List.of(1,2,3).stream()
              .peek(i -> i > 2 )
              .forEach(System.out::print);
```

13. How many stream elements the `peek()` operation prints in the following code?

```
List.of(1,2,3).stream()
              .peek(System.out::println)
              .noneMatch(e -> e == 2);
```

14. What does the `or()` method return when the `Optional` object is empty?

15. What is the value of the s string in the following code?

```
String s = Stream.of("a","X","42")
 .max(Comparator.naturalOrder())
 .orElse("12");
```

16. How many elements does the `IntStream.rangeClosed(42, 42)` stream emit?

17. Name two stateless operations.

18. Name two stateful operations.

15
Reactive Programming

In this chapter, the reader will be introduced to the **Reactive Manifesto** and the world of reactive programming. We start with defining and discussing the main related concepts – asynchronous, non-blocking, and responsive. Using them, we then define and discuss reactive programming, the main reactive frameworks, and talk about **RxJava** in more detail.

The following topics will be covered in this chapter:

- Asynchronous processing
- Non-blocking API
- Reactive – responsive, resilient, elastic, message-driven
- Reactive streams
- RxJava

Asynchronous processing

Asynchronous means that the requestor gets the response immediately, but the result is not there. Instead, the requestor waits until the result is sent to them, or saved in the database, or, for example, presented as an object that allows checking if the result is ready. If the latter is the case, the requestor calls a certain method to this object periodically and, when the result is ready, retrieves it using another method on the same object. The advantage of asynchronous processing is that the requestor can do other things while waiting.

In Chapter 8, *Multithreading and Concurrent Processing*, we have demonstrated how a child thread can be created. Such a child thread then sends a non-asynchronous (blocking) request and waits for its return doing nothing. The main thread, meanwhile, continues executing and periodically calls the child thread object to see whether the result is ready. That is the most basic of asynchronous processing implementations. In fact, we have used it already when we used parallel streams.

The parallel-stream operations behind the scenes that create child threads break the stream into segments, and assign each segment to a dedicated thread for processing, and then aggregate the partial results from all the segments into the final result. In the previous chapter, we have even written functions that did the aggregating job. As a reminder, the function was called a **combiner**.

Let's compare the performance of sequential and parallel streams using an example.

Sequential and parallel streams

To demonstrate the difference between sequential and parallel processing, let's imagine a system that collects data from 10 physical devices (sensors) and calculates an average. The following is the get() method that collects a measurement from a sensor identified by ID:

```
double get(String id){
    try{
        TimeUnit.MILLISECONDS.sleep(100);
    } catch(InterruptedException ex){
        ex.printStackTrace();
    }
    return id * Math.random();
}
```

We have put a delay for 100 milliseconds to imitate the time it takes to collect the measurement from the sensor. As for the resulting measurement value, we use the Math.random() method. We are going to call this get() method using an object of the MeasuringSystem class, where the method belongs.

Then, we are going to calculate an average – to offset the errors and other idiosyncrasies of an individual device:

```
void getAverage(Stream<Integer> ids) {
    LocalTime start = LocalTime.now();
    double a = ids.mapToDouble(id -> new MeasuringSystem().get(id))
                .average()
                .orElse(0);
    System.out.println((Math.round(a * 100.) / 100.) + " in " +
            Duration.between(start, LocalTime.now()).toMillis() + " ms");
}
```

Notice how we convert the stream of IDs into `DoubleStream` using the `mapToDouble()` operation so we can apply the `average()` operation. The `average()` operation returns an `Optional<Double>` object, and we call its `orElse(0)` method that returns either the calculated value or zero (if, for example, the measuring system could not connect to any of its sensors and returned an empty stream).

The last line of the `getAverage()` method prints the result and the time it took to calculate it. In real code, we would return the result and use it for other calculations. But, for demonstration, we just print it.

Now, we can compare the performance of a sequential stream processing with the performance of the parallel processing:

```
List<Integer> ids = IntStream.range(1, 11)
                           .mapToObj(i -> i)
                           .collect(Collectors.toList());
getAverage(ids.stream());              //prints: 2.99 in 1030 ms
getAverage(ids.parallelStream());      //prints: 2.34 in  214 ms
```

The results may be different if you run this example because, as you may recall, we simulate the collected measurements as random values.

As you can see, the processing of a parallel stream is five times faster than the processing of a sequential stream. The results are different because the measurement produces a slightly different result every time.

Although behind the scenes, the parallel stream uses asynchronous processing, this is not what programmers have in mind when talking about the asynchronous processing of requests. From the application's perspective, it is just parallel (also called concurrent) processing. It is faster than sequential processing, but the main thread has to wait until all the calls are made and the data is retrieved. If each call takes at least 100 ms (as it is in our case), then the processing of all the calls cannot be completed in less time.

Of course, we can create a child thread and let it make all the calls and wait until they are complete, while the main thread does something else. We even can create a service that does it, so the application would just tell such a service what has to be done and then continue doing something else. Later, the main thread can call the service again and get the result or pick it up in some agreed location.

That would be the truly asynchronous processing the programmers are talking about. But, before writing such a code, let's look at the `CompletableFuture` class located in the `java.util.concurrent` package. It does everything described, and even more.

Using the CompletableFuture object

Using the `CompletableFuture` object, we can separate sending the request to the measuring system by getting the result from the `CompletableFuture` object. That is exactly the scenario we described while explaining what asynchronous processing is. Let's demonstrate it in the code:

```
List<CompletableFuture<Double>> list =
     ids.stream()
          .map(id -> CompletableFuture.supplyAsync(() ->
                                      new MeasuringSystem().get(id)))
          .collect(Collectors.toList());
```

The `supplyAsync()` method does not wait for the call to the measuring system to return. Instead, it immediately creates a `CompletableFuture` object and returns it, so that a client can use this object any time later to retrieve the values returned by the measuring system:

```
LocalTime start = LocalTime.now();
double a = list.stream()
                .mapToDouble(cf -> cf.join().doubleValue())
                .average()
                .orElse(0);
System.out.println((Math.round(a * 100.) / 100.) + " in " +
     Duration.between(start, LocalTime.now()).toMillis() + " ms");
                                             //prints: 2.92 in 6 ms
```

There are also methods that allow checking whether the value was returned at all, but that is not the point of this demonstration, which is to show how the `CompletableFuture` class can be used to organize asynchronous processing.

The created list of `CompletableFuture` objects can be stored anywhere and processed very quickly (in 6 ms, in our case), provided that the measurements have been received already. Between creating the list of `CompletableFuture` objects and processing them, the system is not blocked and can do something else.

The `CompletableFuture` class has many methods and support from several other classes and interfaces. For example, a fixed-size thread pool can be added to limit the number of threads:

```
ExecutorService pool = Executors.newFixedThreadPool(3);
List<CompletableFuture<Double>> list = ids.stream()
          .map(id -> CompletableFuture.supplyAsync(() ->
                          new MeasuringSystem().get(id), pool))
          .collect(Collectors.toList());
```

There is a variety of such pools for different purposes and different performance. But all that does not change the overall system design, so we omit such details.

As you can see, the power of asynchronous processing is great. There is also a variation of asynchronous API called a **non-blocking API**, which we are going to discuss in the next section.

Non-blocking API

The client of a non-blocking API expects to get the results back quickly, that is, without being blocked for a significant amount of time. So, the notion of a non-blocking API implies a highly responsive application. It can process the request synchronously or asynchronously—it does not matter for the client. In practice, though, this typically means that the application uses asynchronous processing that facilitates an increased throughput and improved performance.

The term **non-blocking** came into use with the `java.nio` package. The **non-blocking input/output** (**NIO**) provides support for intensive input/output operations. It describes how the application is implemented: it does not dedicate an execution thread to each of the requests but provides several lightweight worker threads that do the processing asynchronously and concurrently.

The java.io package versus the java.nio package

Writing and reading data to and from an external memory (a hard drive, for example) is a much slower operation than the processes in the memory only. The already-existing classes and interfaces of the `java.io` package worked well, but once in a while became the performance bottleneck. The new `java.nio` package was created to provide a more effective I/O support.

The `java.io` implementation is based on I/O stream processing. As we have seen in the previous section, it is basically a blocking operation even if some kind of concurrency is happening behind the scenes. To increase the speed, the `java.nio` implementation was introduced based on the reading/writing to/from a buffer in the memory. Such a design allowed it to separate the slow process of filling/emptying the buffer and quickly reading/writing from/to it.

In a way, it is similar to what we have done in our example of `CompletableFuture` usage. The additional advantage of having data in a buffer is that it is possible to inspect the data, going there and back along the buffer, which is impossible while reading sequentially from the stream. It has provided more flexibility during data processing. In addition, the `java.nio` implementation introduced another middleman process called a **channel** for the bulk data transfer to and from a buffer.

The reading thread is getting data from a channel and only receives what is currently available, or nothing at all (if no data is in the channel). If data is not available, the thread, instead of remaining blocked, can do something else – reading/writing to/from other channels, for example, the same way the main thread in our `CompletableFuture` example was free to do whatever had to be done while the measuring system was reading data from its sensors.

This way, instead of dedicating a thread to one I/O process, a few worker threads can serve many I/O processes. Such a solution was called a **non-blocking I/O** and later was applied to other processes, the most prominent being the *event processing in an event loop*, also called a **run loop**.

The event/run loop

Many non-blocking systems are based on the **event** (or **run**) loop – a thread that is continually executed. It receives events (requests, messages) and then dispatches them to the corresponding event handlers (workers). There is nothing special about the event handlers. They are just methods (functions) dedicated by the programmer for the processing of the particular event type.

Such a design is called a **reactor design pattern**. It is constructed around processing concurrent events and service requests. It also gave the name to the **reactive programming** and **reactive systems** that *react* to events and process them concurrently.

Event loop-based design is widely used in operating systems and in graphical user interfaces. It is available in Spring WebFlux in Spring 5 and implemented in JavaScript, and its popular executing environment Node.JS. The last uses an event loop as its processing backbone. The toolkit, Vert.x, is built around the event loop too.

Before the adoption of an event loop, a dedicated thread was assigned to each incoming request – much like in our demonstration of stream processing. Each of the threads required the allocation of a certain amount of resources that were not request-specific, so some of the resources – mostly memory allocation – were wasted. Then, as the number of requests grew, the CPU needed to switch its context from one thread to another more often to allow more-or-less concurrent processing of all the requests. Under the load, the overhead of switching the context is substantial enough to affect the performance of an application.

Implementing an event loop has addressed these two issues. It eliminated the waste of the resources by avoiding the creation of a thread dedicated to each request and keeping it around until the request is processed. With an event loop in place, only much smaller memory allocation is needed for each request to capture its specifics, which makes it possible to keep far more requests in memory so that they can be processed concurrently. The overhead of the CPU context-switching has become much smaller too, because of the diminishing context size.

The non-blocking API is the way the processing of the requests is implemented. It makes the systems able to handle a much bigger load while remaining highly responsive and resilient.

Reactive

The term **reactive** is usually used in the context of reactive programming and reactive systems. The reactive programming (also called Rx programming) is based on asynchronous data streams (also called **reactive streams**). It was introduced as **Reactive Extension (RX)** of Java, also called **RxJava** (http://reactivex.io). Later, the RX support was added to Java 9 in the `java.util.concurrent` package. It allows a `Publisher` to generate a stream of data, to which a `Subscriber` can asynchronously subscribe.

One principal difference between reactive streams and standard streams (also called **Java 8 streams** located in the `java.util.stream` package) is that a source (publisher) of the reactive stream pushes elements to subscribers at its own rate, while in standard streams, a new element is pulled and emitted only after the previous one was processed.

As you have seen, we were able to process data asynchronously even without this new API, by using `CompletableFuture`. But after writing such a code a few times, you notice that most of the code is just plumbing, so you get a feeling there has to be an even simpler and more convenient solution. That's how the Reactive Streams initiative (`http://www.reactive-streams.org`) was born. The scope of the effort was defined as follows:

> *"The scope of Reactive Streams is to find a minimal set of interfaces, methods, and protocols that will describe the necessary operations and entities to achieve the goal – asynchronous streams of data with non-blocking back pressure."*

The term **non-blocking back pressure** refers to one of the problems of asynchronous processing: coordinating of the speed rate of the incoming data with the ability of the system to process them without the need for stopping (blocking) the data input. The solution is to inform the source that the consumer has difficulty in keeping up with the input. Also, the processing should react to the change of the rate of the incoming data in a more flexible manner than just blocking the flow, thus the name *reactive*.

There are several libraries already that implement the Reactive Streams API: RxJava (`http://reactivex.io`), Reactor (`https://projectreactor.io`), Akka Streams (`https://akka.io/docs`), and Vert.x (`https://vertx.io/`) are among the most known. Writing code using RxJava or another library of asynchronous streams constitutes *reactive programming*. It realizes the goal declared in the Reactive Manifesto (`https://www.reactivemanifesto.org`) as building reactive systems that are *responsive, resilient, elastic*, and *message-driven*.

Responsive

It seems that this term is self-explanatory. The ability to respond in a timely manner is one of the primary qualities of any system. There are many ways to achieve it. Even a traditional blocking API supported by enough servers and other infrastructure can achieve decent responsiveness under a growing load.

Reactive programming helps to do it using less hardware. It comes with a price, as reactive code requires changing the way we think about control flow. But after some time, this new way of thinking becomes as natural as any other familiar skill.

We will see quite a few examples of reactive programming in the following sections.

Resilient

Failures are inevitable. The hardware crashes, the software has defects, unexpected data is received, or an untested execution path was taken – any of these events or a combination of them can happen at any time. *Resilience* is the ability of a system to continue delivering the expected results under unexpected circumstances.

It can be achieved using redundancy of the deployable components and hardware, using isolation of parts of the system so the domino effect becomes less probable, designing the system with automatically replaceable parts, raising an alarm so that qualified personnel can interfere, for example. We have also talked about distributed systems as a good example of resilient systems by design.

A distributed architecture eliminates a single point of failure. Also, breaking the system into many specialized components that talk to one another using messages allows better tuning of the duplication of the most critical parts and creates more opportunities for their isolation and the potential failure containment.

Elastic

The ability to sustain the biggest possible load is usually associated with **scalability**. But, the ability to preserve the same performance characteristics under a varying load, not under the growing one only, is called **elasticity**.

A client of an elastic system should not notice any difference between the idle periods and the periods of the peak load. Non-blocking reactive style of implementation facilitates this quality. Also, breaking the program into smaller parts and converting them into services that can be deployed and managed independently allows for the fine-tuning of the resources allocation.

Such small services are called microservices, and many of them together can comprise a reactive system that can be both scalable and elastic. We will talk about such architecture in more details in the following sections and the next chapter.

Message-driven

We have established already that components isolation and system distribution are two aspects that help to keep the system responsive, resilient, and elastic. Loose and flexible connections are important conditions that support these qualities too. And, the asynchronous nature of the reactive system simply does not leave the designer other choices but to build communication between components on messages.

It creates a breathing space around each component without which the system would be a tightly coupled monolith susceptible to all kinds of problems, not to mention a maintenance nightmare.

In the next chapter, we are going to look at the architectural style that can be used to build an application as a collection of loosely-coupled microservices that communicate using messages.

Reactive streams

The Reactive Streams API introduced in Java 9 consists of the following four interfaces:

```
@FunctionalInterface
public static interface Flow.Publisher<T> {
    public void subscribe(Flow.Subscriber<T> subscriber);
}
public static interface Flow.Subscriber<T> {
    public void onSubscribe(Flow.Subscription subscription);
    public void onNext(T item);
    public void onError(Throwable throwable);
    public void onComplete();
}
public static interface Flow.Subscription {
    public void request(long numberOfItems);
    public void cancel();
}
public static interface Flow.Processor<T,R>
            extends Flow.Subscriber<T>, Flow.Publisher<R> {
}
```

A `Flow.Subscriber` object can be passed as a parameter into the `subscribe()` method of `Flow.Publisher<T>`. The publisher then calls the subscriber's `onSubscribe()` method and passes to it as a parameter a `Flow.Subsctiption` object. Now, the subscriber can call `request(long numberOfItems)` on the subscription object to request data from the publisher. That is the way the **pull model** can be implemented, which leaves it to a subscriber to decide when to request another item for processing. The subscriber can unsubscribe from the publisher services by calling the `cancel()` method on subscription.

In return, the publisher can pass to the subscriber a new item by calling the subscriber's `onNext()` method. When no more data will be coming (all the data from the source were emitted) the publisher calls the subscriber's `onComplete()` method. Also, by calling the subscriber's `onError()` method, the publisher can tell the subscriber that it has encountered a problem.

The `Flow.Processor` interface describes an entity that can act as both a subscriber and a publisher. It allows creating chains (pipelines) of such processors, so a subscriber can receive an item from a publisher, transform it, then pass the result to the next subscriber or processor.

In a push model, the publisher can call `onNext()` without any request from the subscriber. If the rate of processing is lower than the rate of the item publishing, the subscriber can use various strategies to relieve the pressure. For example, it can skip the items or create a buffer for temporary storage with the hope that the item production will slow down and the subscriber will be able to catch up.

This is the minimal set of interfaces the Reactive Streams initiative has defined in support of the asynchronous data streams with non-blocking back pressure. As you can see, it allows the subscriber and publisher to talk to each other and coordinate the rate of incoming data, thus making possible a variety of solutions for the back pressure problem we discussed in the *Reactive* section.

There are many ways to implement these interfaces. Currently, in JDK 9, there is only one implementation of one of the interfaces: the `SubmissionPublisher` class implements `Flow.Publisher`. The reason for that is that these interfaces are not supposed to be used by an application developer. It is a **Service Provider Interface (SPI)** that is used by the developers of the reactive streams libraries. If need be, use one of the already-existing toolkits that have implemented the Reactive Streams API we have mentioned already: RxJava, Reactor, Akka Streams, Vert.x, or any other library of your preference.

RxJava

We will use **RxJava 2.2.7** (http://reactivex.io) in our examples. It can be added to the project by the following dependency:

```
<dependency>
    <groupId>io.reactivex.rxjava2</groupId>
    <artifactId>rxjava</artifactId>
    <version>2.2.7</version>
</dependency>
```

Let's first compare two implementations of the same functionality using the java.util.stream package and the io.reactivex package. The sample program is going to be very simple:

- Create a stream of integers 1,2,3,4,5.
- Filter only even numbers (2 and 4).
- Calculate a square root of each of the filtered numbers.
- Calculate the sum of all the square roots.

Here is how it can be implemented using the java.util.stream package:

```
double a = IntStream.rangeClosed(1, 5)
                    .filter(i -> i % 2 == 0)
                    .mapToDouble(Double::valueOf)
                    .map(Math::sqrt)
                    .sum();
System.out.println(a);         //prints: 3.414213562373095
```

And, the same functionality implemented with RxJava looks as follows:

```
Observable.range(1, 5)
          .filter(i -> i % 2 == 0)
          .map(Math::sqrt)
          .reduce((r, d) -> r + d)
          .subscribe(System.out::println);   //prints: 3.414213562373095
```

RxJava is based on the Observable object (which plays the role of Publisher) and Observer that subscribes to the Observable and waits for the data to be emitted.

By contrast, with the Stream functionality, Observable has significantly different capabilities. For example, a stream, once closed, cannot be reopened, while an Observable object can be used again. Here is an example:

```
Observable<Double> observable = Observable.range(1, 5)
        .filter(i -> i % 2 == 0)
        .doOnNext(System.out::println)     //prints 2 and 4 twice
        .map(Math::sqrt);
observable
        .reduce((r, d) -> r + d)
        .subscribe(System.out::println);   //prints: 3.414213562373095
observable
        .reduce((r, d) -> r + d)
        .map(r -> r / 2)
        .subscribe(System.out::println);   //prints: 1.7071067811865475
```

In the preceding example, as you can see from the comments, the doOnNext() operation was called twice, which means the observable object emitted values twice, once for each processing pipeline:

```
2
4
3.414213562373095
2
4
1.7071067811865475
```

If we do not want Observable running twice, we can cache its data, by adding the cache() operation:

```
Observable<Double> observable = Observable.range(1,5)
        .filter(i -> i % 2 == 0)
        .doOnNext(System.out::println)   //prints 2 and 4 only once
        .map(Math::sqrt)
        .cache();
observable
        .reduce((r, d) -> r + d)
        .subscribe(System.out::println); //prints: 3.414213562373095
observable
        .reduce((r, d) -> r + d)
        .map(r -> r / 2)
        .subscribe(System.out::println);   //prints: 1.7071067811865475
```

As you can see, the second usage of the same `Observable` took advantage of the cached data, thus allowing for better performance:

```
2
4
3.414213562373095
1.7071067811865475
```

RxJava provides such a rich functionality that there is no way we can review it all in detail in this book. Instead, we will try to cover the most popular API. The API describes the methods available for invocation using an `Observable` object. Such methods are often also called **operations** (as in the case with the standard Java 8 streams too) or **operators** (mostly used in connection to reactive streams). We will use these three terms, methods, operations, and operators, interchangeably as synonyms.

Observable types

Talking about RxJava 2 API (notice that is it quite different than RxJava 1), we will use the online documentation that can be found at `http://reactivex.io/RxJava/2.x/javadoc/index.html`.

An observer subscribes to receive values from an observable object, which can behave as one of the following types:

- **Blocking**: Waiting until the result is returned
- **Non-blocking**: Processing the emitted elements asynchronously
- **Cold**: Emitting an element at the observer's request
- **Hot**: Emitting elements whether an observer has subscribed or not

An observable object can be an object of one of the following classes of the `io.reactivex` package:

- `Observable<T>`: Can emit none, one, or many elements; does not support backpressure.
- `Flowable<T>`: Can emit none, one, or many elements; supports backpressure.
- `Single<T>`: Can emit either one element or error; the notion of backpressure does not apply.

- `Maybe<T>`: Represents a deferred computation; can emit either no value, one value, or error; the notion of backpressure does not apply.
- `Completable`: Represents a deferred computation without any value; indicates completion of the task or error; the notion of backpressure does not apply.

An object of each of these classes can behave as blocking, non-blocking, cold, or a hot observable. They are different because of the number of values that can be emitted, an ability to defer the returning of the result, or returning the flag of the task completion only, and because of their ability to handle backpressure.

Blocking versus non-blocking

To demonstrate this behavior, we create an observable that emits five sequential integers, starting with 1:

```
Observable<Integer> obs = Observable.range(1,5);
```

All the blocking methods (operators) of `Observable` starts with the "blocking", so the `blockingLast()` is one of the blocking operators, which blocks the pipeline until the last of the elements is emitted:

```
Double d2 = obs.filter(i -> i % 2 == 0)
               .doOnNext(System.out::println)   //prints 2 and 4
               .map(Math::sqrt)
               .delay(100, TimeUnit.MILLISECONDS)
               .blockingLast();
System.out.println(d2);                          //prints: 2.0
```

In this example, we select only even numbers, print the selected element, then calculate the square root and wait for 100 ms (imitating a long-running calculation). The result of this example looks as follows:

The non-blocking version of the same functionality looks as follows:

```
List<Double> list = new ArrayList<>();
obs.filter(i -> i % 2 == 0)
   .doOnNext(System.out::println)   //prints 2 and 4
   .map(Math::sqrt)
   .delay(100, TimeUnit.MILLISECONDS)
```

```
    .subscribe(d -> {
        if(list.size() == 1){
            list.remove(0);
        }
        list.add(d);
    });
System.out.println(list);              //prints: []
```

We use the `List` object to capture the result because, as you may remember, the lambda expression does not allow using the non-final variables.

As you can see, the resulting list is empty. That is because the pipeline calculations are performed without blocking (asynchronously). So, because of the delay of 100 ms, the control meanwhile went down to the last line that prints the list content, which is still empty. We can set a delay in front of the last line:

```
try {
    TimeUnit.MILLISECONDS.sleep(200);
} catch (InterruptedException e) {
    e.printStackTrace();
}
System.out.println(list);    //prints: [2.0]
```

The delay has to be 200 ms at least because the pipeline processes two elements, each with 100 ms delay. Now, you can see, the list contains an expected value of 2.0.

That is basically the difference between blocking and non-blocking operators. Other classes, that represent an `observable`, have similar blocking operators. Here are examples of blocking `Flowable`, `Single`, and `Maybe`:

```
Flowable<Integer> obs = Flowable.range(1,5);
Double d2 = obs.filter(i -> i % 2 == 0)
        .doOnNext(System.out::println)   //prints 2 and 4
        .map(Math::sqrt)
        .delay(100, TimeUnit.MILLISECONDS)
        .blockingLast();
System.out.println(d2);                  //prints: 2.0

Single<Integer> obs2 = Single.just(42);
int i2 = obs2.delay(100, TimeUnit.MILLISECONDS).blockingGet();
System.out.println(i2);                  //prints: 42

Maybe<Integer> obs3 = Maybe.just(42);
int i3 = obs3.delay(100, TimeUnit.MILLISECONDS).blockingGet();
System.out.println(i3);                  //prints: 42
```

The `Completable` class has blocking operators that allow setting a timeout:

```
(1)  Completable obs = Completable.fromRunnable(() -> {
            System.out.println("Running...");       //prints: Running...
            try {
                TimeUnit.MILLISECONDS.sleep(200);
            } catch (InterruptedException e) {
                e.printStackTrace();
            }
      });
(2)  Throwable ex = obs.blockingGet();
(3)  System.out.println(ex);                        //prints: null

//(4)  ex = obs.blockingGet(15, TimeUnit.MILLISECONDS);
//                              java.util.concurrent.TimeoutException:
//              The source did not signal an event for 15 milliseconds.

(5)  ex = obs.blockingGet(150, TimeUnit.MILLISECONDS);
(6)  System.out.println(ex);                        //prints: null

(7)  obs.blockingAwait();
(8)  obs.blockingAwait(15, TimeUnit.MILLISECONDS);
```

The result of the preceding code is presented in the following screenshot:

The first **Run** message comes from line 2 in response to the call of the blocking `blockingGet()` method. The first **null** message comes from line 3. Line 4 throws an exception because the timeout was set to 15 ms, while the actual processing is set to a 100 ms delay. The second **Run** message comes from line 5 in response to the `blockingGet()` method call. This time, the timeout is set to 150 ms, which is more than 100 ms, and the method was able to return before the timeout was up.

The last two lines, 7 and 8, demonstrate the usage of the `blockingAwait()` method with and without a timeout. This method does not return a value but allows the observable pipeline to run its course. It is interesting to notice that it does not break with an exception even when the timeout is set to a smaller value than the time the pipelines takes to finish. Apparently, it starts its waiting after the pipeline has finished processing unless it is a defect that will be fixed later (the documentation is not clear on this point).

Although the blocking operations exist (and we will review more of such while talking about each observable type in the following sections), they are and should be used only in the cases when it is not possible to implement the required functionality using non-blocking operations only. The main thrust of reactive programming is to strive to process all requests asynchronously in a non-blocking style.

Cold versus hot

All the examples we have seen so far demonstrated only a `cold` observable, those that provide the next value only at the request of the processing pipeline after the previous value has been processed already. Here is another example:

```
Observable<Long> cold = Observable.interval(10, TimeUnit.MILLISECONDS);
cold.subscribe(i -> System.out.println("First: " + i));
pauseMs(25);
cold.subscribe(i -> System.out.println("Second: " + i));
pauseMs(55);
```

We have used the method `interval()` to create an `Observable` object that represents a stream of sequential numbers emitted every specified interval (in our case, every 10 ms). Then, we subscribed to the created object, wait 25 ms, subscribe again, and wait another 55 ms. The `pauseMs()` method looks as follows:

```
void pauseMs(long ms){
    try {
        TimeUnit.MILLISECONDS.sleep(ms);
    } catch (InterruptedException e) {
        e.printStackTrace();
    }
}
```

If we run the preceding example, the output will be as follows:

```
First: 0
First: 1
First: 2
First: 3
Second: 0
Second: 1
First: 4
Second: 2
First: 5
First: 6
Second: 3
Second: 4
First: 7
```

As you can see, each of the pipelines processed every value emitted by the cold observable.

To convert the *cold* observable into a *hot* one, we use the `publish()` method that converts the observable into a `ConnectableObservable` object that extends the `Observable`:

```
ConnectableObservable<Long> hot =
        Observable.interval(10, TimeUnit.MILLISECONDS).publish();
hot.connect();
hot.subscribe(i -> System.out.println("First: " + i));
pauseMs(25);
hot.subscribe(i -> System.out.println("Second: " + i));
pauseMs(55);
```

As you can see, we have to call the `connect()` method in order that the `ConnectableObservable` object starts emitting values. The output looks like the following:

```
First: 0
First: 1
First: 2
First: 3
Second: 3
First: 4
Second: 4
First: 5
Second: 5
First: 6
Second: 6
First: 7
Second: 7
```

The output shows that the second pipeline did not receive the first three values because it has subscribed to the observable later. So, the observable emits values independent of the ability of the observers to process them. If the processing falls behind, and new values keep coming while previous ones are not fully processed yet, the `Observable` class puts them into a buffer. If this buffer grows big enough, the JVM can run out of memory because, as we have mentioned already, the `Observable` class does not have an ability of backpressure management.

For such cases, the `Flowable` class is a better candidate for the observable because it does have an ability to handle the backpressure. Here is an example:

```
PublishProcessor<Integer> hot = PublishProcessor.create();
hot.observeOn(Schedulers.io(), true)
    .subscribe(System.out::println, Throwable::printStackTrace);
```

```
for (int i = 0; i < 1_000_000; i++) {
    hot.onNext(i);
}
```

The `PublishProcessor` class extends `Flowable` and has the `onNext(Object o)` method that forces it to emit the passed-in object. Before calling it, we have subscribed to the observable using the `Schedulers.io()` thread. We will talk about the schedulers in the *Multithreading (Scheduler)* section.

The `subscribe()` method has several overloaded versions. We decided to use the one that accepts two `Consumer` functions: the first one processes the passed-in value, the second processes an exception if it was thrown by any of the pipeline operations (similar to a `Catch` block).

If we run the preceding example, it will print successfully the first 127 values and then will throw `MissingBackpressureException`, as shown on the following screenshot:

```
126
127
io.reactivex.exceptions.MissingBackpressureException: Could not emit value due to lack of requests
    at io.reactivex.processors.PublishProcessor$PublishSubscription.onNext(PublishProcessor.java:364)
    at io.reactivex.processors.PublishProcessor.onNext(PublishProcessor.java:243)
    at com.packt.learnjava.ch15_reactive.HotObservable.hot(HotObservable.java:31)
    at com.packt.learnjava.ch15_reactive.HotObservable.main(HotObservable.java:14)
```

The message in the exception provides a clue: `Could not emit value due to lack of requests`. Apparently, the rate of emitting the values is higher than the rate of consuming them and an internal buffer can keep only 128 elements. If we add a delay (simulating a longer processing time), the result will be even worse:

```
PublishProcessor<Integer> hot = PublishProcessor.create();
hot.observeOn(Schedulers.io(), true)
    .delay(10, TimeUnit.MILLISECONDS)
    .subscribe(System.out::println, Throwable::printStackTrace);
for (int i = 0; i < 1_000_000; i++) {
    hot.onNext(i);
}
```

Even the first 128 elements will not get through and the output will have only `MissingBackpressureException`.

To address the issue, a backpressure strategy has to be set. For example, let's drop every value that the pipeline did not manage to process:

```
PublishProcessor<Integer> hot = PublishProcessor.create();
hot.onBackpressureDrop(v -> System.out.println("Dropped: "+ v))
```

```
        .observeOn(Schedulers.io(), true)
        .subscribe(System.out::println, Throwable::printStackTrace);
    for (int i = 0; i < 1_000_000; i++) {
        hot.onNext(i);
    }
}
```

Notice that the strategy has to be set before the `observeOn()` operation, so it will be picked up by the created `Schedulers.io()` thread.

The output shows that many of the emitted values were dropped. Here is an output fragment:

We will talk about other backpressure strategies in the *Operators* section, in the overview of the corresponding operators.

Disposable

Please notice that a `subscribe()` method actually returns a `Disposable` object that can be queried to check whether the pipeline processing has been completed (and disposed of):

```
Observable<Integer> obs = Observable.range(1,5);
List<Double> list = new ArrayList<>();
Disposable disposable =
    obs.filter(i -> i % 2 == 0)
        .doOnNext(System.out::println)      //prints 2 and 4
        .map(Math::sqrt)
        .delay(100, TimeUnit.MILLISECONDS)
        .subscribe(d -> {
            if(list.size() == 1){
                list.remove(0);
            }
            list.add(d);
        });
System.out.println(list);                   //prints: []
```

```
System.out.println(disposable.isDisposed()); //prints: false
try {
    TimeUnit.MILLISECONDS.sleep(200);
} catch (InterruptedException e) {
    e.printStackTrace();
}
System.out.println(disposable.isDisposed());   //prints: true
System.out.println(list);                      //prints: [2.0]
```

It is also possible to enforce the disposing of a pipeline, thus effectively canceling the processing:

```
Observable<Integer> obs = Observable.range(1,5);
List<Double> list = new ArrayList<>();
Disposable disposable =
    obs.filter(i -> i % 2 == 0)
        .doOnNext(System.out::println)        //prints 2 and 4
        .map(Math::sqrt)
        .delay(100, TimeUnit.MILLISECONDS)
        .subscribe(d -> {
            if(list.size() == 1){
                list.remove(0);
            }
            list.add(d);
        });
System.out.println(list);                      //prints: []
System.out.println(disposable.isDisposed()); //prints: false
disposable.dispose();
try {
    TimeUnit.MILLISECONDS.sleep(200);
} catch (InterruptedException e) {
    e.printStackTrace();
}
System.out.println(disposable.isDisposed()); //prints: true
System.out.println(list);                      //prints: []
```

As you can see, by adding the call to disposable.dispose(), we have stopped processing: the list content, even after a 200 ms delay, remains empty (the last line of the preceding example).

This method of forced disposal can be used to make sure that there are no run-away threads. Each created Disposable object can be disposed of in the same way the resources are released in a finally block. The CompositeDisposable class helps to handle multiple Disposable objects in a coordinated manner.

When an onComplete or onError event happens, the pipeline is disposed of automatically.

For example, you can use the `add()` method and add a newly created `Disposable` object to the `CompositeDisposable` object. Then, when necessary, the `clear()` method can be invoked on the `CompositeDisposable` object. It will remove the collected `Disposable` objects and call the `dispose()` method on each of them.

Creating an observable

You have seen already a few methods of creating an observable in our examples. There are many other factory methods in `Observable`, `Flowable`, `Single`, `Maybe`, and `Completable`. Not all of the following methods are available in each of these interfaces though (see the comments; *all* means that all of the listed interfaces have it):

- `create()`: Creates an `Observable` object by providing the full implementation (all)
- `defer()`: Creates a new `Observable` object every time a new `Observer` subscribes (all)
- `empty()`: Creates an empty `Observable` object that completes immediately upon subscription (all, except `Single`)
- `never()`: Creates an `Observable` object that does not emit anything and does nothing at all; does not even complete (all)
- `error()`: Creates an `Observable` object that emits an exception immediately upon subscription (all)
- `fromXXX()`: Creates an `Observable` object, where XXX can be *Callable*, *Future* (all), *Iterable, Array, Publisher* (`Observable` and `Flowable`), *Action*, *Runnable* (`Maybe` and `Completable`); which means it creates an `Observable` object based on the provided function or object
- `generate()`: Creates a cold `Observable` object that generates values based on the provided function or object (`Observable` and `Flowable` only)
- `range()`, `rangeLong()`, `interval()`, `intervalRange()`: Creates an `Observable` object that emits sequential `int` or `long` values limited or not by the specified range and spaced by the specified time interval (`Observable` and `Flowable` only)
- `just()`: Creates an `Observable` object based on the provided object or a set of objects (all, except `Completable`)
- `timer()`: Creates an `Observable` object that, after the specified time, emits an 0L signal (all) and then completes for `Observable` and `Flowable`

There are also many other helpful methods, such as `repeat()`, `startWith()`, and similar. We just do not have enough space to list all of them. Refer to the online documentation (`http://reactivex.io/RxJava/2.x/javadoc/index.html`).

Let's look at an example of the `create()` method usage. The `create()` method of `Observable` looks as follows:

```
public static Observable<T> create(ObservableOnSubscribe<T> source)
```

The passed-in object has to be an implementation of the functional interface `ObservableOnSubscribe<T>` that has only one abstract method, `subscribe()`:

```
void subscribe(ObservableEmitter<T> emitter)
```

The `ObservableEmitter<T>` interface contains the following methods:

- `boolean isDisposed()`: Returns `true` if the processing pipeline was disposed or the emitter was terminated
- `ObservableEmitter<T> serialize()`: Provides the serialization algorithm used by the calls to `onNext()`, `onError()`, and `onComplete()`, located in the base class `Emitter`
- `void setCancellable(Cancellable c)`: Sets on this emitter a `Cancellable` implementation (a functional interface that has only one method, `cancel()`)
- `void setDisposable(Disposable d)`: Sets on this emitter a `Disposable` implementation (an interface that has two methods: `isDispose()` and `dispose()`)
- `boolean tryOnError(Throwable t)`: Handles the error condition, attempts to emit the provided exception, and returns `false` if the emission is not allowed

To create an observable, all the preceding interfaces can be implemented as follows:

```
ObservableOnSubscribe<String> source = emitter -> {
    emitter.onNext("One");
    emitter.onNext("Two");
    emitter.onComplete();
};
Observable.create(source)
        .filter(s -> s.contains("w"))
        .subscribe(v -> System.out.println(v),
                e -> e.printStackTrace(),
                () -> System.out.println("Completed"));
pauseMs(100);
```

Let's look at the preceding example closer. We have created an `ObservableOnSubscribe` function `source` and implemented the emitter: we told the emitter to emit **One** at the first call to `onNext()`, to emit **Two** at the second call to `onNext()`, and then to call `onComplete()`. We have passed the `source` function into the `create()` method and built the pipeline to process all the emitted values.

To make it more interesting, we have added the `filter()` operator that allows propagating further only the values with the *w* character. We have also chosen the `subscribe()` method version with three parameters: the functions `Consumer onNext`, `Consumer onError`, and `Action onComplete`. The first is called every time a next value reached the method, the second is called when an exception was emitted, and the third is called when the source emits an `onComplete()` signal. After creating the pipeline, we have paused for 100 ms to give the asynchronous process a chance to finish. The result looks as follows:

If we remove the line `emitter.onComplete()` from the emitter implementation, only message **Two** will be displayed.

Those are the basics of how the `create()` method can be used. As you can see, it allows for a full customization. In practice, it is rarely used because there are many simpler ways to create an observable. We review them in the following sections.

You will see examples of other factory methods used in our examples throughout other sections of this chapter.

Operators

There are literally hundreds (if we count all the overloaded versions) of operators available in each of the observable interfaces, `Observable`, `Flowable`, `Single`, `Maybe`, or `Completable`.

In `Observable` and `Flowable`, the number of methods goes beyond 500. That is why in this section we are going to provide just an overview and a few examples that will help the reader to navigate the maze of possible options.

To help see the big picture, we have grouped all the operators into ten categories: transforming, filtering, combining, converting from XXX, exceptions handling, life cycle events handling, utilities, conditional and boolean, backpressure, and connectable.

Please notice, these are not all the operators available. You can see more in the online documentation (http://reactivex.io/RxJava/2.x/javadoc/index.html).

Transforming

These operators transform the values emitted by an observable:

- buffer(): Collects the emitted values into bundles according to the provided parameters or using the provided functions, and emits these bundles periodically one at a time
- flatMap(): Produces observables based on the current observable and inserts them into the current flow, one of the most popular operators
- groupBy(): Divides the current Observable into groups of observables (GroupedObservables objects)
- map(): Transforms the emitted value using the provided function
- scan(): Applies the provided function to each value in combination with the value produced as the result of the previous application of the same function to the previous value
- window(): Emits groups of values similar to buffer() but as observables, each of which emits a subset of values from the original observable and then terminates with an onCompleted()

The following code demonstrates the use of map(), flatMap(), and groupBy():

```
Observable<String> obs = Observable.fromArray("one", "two");

obs.map(s -> s.contains("w") ? 1 : 0)
    .forEach(System.out::print);                    //prints: 01

List<String> os = new ArrayList<>();
List<String> noto = new ArrayList<>();
obs.flatMap(s -> Observable.fromArray(s.split("")))
        .groupBy(s -> "o".equals(s) ? "o" : "noto")
        .subscribe(g -> g.subscribe(s -> {
            if (g.getKey().equals("o")) {
                os.add(s);
            } else {
                noto.add(s);
            }
        }));
```

```
System.out.println(os);                              //prints: [o, o]
System.out.println(noto);                            //prints: [n, e, t, w]
```

Filtering

The following operators (and their multiple overloaded versions) select which of the values will continue to flow through the pipeline:

- `debounce()`: Emits a value only when a specified time-span has passed without the observable emitting another value
- `distinct()`: Selects only unique values
- `elementAt(long n)`: Emits only one value with the specified n position in the stream
- `filter()`: Emits only the values that match the specified criteria
- `firstElement()`: Emits only the first value
- `ignoreElements()`: Does not emit values; only the `onComplete()` signal goes through
- `lastElement()`: Emits only the last value
- `sample()`: Emits the most recent value emitted within the specified time interval
- `skip(long n)`: Skips the first n values
- `take(long n)`: Emits only the first n values

The following are examples of some of the just-listed operators' uses:

```
Observable<String> obs = Observable.just("onetwo")
        .flatMap(s -> Observable.fromArray(s.split("")));
// obs emits "onetwo" as characters
obs.map(s -> {
            if("t".equals(s)){
                NonBlockingOperators.pauseMs(15);
            }
            return s;
        })
        .debounce(10, TimeUnit.MILLISECONDS)
        .forEach(System.out::print);                 //prints: eo
obs.distinct().forEach(System.out::print);           //prints: onetw
obs.elementAt(3).subscribe(System.out::println);     //prints: t
obs.filter(s -> s.equals("o"))
    .forEach(System.out::print);                     //prints: oo
obs.firstElement().subscribe(System.out::println);   //prints: o
obs.ignoreElements().subscribe(() ->
        System.out.println("Completed!"));           //prints: Completed!
```

```
Observable.interval(5, TimeUnit.MILLISECONDS)
    .sample(10, TimeUnit.MILLISECONDS)
    .subscribe(v -> System.out.print(v + " "));      //prints: 1 3 4 6 8
pauseMs(50);
```

Combining

The following operators (and their multiple overloaded versions) create a new observable using multiple source observables:

- `concat(src1, src2)`: Creates an `Observable` that emits all values of `src1`, then all values of `src2`

- `combineLatest(src1, src2, combiner)`: Creates an `Observable` that emits a value emitted by either of two sources combined with the latest value emitted by each source using the provided function combiner

- `join(src2, leftWin, rightWin, combiner)`: Combines values emitted by two observables during the `leftWin` and `rightWin` time windows according to the `combiner` function

- `merge()`: Combines multiple observables into one; in contrast to `concat()`, it may interleave them, whereas `concat()` never interleaves the emitted values from different observables

- `startWith(T item)`: Adds the specified value before emitting values from the source observable

- `startWith(Observable<T> other)`: Adds the values from the specified observable before emitting values from the source observable

- `switchOnNext(Observable<Observable> observables)`: Creates a new `Observable` that emits the most-recently emitted values of the specified observables

- `zip()`: Combines the values of the specified observables using the provided function

The following code demonstrates the use of some of these operators:

```
Observable<String> obs1 = Observable.just("one")
                      .flatMap(s -> Observable.fromArray(s.split("")));
Observable<String> obs2 = Observable.just("two")
                      .flatMap(s -> Observable.fromArray(s.split("")));
Observable.concat(obs2, obs1, obs2)
        .subscribe(System.out::print);                //prints: twoonetwo
Observable.combineLatest(obs2, obs1, (x,y) -> "("+x+y+")")
        .subscribe(System.out::print);                //prints: (oo)(on)(oe)
```

```
System.out.println();
obs1.join(obs2, i -> Observable.timer(5, TimeUnit.MILLISECONDS),
                i -> Observable.timer(5, TimeUnit.MILLISECONDS),
          (x,y) -> "("+x+y+")").subscribe(System.out::print);
                    //prints: (ot)(nt)(et)(ow)(nw)(ew)(oo)(no)(eo)
Observable.merge(obs2, obs1, obs2)
        .subscribe(System.out::print);              //prints: twoonetwo
obs1.startWith("42")
    .subscribe(System.out::print); //prints: 42one
Observable.zip(obs1, obs2, obs1,  (x,y,z) -> "("+x+y+z+")")
        .subscribe(System.out::print);       //prints: (oto)(nwn)(eoe)
```

Converting from XXX

These operators are pretty straightforward. Here is the list of from-XXX operators of the `Observable` class:

- `fromArray(T... items)`: Creates an `Observable` from a varargs
- `fromCallable(Callable<T> supplier)`: Creates an `Observable` from a `Callable` function
- `fromFuture(Future<T> future)`: Creates an `Observable` from a `Future` object
- `fromFuture(Future<T> future, long timeout, TimeUnit unit)`: Creates an `Observable` from a `Future` object with the timeout parameters applied to the `future`
- `fromFuture(Future<T> future, long timeout, TimeUnit unit, Scheduler scheduler)`: Creates an `Observable` from a `Future` object with the timeout parameters applied to the `future` and the scheduler (`Schedulers.io()` is recommended, see the *Multithreading (Scheduler)* section)
- `fromFuture(Future<T> future, Scheduler scheduler)`: Creates an `Observable` from a `Future` object on the specified scheduler (`Schedulers.io()` is recommended, see the *Multithreading (Scheduler)* section)
- `fromIterable(Iterable<T> source)`: Creates an `Observable` from an iterable object (`List`, for example)
- `fromPublisher(Publisher<T> publisher)`: Creates an `Observable` from a `Publisher` object

Exceptions handling

The subscribe() operator has an overloaded version that accepts the Consumer<Throwable> function that handles any exception raised anywhere in the pipeline. It works similar to the all-embracing try-catch block. If you have this function passed into the subscribe() operator, you can be sure that is the only place where all exceptions will end up.

But, if you need to handle the exceptions in the middle of the pipeline, so the values flow can be recovered and processed by the rest of the operators, after the operator that has thrown the exception, the following operators (and their multiple overloaded versions) can help with that:

- onErrorXXX(): Resumes the provided sequence when an exception was caught; XXX indicates what the operator does: onErrorResumeNext(), onErrorReturn(), or onErrorReturnItem()
- retry(): Creates an Observable that repeats the emissions emitted from the source; re-subscribes to the source Observable if it calls onError()

The demo code looks as follows:

```
Observable<String> obs = Observable.just("one")
                    .flatMap(s -> Observable.fromArray(s.split("")));
Observable.error(new RuntimeException("MyException"))
        .flatMap(x -> Observable.fromArray("two".split("")))
        .subscribe(System.out::print,
         e -> System.out.println(e.getMessage())//prints: MyException
        );
Observable.error(new RuntimeException("MyException"))
        .flatMap(y -> Observable.fromArray("two".split("")))
        .onErrorResumeNext(obs)
        .subscribe(System.out::print);          //prints: one
Observable.error(new RuntimeException("MyException"))
        .flatMap(z -> Observable.fromArray("two".split("")))
        .onErrorReturnItem("42")
        .subscribe(System.out::print);          //prints: 42
```

Life cycle events handling

These operators are invoked each on a certain event that happened anywhere in the pipeline. They work similarly to the operators described in the *Exceptions handling* section.

The format of these operators is doXXX(), where XXX is the name of the event: onComplete, onNext, onError, and similar. Not all of them are available in all the classes, and some of them are slightly different in Observable, Flowable, Single, Maybe, or Completable. But, we do not have space to list all the variations of all these classes and will limit our overview to a few examples of the life cycle events-handling operators of the Observable class:

- doOnSubscribe(Consumer<Disposable> onSubscribe): Executes when an observer subscribes
- doOnNext(Consumer<T> onNext): Applies the provided Consumer function when the source observable calls onNext
- doAfterNext(Consumer<T> onAfterNext): Applies the provided Consumer function to the current value after it is pushed downstream
- doOnEach(Consumer<Notification<T>> onNotification): Executes the Consumer function for each emitted value
- doOnEach(Observer<T> observer): Notifies an Observer for each emitted value and the terminal event it emits
- doOnComplete(Action onComplete): Executes the provided Action function after the source observable generates the onComplete event
- doOnDispose(Action onDispose): Executes the provided Action function after the pipeline was disposed of by the downstream
- doOnError(Consumer<Throwable> onError): Executes when the onError event is sent
- doOnLifecycle(Consumer<Disposable> onSubscribe, Action onDispose): Calls the corresponding onSubscribe or onDispose function for the corresponding event
- doOnTerminate(Action onTerminate): Executes the provided Action function when the source observable generates the onComplete event or an exception (onError event) is raised
- doAfterTerminate(Action onFinally): Executes the provided Action function after the source observable generates the onComplete event or an exception (onError event) is raised
- doFinally(Action onFinally): Executes the provided Action function after the source observable generates the onComplete event or an exception (onError event) is raised, or the pipeline was disposed of by the downstream

Here is a demo code:

```
Observable<String> obs = Observable.just("one")
            .flatMap(s -> Observable.fromArray(s.split("")));

obs.doOnComplete(() -> System.out.println("Completed!"))
        .subscribe(v -> {
            System.out.println("Subscribe onComplete: " + v);
        });
pauseMs(25);
```

If we run this code, the output will be as follows:

```
Subscribe onNext: o
Subscribe onNext: n
Subscribe onNext: e
Completed!
```

You will also see other examples of these operators' usage in the *Multithreading (scheduler)* section.

Utilities

Various useful operators (and their multiple overloaded versions) can be used for controlling the pipeline behavior:

- `delay()`: Delays the emission by the specified period of time
- `materialize()`: Creates an `Observable` that represents both the emitted values and the notifications sent
- `dematerialize()`: Reverses the result of the `materialize()` operator
- `observeOn()`: Specifies the `Scheduler` (thread) on which the `Observer` should observe the `Observable` (see the *Multithreading (scheduler)* section)
- `serialize()`: Forces serialization of the emitted values and notifications
    - `subscribe()`: Subscribes to the emissions and notifications from an observable; various overloaded versions accept callbacks used for a variety of events, including `onComplete` and `onError`; only after `subscribe()` is invoked the values start flowing through the pipeline

- subscribeOn(): Subscribes the Observer to the Observable asynchronously using the specified Scheduler (see the *Multithreading (scheduler)* section)
- timeInterval(), timestamp(): Converts an Observable<T> that emits values into Observable<Timed<T>>, which, in turn, emits the amount of time elapsed between the emissions or the timestamp correspondingly
- timeout(): Repeats the emissions of the source Observable; generates an error if no emissions happen after the specified period of time
- using(): Creates a resource that is disposed of automatically along with the Observable; works similarly to the try-with-resources construct

The following code contains a few examples of some of these operators used in a pipeline:

```
Observable<String> obs = Observable.just("one")
                    .flatMap(s -> Observable.fromArray(s.split("")));
obs.delay(5, TimeUnit.MILLISECONDS)
    .subscribe(System.out::print);                          //prints: one
pauseMs(10);
System.out.println(); //used here just to break the line
Observable source = Observable.range(1,5);
Disposable disposable = source.subscribe();
Observable.using(
   () -> disposable,
   x -> source,
   y -> System.out.println("Disposed: " + y) //prints: Disposed: DISPOSED
)
.delay(10, TimeUnit.MILLISECONDS)
.subscribe(System.out::print);                              //prints: 12345
pauseMs(25);
```

If we run all these examples, the output will look as follows:

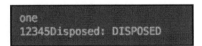

As you can see, the pipeline, when completed, sends the **DISPOSED** signal to the using operator (the third parameter), so the Consumer function we pass as the third parameter can dispose of the resource used by the pipeline.

Conditional and Boolean

The following operators (and their multiple overloaded versions) allow the evaluating of one or more observables or emitted values and change the logic of the processing accordingly:

- `all(Predicate criteria)`: Returns `Single<Boolean>` with a `true` value, if all the emitted values match the provided criteria
- `amb()`: Accepts two or more source observables and emits values from only the first of them that starts emitting
- `contains(Object value)`: Returns `Single<Boolean>` with `true`, if the observable emits the provided value
- `defaultIfEmpty(T value)`: Emits the provided value if the source `Observable` does not emit anything
- `sequenceEqual()`: Returns `Single<Boolean>` with `true`, if the provided sources emit the same sequence; an overloaded version allows to provide the equality function used for comparison
- `skipUntil(Observable other)`: Discards emitted values until the provided `Observable other` emits a value
- `skipWhile(Predicate condition)`: Discards emitted values as long as the provided condition remains `true`
- `takeUntil(Observable other)`: Discards emitted values after the provided `Observable other` emits a value
- `takeWhile(Predicate condition)`: Discards emitted values after the provided condition became `false`

This code contains a few demo examples:

```
Observable<String> obs = Observable.just("one")
                .flatMap(s -> Observable.fromArray(s.split("")));
Single<Boolean> cont = obs.contains("n");
System.out.println(cont.blockingGet());           //prints: true
obs.defaultIfEmpty("two")
   .subscribe(System.out::print);                 //prints: one
Observable.empty().defaultIfEmpty("two")
        .subscribe(System.out::print);            //prints: two

Single<Boolean> equal = Observable.sequenceEqual(obs,
                               Observable.just("one"));
System.out.println(equal.blockingGet());          //prints: false

equal = Observable.sequenceEqual(Observable.just("one"),
```

```
                            Observable.just("one"));
System.out.println(equal.blockingGet());          //prints: true

equal = Observable.sequenceEqual(Observable.just("one"),
                            Observable.just("two"));
System.out.println(equal.blockingGet());          //prints: false
```

Backpressure

We have discussed and demonstrated the **backpressure** effect and the possible drop strategy in *Cold versus hot* section. The other strategy may be as follows:

```
Flowable<Double> obs = Flowable.fromArray(1.,2.,3.);
obs.onBackpressureBuffer().subscribe();
//or
obs.onBackpressureLatest().subscribe();
```

The buffering strategy allows defining the buffer size and providing a function that can be executed if the buffer overflows. The latest strategy tells the values producer to pause (when the consumer cannot process the emitted values on time) and emit the next value on request.

The backpressure operators are available only in the `Flowable` class.

Connectable

The operators of this category allow connecting observables and thus achieve more precisely-controlled subscription dynamics:

- `publish()`: Converts an `Observable` object into a `ConnectableObservable` object
- `replay()`: Returns a `ConnectableObservable` object that repeats all the emitted values and notifications every time a new `Observer` subscribes
- `connect()`: Instructs a `ConnectableObservable` to begin emitting values to the subscribers
- `refCount()`: Converts a `ConnectableObservable` to an `Observable`

We have demonstrated how `ConnectableObservable` works in the *Cold versus hot* section. One principal difference between `ConnectableObservable` and `Observable` is that `ConnectableObservable` does not start emitting value until its `connect` operator is called.

Multithreading (scheduler)

RxJava is single-threaded by default. This means that the source observable and all its operators notify the observers on the same thread on which the subscribe() operator is called.

There are two operators, observeOn() and subscribeOn(), that allow moving the executing of individual actions to a different thread. These methods take as an argument a Scheduler object that, well, schedules the individual actions to be executed on a different thread.

 The subscribeOn() operator declares which scheduler should emit the values.
The observeOn() operator declares which scheduler should observe and process values.

The Schedulers class contains factory methods that create Scheduler objects with different life cycles and performance configuration:

- computation(): Creates a scheduler based on a bounded thread pool with a size up to the number of available processors; it should be used for CPU-intensive computations;
  use Runtime.getRuntime().availableProcessors() to avoid using more this type of schedulers than available processors; otherwise, the performance may degrade because of the overhead of the thread-context switching
- io(): Creates a scheduler based on an unbounded thread pool used for I/O-related work, such as working with files and databases in general when the interaction with the source is blocking by nature; avoid using it otherwise, because it may spin too many threads and negatively affect performance and memory usage
- newThread(): Creates a new thread every time and does not use any pool; it is an expensive way to create a thread, so you are expected to know exactly what is the reason for using it
- single(): Creates a scheduler based on a single thread that executes all the tasks sequentially; useful when the sequence of the execution matters

- `trampoline()`: Creates a scheduler that executes tasks in a first-in-first-out manner; useful for executing recursive algorithms
- `from(Executor executor)`: Creates a scheduler based on the provided executor (thread pool), which allows for better controlling the max number of threads and their life cycles. We talked about thread pools in Chapter 8, *Multithreading and Concurrent Processing*. To remind you, here are the pools we have discussed:

```
Executors.newCachedThreadPool();
Executors.newSingleThreadExecutor();
Executors.newFixedThreadPool(int nThreads);
Executors.newScheduledThreadPool(int poolSize);
Executors.newWorkStealingPool(int parallelism);
```

As you can see, some of the other factory methods of the `Schedulers` class are backed by one of these thread pools and serves just as a simpler and shorter expression of a thread pool declaration. To make the examples simpler and comparable, we are going to use only a `computation()` scheduler. Let's look at the basics of parallel/concurrent processing in RxJava.

The following code is an example of delegating CPU-intensive calculations to dedicated threads:

```
Observable.fromArray("one","two","three")
        .doAfterNext(s -> System.out.println("1: " +
            Thread.currentThread().getName() + " => " + s))
        .flatMap(w -> Observable.fromArray(w.split(""))
                        .observeOn(Schedulers.computation())
        //.flatMap(s -> {
        //      CPU-intensive calculations go here
        // }
            .doAfterNext(s -> System.out.println("2: " +
                    Thread.currentThread().getName() + " => "
+ s))
        )
        .subscribe(s -> System.out.println("3: " + s));
pauseMs(100);
```

In this example, we decided to create a sub-flow of characters from each emitted word and let a dedicated thread process characters of each word. The output of this example looks as follows:

```
3: o
1: main => one
2: RxComputationThreadPool-1 => o
3: n
2: RxComputationThreadPool-1 => n
3: e
2: RxComputationThreadPool-1 => e
1: main => two
3: t
2: RxComputationThreadPool-2 => t
3: w
2: RxComputationThreadPool-2 => w
3: o
2: RxComputationThreadPool-2 => o
1: main => three
3: t
2: RxComputationThreadPool-3 => t
3: h
2: RxComputationThreadPool-3 => h
3: r
2: RxComputationThreadPool-3 => r
3: e
2: RxComputationThreadPool-3 => e
3: e
2: RxComputationThreadPool-3 => e
```

As you can see, the main thread was used to emit the words, and the characters of each word were processed by a dedicated thread. Please notice that although in this example the sequence of the results coming to the subscribe() operation corresponds to the sequence the words and characters were emitted, in real-life cases, the calculation time of each value will not be the same, so there is no guarantee that the results will come in the same sequence.

If need be, we can put each word emission on a dedicated non-main thread too, so the main thread can be free to do what else can be done. For example, note the following:

```
Observable.fromArray("one","two","three")
        .observeOn(Schedulers.computation())
        .doAfterNext(s -> System.out.println("1: " +
                        Thread.currentThread().getName() + " => " + s))
        .flatMap(w -> Observable.fromArray(w.split(""))
                .observeOn(Schedulers.computation())
                .doAfterNext(s -> System.out.println("2: " +
                        Thread.currentThread().getName() + " => " + s))
        )
        .subscribe(s -> System.out.println("3: " + s));
pauseMs(100);
```

The output of this example is as follows:

```
3: o
2: RxComputationThreadPool-2 => o
1: RxComputationThreadPool-1 => one
3: n
2: RxComputationThreadPool-2 => n
3: e
2: RxComputationThreadPool-2 => e
1: RxComputationThreadPool-1 => two
3: t
2: RxComputationThreadPool-3 => t
3: w
2: RxComputationThreadPool-3 => w
1: RxComputationThreadPool-1 => three
3: o
2: RxComputationThreadPool-3 => o
3: t
2: RxComputationThreadPool-4 => t
3: h
2: RxComputationThreadPool-4 => h
3: r
2: RxComputationThreadPool-4 => r
3: e
2: RxComputationThreadPool-4 => e
3: e
2: RxComputationThreadPool-4 => e
```

As you can see, the main thread does not emit the words anymore.

In RxJava 2.0.5, a new simpler way of parallel processing was introduced, similar to the parallel processing in the standard Java 8 streams. Using `ParallelFlowable`, the same functionality can be achieved as follows:

```
ParallelFlowable src =
                    Flowable.fromArray("one","two","three").parallel();
src.runOn(Schedulers.computation())
   .doAfterNext(s -> System.out.println("1: " +
                         Thread.currentThread().getName() + " => " + s))
   .flatMap(w -> Flowable.fromArray(((String)w).split("")))
   .runOn(Schedulers.computation())
   .doAfterNext(s -> System.out.println("2: " +
                         Thread.currentThread().getName() + " => " + s))
   .sequential()
   .subscribe(s -> System.out.println("3: " + s));
pauseMs(100);
```

As you can see, the `ParallelFlowable` object is created by applying
the `parallel()` operator to the regular `Flowable`. Then, the `runOn()` operator tells the
created observable to use the `computation()` scheduler for emitting the values. Please
notice that there is no need anymore to set another scheduler (for processing the characters)
inside the `flatMap()` operator. It can be set outside it – just in the main pipeline, which
makes the code simpler. The result looks like this:

```
1: RxComputationThreadPool-2 => two
1: RxComputationThreadPool-3 => three
1: RxComputationThreadPool-1 => one
2: RxComputationThreadPool-3 => t
2: RxComputationThreadPool-1 => o
2: RxComputationThreadPool-3 => h
2: RxComputationThreadPool-1 => n
2: RxComputationThreadPool-3 => r
2: RxComputationThreadPool-1 => e
2: RxComputationThreadPool-3 => e
2: RxComputationThreadPool-3 => e
3: t
3: o
3: t
3: n
3: h
3: e
3: r
3: e
3: e
2: RxComputationThreadPool-2 => t
3: w
2: RxComputationThreadPool-2 => w
3: o
2: RxComputationThreadPool-2 => o
```

As for the `subscribeOn()` operator, its location in the pipeline does not play any role.
Wherever it is placed, it still tells the observable which scheduler should emit the values.
Here is an example:

```
Observable.just("a", "b", "c")
        .doAfterNext(s -> System.out.println("1: " +
                    Thread.currentThread().getName() + " => " + s))
        .subscribeOn(Schedulers.computation())
        .subscribe(s -> System.out.println("2: " +
                    Thread.currentThread().getName() + " => " + s));
pauseMs(100);
```

The result looks like this:

```
2: RxComputationThreadPool-1 => a
1: RxComputationThreadPool-1 => a
2: RxComputationThreadPool-1 => b
1: RxComputationThreadPool-1 => b
2: RxComputationThreadPool-1 => c
1: RxComputationThreadPool-1 => c
```

Even if we change the location of the subscribeOn() operator as in the following example, the result does not change:

```
Observable.just("a", "b", "c")
        .subscribeOn(Schedulers.computation())
        .doAfterNext(s -> System.out.println("1: " +
                    Thread.currentThread().getName() + " => " + s))
        .subscribe(s -> System.out.println("2: " +
                    Thread.currentThread().getName() + " => " + s));
pauseMs(100);
```

And, finally, here is the example with both operators:

```
Observable.just("a", "b", "c")
        .subscribeOn(Schedulers.computation())
        .doAfterNext(s -> System.out.println("1: " +
                    Thread.currentThread().getName() + " => " + s))
        .observeOn(Schedulers.computation())
        .subscribe(s -> System.out.println("2: " +
                    Thread.currentThread().getName() + " => " + s));
pauseMs(100);
```

The result now shows that two threads are used: one for subscribing and another for observing:

```
1: RxComputationThreadPool-1 => a
2: RxComputationThreadPool-2 => a
1: RxComputationThreadPool-1 => b
2: RxComputationThreadPool-2 => b
1: RxComputationThreadPool-1 => c
2: RxComputationThreadPool-2 => c
```

This concludes our short overview of RxJava, which is a big and still-growing library with a lot of possibilities, many of which we just did not have space in this book to review. We encourage you to try and learn it because it seems that reactive programming is the way modern data processing is heading.

Summary

In this chapter, the reader has learned what reactive programming is and what its main concepts are: asynchronous, non-blocking, responsive, and so on. The reactive streams were introduced and explained in simple terms, as well as the RxJava library, the first solid implementation that supports reactive programming principles.

In the next chapter, we will talk about microservices as the foundation for creating reactive systems and will review another library that successfully supports reactive programming: **Vert.x**. We will use it to demonstrate how various microservices can be built.

Quiz

1. Select all the correct statements:

   a. Asynchronous processing always provides results later.
   b. Asynchronous processing always provides responses quickly.
   c. Asynchronous processing can use parallel processing.
   d. Asynchronous processing always provides results faster than a blocking call.

2. Can `CompletableFuture` be used without using a thread pool?
3. What does *nio* in `java.nio` stand for?
4. Is an `event` loop the only design that supports a non-blocking API?
5. What does the *Rx* in RxJava stand for?
6. Which Java package of **Java Class Library** (**JCL**) supports reactive streams?
7. Select all classes from the following list that can represent an observable in a reactive stream:
   a. `Flowable`
   b. `Probably`
   c. `CompletableFuture`
   d. `Single`

8. How do you know that the particular method (operator) of the `Observable` class is blocking?

9. What is the difference between a cold and a hot observable?

10. The `subscribe()` method of `Observable` returns the `Disposable` object. What happens when the `dispose()` method is called on this object?

11. Select all the names of the methods that create an `Observable` object:

       a. `interval()`

       b. `new()`

       c. `generate()`

       d. `defer()`

12. Name two transforming `Observable` operators.

13. Name two filtering `Observable` operators.

14. Name two backpressure-processing strategies.

15. Name two `Observable` operators that allow adding threads to the pipeline processing.

16
Microservices

In this chapter, you will learn what microservices are, how they are different from other architectural styles, and how existing microservice frameworks support message-driven architecture. We will also help you to decide on the size of a microservice and discuss whether the service size plays any role in identifying it as a microservice or not. By the end of the chapter, you will understand how to build microservices and use them as the foundational component for creating a reactive system. We will support the discussion with a detailed code demonstration of a small reactive system built using the Vert.x toolkit.

The following topics will be covered in this chapter:

- What is a microservice?
- The size of a microservice
- How microservices talk to each other
- An example of a reactive system of microservices

What is a microservice?

With processing loads constantly increasing, the conventional way of addressing the issue is to add more servers with the same `.ear` or `.war` file deployed and then join them all together in a cluster. This way, a failed server can be automatically replaced with another one, and the system will experience no decrease in its performance. The database that backs all the clustered servers is typically clustered too.

Increasing the number of clustered servers, however, is far too coarse-grained a solution for scalability, especially if the processing bottleneck is localized in only one of many procedures that are running in the application. Imagine that one particular CPU- or I/O-intensive process slows down the whole application; adding another server just to mitigate the problem of only one part of the application may carry too much of an overhead.

One way to decrease the overhead is to split the application into tiers: frontend (or a web tier), middle tier (or an app tier), and backend (or a backend tier). Each tier can be deployed independently with its own cluster of servers so that each tier can grow horizontally and remain independent of the other tiers. Such a solution makes scalability more flexible; however, at the same, this complicates the deployment process as more deployable units need to be taken care of.

Another way to guarantee the smooth deployment of each tier could be to deploy new code in one server at a time – especially if the new code is designed and implemented with backward compatibility in mind. This approach works fine for the front and middle tiers, but may not be as smooth for the backend. Added to this are unexpected outages during the deployment process that are caused by human error, defects in the code, pure accident, or a combination of all of these – thus, it is easy to understand why very few people look forward to the deployment process of a major release during production.

Yet, breaking the application into tiers may still be too coarse. In this case, some of the critical parts of the application, especially those that require more scaling than others, can be deployed in their own cluster of servers, and simply provide *services* to other parts of the system.

In fact, this is how **Service-Oriented Architecture (SOA)** was born. The complication that arose from increasing the number of deployable units was partially offset when independently deployed services were identified not only by their need for scalability but also by how often the code in them was changed. Early identification of this during the design process simplifies the deployment because only a few parts need to be changed and redeployed more often than the other parts of the system. Unfortunately, it is not easy to predict how the future system is going to evolve. That is why an independent deployment unit is often identified as a precaution because it is easier to do this during design time as opposed to later on. And this, in turn, leads to a continual decrease in the size of the deployable units.

Unfortunately, maintaining and coordinating a loose system of services comes at a price. Each participant has to be responsible for maintaining its API not only in formal terms (such as names and types) but also in spirit: the results produced by a new version of the same service have to be the same in terms of scale. Keeping the same value by type but then making it bigger or smaller in terms of scale would probably be unacceptable to the clients of the service. So, despite the declared independence, the service authors have to be more aware of who their clients are and what their needs are.

Fortunately, splitting an application into independently deployable units has brought several unexpected benefits that have increased the motivation for breaking a system into smaller services. The physical isolation allows more flexibility in choosing a programming language and the implementation platform. It also helps you to select technology that is the best for the job and to hire specialists who are able to implement it. By doing so, you are not bound by the technology choices made for the other parts of the system. This has also helped recruiters be more flexible when it comes to finding necessary talent, which is a big advantage as the demand for work continues to outpace the inflow of specialists into the job market.

Each independent part (service) is able to evolve at its own pace and become more sophisticated as long as the contract with the other parts of the system does not change or is introduced in a well-coordinated manner. This is how microservices came into being, and they have since been put to work by the giants of data processing, such as Netflix, Google, Twitter, eBay, Amazon, and Uber. Now let's talk about the results of this effort and the lessons learned.

The size of a microservice

There is no universal answer to the question *How small does a microservice have to be?* The general consensus aligns itself with the following characteristics of a microservice (in no particular order):

- The size of the source code should be smaller than that of the service in SOA architecture.
- One development team should be able to support several microservices, and the size of the team should be such that two pizzas are enough to provide lunch for the whole team.
- It has to be deployable and independent of other microservices, assuming there is no change in the contract (that is, the API).
- Each microservice has to have its own database (or schema, or set of tables, at least) – although, this is a subject of debate, especially in cases where several microservices are able to modify the same dataset; if the same team maintains all of them, it is easier to avoid a conflict while modifying the same data concurrently.
- It has to be stateless and idempotent; if one instance of the microservice has failed, another one should be able to accomplish what was expected from the failed microservice.

- It should provide a way to check its *health*, which proves that the service is up and running, has all the necessary resources, and is ready to do the job.

The sharing of resources needs to be considered during the design process, development, and after deployment and monitored for the validation of the assumptions about the degree of interference (blocking, for example) while accessing the same resource from different processes. Special care also needs to be taken during the modification process of the same persistent data whether shared across databases, schemas, or just tables within the same schema. If *eventual consistency* is acceptable (which is often the case for larger sets of data used for statistical purposes), then special measures are necessary. But the need for transactional integrity poses an often difficult problem.

One way to support a transaction across several microservices is to create a service that will play the role of a **Distributed Transaction Manager (DTM)**. In this way, other services can pass requests for data modification to it. The DTM service can keep the concurrently modified data in its own database table and move the results into the target table(s) in one transaction only after the data becomes consistent. For example, money can be added to an account by one microservice only when the corresponding amount is added to a ledger by another microservice.

If the time taken to access the data is an issue, or if you need to protect the database from an excessive number of concurrent connections, dedicating a database to a microservice may be the solution. Alternatively, a memory cache may be the way to go. Adding a service that provides access to the cache increases the isolation of the services, but requires synchronization between the peers that are managing the same cache (which is sometimes difficult).

After reviewing all the listed points and possible solutions, the size of each microservice should depend on the result of these considerations, and not as a blank statement of size imposed on all the services. This helps to avoid unproductive discussions and produce a result that is tailored to address a particular project and its needs.

How microservices talk to each other

There are more than a dozen frameworks that are currently used for building microservices. Two of the most popular are Spring Boot (`https://spring.io/projects/spring-boot`) and MicroProfile (`https://microprofile.io`) with the declared goal of optimizing Enterprise Java for a microservices-based architecture. The lightweight open source microservice framework, KumuluzEE (`https://ee.kumuluz.com`) is compliant with MicroProfile.

Here is a list of other frameworks (in alphabetical order):

- **Akka**: This is a toolkit for building highly concurrent, distributed, and resilient, message-driven applications for Java and Scala (`akka.io`).
- **Bootique**: This is a minimally opinionated framework for runnable Java apps (`bootique.io`).
- **Dropwizard**: This is a Java framework for developing operations-friendly, high-performance, and RESTful web services (`www.dropwizard.io`).
- **Jodd**: This is a set of Java microframeworks, tools, and utilities under 1.7 MB (`jodd.org`).
- **Lightbend Lagom**: This is an opinionated microservice framework built on Akka and Play (`www.lightbend.com`).
- **Ninja**: This is a full-stack framework for Java (`https://www.ninjaframework.org/`).
- **Spotify Apollo**: This is a set of Java libraries used by Spotify for writing microservices (Spotify/Apollo).
- **Vert.x**: This is a toolkit for building reactive applications on a JVM (`vertx.io`).

All of these frameworks support REST-based communication between microservices; some of them also have an additional way of sending messages.

To demonstrate the alternative versus traditional methods of communication, we will use Vert.x, which is an event-driven non-blocking lightweight polyglot toolkit. It allows you to write components in Java, JavaScript, Groovy, Ruby, Scala, Kotlin, and Ceylon. It supports an asynchronous programming model and a distributed event bus that reaches into in-browser JavaScript, thus allowing the creation of real-time web applications. However, because of the focus of this book, we are going to use Java only.

Vert.x API has two source trees: the first starts with `io.vertx.core`, and the second starts with `io.vertx.rxjava.core`. The second source tree is a reactive version of the `io.vertx.core` class. In fact, the reactive source tree is based on non-reactive sources, so these two source trees are not incompatible. On the contrary, the reactive version is provided in *addition* to the non-reactive implementation. Since our discussion is focused around reactive programming, we will mainly use the classes and interfaces of the `io.vertx.rxjava` source tree, also called the **rxfied Vert.x API**.

To begin, we will add the following dependency to the `pom.xml` file, as follows:

```
<dependency>
    <groupId>io.vertx</groupId>
    <artifactId>vertx-web</artifactId>
    <version>3.6.3</version>
</dependency>
<dependency>
    <groupId>io.vertx</groupId>
    <artifactId>vertx-rx-java</artifactId>
    <version>3.6.3</version>
</dependency>
```

A class that implements the `io.vertx.core.Verticle` interface serves as a building block of a Vert.x-based application. The `io.vertx.core.Verticle` interface has four abstract methods:

```
Vertx getVertx();
void init(Vertx var1, Context var2);
void start(Future<Void> var1) throws Exception;
void stop(Future<Void> var1) throws Exception;
```

To make the coding easier in practice, there is an abstract `io.vertx.rxjava.core.AbstractVerticle` class that implements all of the methods, but they are empty and do not do anything. It allows creating a verticle by extending the `AbstractVerticle` class and implementing only those methods of the `Verticle` interface that are needed for the application. In most cases, implementing the `start()` method only is enough.

Vert.x has its own system of exchanging messages (or events) through an event bus. By using the `rxSend(String address, Object msg)` method of the `io.vertx.rxjava.core.eventBus.EventBus` class, any verticle can send a message to any address (which is just a string):

```
Single<Message<String>> reply = vertx.eventBus().rxSend(address, msg);
```

The `vertx` object (which is the protected property of `AbstractVerticle` and is available for every verticle) allows access to the event bus and the `rxSend()` call method. The `Single<Message<String>>` return value represents a reply that can be returned in response to the message; you can subscribe to it, or process it in any other way.

A verticle can also register as a message receiver (consumer) for a certain address:

```
vertx.eventBus().consumer(address);
```

If several verticles are registered as consumers for the same address, then the `rxSend()` method delivers the message only to one of these consumers using a round-robin algorithm.

Alternatively, the `publish()` method can be used to deliver a message to all consumers that are registered with the same address:

```
EventBus eb = vertx.eventBus().publish(address, msg);
```

The returned object is the `EventBus` object, which allows you to add other `EventBus` methods invocations, if necessary.

As you may recall, message-driven asynchronous processing is a foundation for the elasticity, responsiveness, and resilience of a reactive system composed of microservices. That is why, in the next section, we will demonstrate how you can build a reactive system that uses both REST-based communication and Vert.x `EventBus`-based messages.

The reactive system of microservices

To demonstrate how a reactive system of microservices may look if implemented using Vert.x, we are going to create an HTTP server that can accept a REST-based request to the system, send an `EventBus`-based message to another verticle, receive a reply, and send the response back to the original request.

To demonstrate how it all works, we will also write a program that generates HTTP requests to the system and allows you to test the system from outside.

The HTTP server

Let's assume that the entry point into the reactive system demonstration is going to be an HTTP call. This means that we need to create a verticle that acts as an HTTP server. Vert.x makes this really easy; the following three lines in a verticle will do the trick:

```
HttpServer server = vertx.createHttpServer();
server.requestStream().toObservable()
      .subscribe(request -> request.response()
            .setStatusCode(200)
            .end("Hello from " + name + "!\n")
      );
server.rxListen(port).subscribe();
```

As you can see, the created server listens to the specified port and responds with **Hello...** to each incoming request. By default, the hostname is `localhost`. If necessary, another address for the host can be specified using an overloaded version of the same method:

```
server.rxListen(port, hostname).subscribe();
```

Here is the entire code of the verticle that we have created:

```
package com.packt.learnjava.ch16_microservices;
import io.vertx.core.Future;
import io.vertx.rxjava.core.AbstractVerticle;
import io.vertx.rxjava.core.http.HttpServer;
public class HttpServerVert extends AbstractVerticle {
    private int port;
    public HttpServerVert(int port) { this.port = port; }
    public void start(Future<Void> startFuture) {
        String name = this.getClass().getSimpleName() +
                    "(" + Thread.currentThread().getName() +
                                ", localhost:" + port + ")";
        HttpServer server = vertx.createHttpServer();
        server.requestStream().toObservable()
            .subscribe(request -> request.response()
                    .setStatusCode(200)
                    .end("Hello from " + name + "!\n")
            );
        server.rxListen(port).subscribe();
        System.out.println(name + " is waiting...");
    }
}
```

We can deploy this server using the following code:

```
Vertx vertx = Vertx.vertx();
RxHelper.deployVerticle(vertx, new HttpServerVert(8082));
```

The result will look as follows:

```
HttpServerVert(vert.x-eventloop-thread-0, localhost:8082) is waiting...
```

Notice that the **...is waiting...** message appears immediately, even before any request has come in – that is the asynchronous nature of this server. The `name` prefix is constructed to contain the class name, thread name, hostname, and port. Notice that the thread name tells us that the server listens on the event loop thread, `0`.

Now we can issue a request to the deployed server using the `curl` command; the response will be as follows:

```
demo> curl localhost:8082
Hello from HttpServerVert(vert.x-eventloop-thread-0, localhost:8082)!
demo>
```

As you can see, we have issued the HTTP GET (default) request and got back the expected **Hello...** message with the expected name.

The following code is a more realistic version of the `start()` method:

```
Router router = Router.router(vertx);
router.get("/some/path/:name/:address/:anotherParam")
      .handler(this::processGetSomePath);
router.post("/some/path/send")
      .handler(this::processPostSomePathSend);
router.post("/some/path/publish")
      .handler(this::processPostSomePathPublish);
vertx.createHttpServer()
      .requestHandler(router::handle)
      .rxListen(port)
      .subscribe();
System.out.println(name + " is waiting...");
```

Now we use the `Router` class and send requests to different handlers depending on the HTTP method (GET or POST) and the path. It requires you to add the following dependency to the `pom.xml` file:

```
<dependency>
    <groupId>io.vertx</groupId>
    <artifactId>vertx-web</artifactId>
    <version>3.6.3</version>
</dependency>
```

The first route has the `/some/path/:name/:address/:anotherParam` path, which includes three parameters (`name`, `address`, and `anotherParam`). The HTTP request is passed inside the `RoutingContext` object to the following handler:

```
private void processGetSomePath(RoutingContext ctx){
    ctx.response()
       .setStatusCode(200)
       .end("Got into processGetSomePath using " +
                                ctx.normalisedPath() + "\n");
}
```

The handler simply returns an HTTP code of `200` and a hardcoded message that is set on the HTTP response object and is returned by the `response()` method. Behind the scenes, the HTTP response object comes from the HTTP request. We have made the first implementation of the handlers simple for clarity. Later, we will reimplement them in a more realistic way.

The second route has the `/some/path/send` path with the following handler:

```
private void processPostSomePathSend(RoutingContext ctx){
    ctx.response()
       .setStatusCode(200)
       .end("Got into processPostSomePathSend using " +
                                   ctx.normalisedPath() + "\n");
```

The third route has the `/some/path/publish` path with the following handler:

```
private void processPostSomePathPublish(RoutingContext ctx){
    ctx.response()
       .setStatusCode(200)
       .end("Got into processPostSomePathPublish using " +
                                   ctx.normalisedPath() + "\n");
}
```

If we deploy our server again and issue HTTP requests to hit each of the routes, we will see the following screenshot:

```
demo>
demo> curl localhost:8082/some/path
Got into processGetSomePath using /some/path
demo> curl localhost:8082/some/path/send
<html><body><h1>Resource not found</h1></body></html>demo>
demo> curl -X POST localhost:8082/some/path/send
Got into processPostSomePathSend using /some/path/send
demo> curl -X POST localhost:8082/some/path/publish
Got into processPostSomePathPublish using /some/path/publish
demo>
```

The preceding screenshot illustrates that we sent the expected message to the first HTTP GET request, but received **Resource not found** in response to the second HTTP GET request. This is because there is no `/some/path/send` route for the HTTP GET request in our server. We then switched to the HTTP POST request and received the expected messages for both POST requests.

From the name of the paths, you could guess that we are going to use the /some/path/send route to send the EventBus message, and the /some/path/publish route to publish the EventBus message. But before implementing the corresponding route handlers, let's create a verticle that is going to receive the EventBus messages.

The EventBus message receiver

The implementation of the message receiver is pretty straightforward:

```
vertx.eventBus()
    .consumer(address)
    .toObservable()
    .subscribe(msgObj -> {
            String body = msgObj.body().toString();
            String msg = name + " got message '" + body + "'.";
            System.out.println(msg);
            String reply = msg + " Thank you.";
            msgObj.reply(reply);
    }, Throwable::printStackTrace );
```

The EventBus object can be accessed via the vertx object. The consumer(address) method of the EventBus class allows you to set the address that is associated with this message receiver and returns MessageConsumer<Object>. We then convert this object to Observable and subscribe to it, waiting for the message to be received asynchronously. The subscribe() method has several overloaded versions. We have selected one that accepts two functions: the first is called for each of the emitted values (for each received message, in our case); the second is called when an exception is thrown anywhere in the pipeline (that is, it acts like the all-embracing try...catch block). The MessageConsumer<Object> class indicates that, in principle, the message can be represented by an object of any class. As you can see, we have decided that we are going to send a string, so we just cast the message body to String.
The MessageConsumer<Object> class also has a reply(Object) method that allows you to send a message back to the sender.

The full implementation of the message-receiving verticle is as follows:

```
package com.packt.learnjava.ch16_microservices;
import io.vertx.core.Future;
import io.vertx.rxjava.core.AbstractVerticle;
public class MessageRcvVert extends AbstractVerticle {
    private String id, address;
    public MessageRcvVert(String id, String address) {
```

```
            this.id = id;
            this.address = address;
        }
        public void start(Future<Void> startFuture) {
            String name = this.getClass().getSimpleName() +
                    "(" + Thread.currentThread().getName() +
                            ", " + id + ", " + address + ")";
            vertx.eventBus()
                .consumer(address)
                .toObservable()
                .subscribe(msgObj -> {
                        String body = msgObj.body().toString();
                        String msg = name + " got message '" + body + "'.";
                        System.out.println(msg);
                        String reply = msg + " Thank you.";
                        msgObj.reply(reply);
                }, Throwable::printStackTrace );
            System.out.println(name + " is waiting...");
        }
    }
```

We can deploy this verticle in the same way that we deployed the `HttpServerVert` verticle:

```
String address = "One";
Vertx vertx = Vertx.vertx();
RxHelper.deployVerticle(vertx, new MessageRcvVert("1", address));
```

If we run this code, the following message will be displayed:

```
MessageRcvVert(vert.x-eventloop-thread-0, 1, One) is waiting...
```

As you can see, the last line of `MessageRcvVert` is reached and executed, while the created pipeline and the functions we have passed to its operators are waiting for the message to be sent. So, let's go ahead and do that now.

The EventBus message senders

As we have promised, we will now reimplement the handlers of the `HttpServerVert` verticle in a more realistic manner. The GET method handler now looks like the following code block:

```
private void processGetSomePath(RoutingContext ctx){
    String caller = ctx.pathParam("name");
    String address = ctx.pathParam("address");
    String value = ctx.pathParam("anotherParam");
    System.out.println("\n" + name + ": " + caller + " called.");
    vertx.eventBus()
        .rxSend(address, caller + " called with value " + value)
        .toObservable()
        .subscribe(reply -> {
            System.out.println(name +
                        ": got message\n      " + reply.body());
            ctx.response()
                .setStatusCode(200)
                .end(reply.body().toString() + "\n");
        }, Throwable::printStackTrace);
}
```

As you can see, the `RoutingContext` class provides the `pathParam()` method, which extracts parameters from the path (if they are marked with `:`, as in our example). Then, again, we use the `EventBus` object to send a message asynchronously to the address provided as a parameter. The `subscribe()` method uses the provided function to process the reply from the message receiver and to send the response back to the original request to the HTTP server.

Let's now deploy both verticles – the `HttpServerVert` and the `MessageRcvVert` verticles:

```
String address = "One";
Vertx vertx = Vertx.vertx();
RxHelper.deployVerticle(vertx, new MessageRcvVert("1", address));
RxHelper.deployVerticle(vertx, new HttpServerVert(8082));
```

When we run the preceding code, the screen displays the following messages:

```
MessageRcvVert(vert.x-eventloop-thread-0, 1, One) is waiting...
HttpServerVert(vert.x-eventloop-thread-1, localhost:8082) is waiting...
```

Notice that each of the verticles is run on its own thread. Now we can submit the HTTP GET request using the `curl` command; the result is as follows:

```
demo> curl localhost:8082/some/path/Nick/One/someValue
MessageRcvVert(vert.x-eventloop-thread-0, 1, One) got message 'Nick called with value someValue'. Thank you.
demo>
```

This is how the interaction is viewed from outside our demonstration system. Inside, we can also see the following messages, which allow us to trace how our verticles interact and send messages to each other:

```
HttpServerVert(vert.x-eventloop-thread-1, localhost:8082): Nick called.
MessageRcvVert(vert.x-eventloop-thread-0, 1, One) got message 'Nick called with value someValue'.
HttpServerVert(vert.x-eventloop-thread-1, localhost:8082): got message
    MessageRcvVert(vert.x-eventloop-thread-0, 1, One) got message 'Nick called with value someValue'. Thank you.
```

The result looks exactly as expected.

Now, the handler for the `/some/path/send` path is as follows:

```java
private void processPostSomePathSend(RoutingContext ctx){
    ctx.request().bodyHandler(buffer -> {
        System.out.println("\n" + name + ": got payload\n    " + buffer);
        JsonObject payload = new JsonObject(buffer.toString());
        String caller = payload.getString("name");
        String address = payload.getString("address");
        String value = payload.getString("anotherParam");
        vertx.eventBus()
            .rxSend(address, caller + " called with value " + value)
            .toObservable()
            .subscribe(reply -> {
                System.out.println(name +
                            ": got message\n    " + reply.body());
                ctx.response()
                   .setStatusCode(200)
                   .end(reply.body().toString() + "\n");
            }, Throwable::printStackTrace);
    });
}
```

For an HTTP POST request, we expect a payload in the JSON format to be sent with the same values that we sent as the parameters for the HTTP GET request. The rest of the method is very similar to the `processGetSomePath()` implementation. Let's deploy the `HttpServerVert` and `MessageRcvVert` verticles again, and then issue the HTTP POST request with a payload; the result will be as follows:

```
demo>
demo> curl -X POST localhost:8082/some/path/send -d '{"name":"Nick","address":"One","anotherParam":"someValue"}'
MessageRcvVert(vert.x-eventloop-thread-0, 1, One) got message 'Nick called with value someValue'. Thank you.
demo>
```

This looks exactly like the result of the HTTP GET request, as was designed. At the backend, the following messages are displayed:

```
HttpServerVert(vert.x-eventloop-thread-1, localhost:8082): got payload
    {"name":"Nick","address":"One","anotherParam":"someValue"}
MessageRcvVert(vert.x-eventloop-thread-0, 1, One) got message 'Nick called with value someValue'.
HttpServerVert(vert.x-eventloop-thread-1, localhost:8082): got message
    MessageRcvVert(vert.x-eventloop-thread-0, 1, One) got message 'Nick called with value someValue'. Thank you.
```

There is nothing new in these messages either, except that the JSON format is displayed.

Finally, let's take a look at the handler of the HTTP POST request for the `/some/path/publish` path:

```java
private void processPostSomePathPublish(RoutingContext ctx){
    ctx.request().bodyHandler(buffer -> {
        System.out.println("\n" + name + ": got payload\n    " + buffer);
        JsonObject payload = new JsonObject(buffer.toString());
        String caller = payload.getString("name");
        String address = payload.getString("address");
        String value = payload.getString("anotherParam");
        vertx.eventBus()
            .publish(address, caller + " called with value " + value);
        ctx.response()
            .setStatusCode(202)
            .end("The message was published to address " +
                                            address + ".\n");
    });
}
```

This time, we have used the publish() method to send the message. Notice that this method does not have the ability to receive a reply. That is because, as we have mentioned already, the publish() method sends the message to all receivers that are registered with this address. If we issue an HTTP POST request with the /some/path/publish path, the result looks slightly different:

```
demo>
demo> curl -X POST localhost:8082/some/path/publish -d '{"name":"Nick","address":"One","anotherParam":"someValue"}'
The message was published to address One.
demo>
```

Additionally, the messages on the backend look different too:

```
HttpServerVert(vert.x-eventloop-thread-1, localhost:8082): got payload
    {"name":"Nick","address":"One","anotherParam":"someValue"}
MessageRcvVert(vert.x-eventloop-thread-0, 1, One) got message 'Nick called with value someValue'.
```

All of these differences are related to the fact that the server cannot get back a reply, even though the receiver sends it in exactly the same way that it does in response to the message sent by the rxSend() method.

In the next section, we will deploy several instances of the sender and receiver and examine the difference between the message distribution by the rxSend() and publish() methods.

The reactive system demonstration

Let's now assemble and deploy a small reactive system using the verticles created in the previous section:

```java
package com.packt.learnjava.ch16_microservices;
import io.vertx.rxjava.core.RxHelper;
import io.vertx.rxjava.core.Vertx;
public class ReactiveSystemDemo {
    public static void main(String... args) {
        String address = "One";
        Vertx vertx = Vertx.vertx();
        RxHelper.deployVerticle(vertx, new MessageRcvVert("1", address));
        RxHelper.deployVerticle(vertx, new MessageRcvVert("2", address));
        RxHelper.deployVerticle(vertx, new MessageRcvVert("3", "Two"));
        RxHelper.deployVerticle(vertx, new HttpServerVert(8082));
    }
}
```

As you can see, we are going to deploy two verticles that use the same `One` address to receive messages and one verticle that uses the `Two` address. If we run the preceding program, the screen will display the following messages:

```
MessageRcvVert(vert.x-eventloop-thread-2, 3, Two) is waiting...
MessageRcvVert(vert.x-eventloop-thread-1, 2, One) is waiting...
MessageRcvVert(vert.x-eventloop-thread-0, 1, One) is waiting...
HttpServerVert(vert.x-eventloop-thread-3, localhost:8082) is waiting...
```

Let's now start sending HTTP requests to our system. First, let's send the same HTTP `GET` request three times:

```
demo>
demo> curl localhost:8082/some/path/Nick/One/someValue
MessageRcvVert(vert.x-eventloop-thread-0, 1, One) got message 'Nick called with value someValue'. Thank you.
demo>
demo> curl localhost:8082/some/path/Nick/One/someValue
MessageRcvVert(vert.x-eventloop-thread-1, 2, One) got message 'Nick called with value someValue'. Thank you.
demo>
demo> curl localhost:8082/some/path/Nick/One/someValue
MessageRcvVert(vert.x-eventloop-thread-0, 1, One) got message 'Nick called with value someValue'. Thank you.
demo>
```

As we have mentioned already, if there are several verticles registered with the same address, the `rxSend()` method uses a round-robin algorithm to select the verticle that should receive the next message. The first request went to the receiver with `ID="1"`, the second request went to the receiver with `ID="2"`, and the third request went to the receiver with `ID="1"` again.

We get the same results using the HTTP `POST` request for the `/some/path/send` path:

```
demo>
demo> curl -X POST localhost:8082/some/path/send -d '{"name":"Nick","address":"One","anotherParam":"someValue"}'
MessageRcvVert(vert.x-eventloop-thread-1, 2, One) got message 'Nick called with value someValue'. Thank you.
demo>
demo> curl -X POST localhost:8082/some/path/send -d '{"name":"Nick","address":"One","anotherParam":"someValue"}'
MessageRcvVert(vert.x-eventloop-thread-0, 1, One) got message 'Nick called with value someValue'. Thank you.
demo>
demo> curl -X POST localhost:8082/some/path/send -d '{"name":"Nick","address":"One","anotherParam":"someValue"}'
MessageRcvVert(vert.x-eventloop-thread-1, 2, One) got message 'Nick called with value someValue'. Thank you.
demo>
```

Again, the receiver of the message is rotated using the round-robin algorithm.

Now, let's publish a message to our system twice:

```
demo>
demo> curl -X POST localhost:8082/some/path/publish -d '{"name":"Nick","address":"One","anotherParam":"someValue"}'
The message was published to address One.
demo>
demo> curl -X POST localhost:8082/some/path/publish -d '{"name":"Nick","address":"One","anotherParam":"someValue"}'
The message was published to address One.
demo>
```

Since the receiver's reply cannot propagate back to the system user, we need to take a look at the messages that are logged on the backend:

```
HttpServerVert(vert.x-eventloop-thread-3, localhost:8082): got payload
    {"name":"Nick","address":"One","anotherParam":"someValue"}
MessageRcvVert(vert.x-eventloop-thread-0, 1, One) got message 'Nick called with value someValue'.
MessageRcvVert(vert.x-eventloop-thread-1, 2, One) got message 'Nick called with value someValue'.

HttpServerVert(vert.x-eventloop-thread-3, localhost:8082): got payload
    {"name":"Nick","address":"One","anotherParam":"someValue"}
MessageRcvVert(vert.x-eventloop-thread-1, 2, One) got message 'Nick called with value someValue'.
MessageRcvVert(vert.x-eventloop-thread-0, 1, One) got message 'Nick called with value someValue'.
```

As you can see, the `publish()` method sends the message to all verticles that are registered to the specified address. And note that the verticle with ID=`"3"` (registered with the Two address) never received a message.

Before we wrap up this reactive system demonstration, it is worth mentioning that Vert.x allows you to easily cluster verticles. You can read about this feature in the Vert.x documentation (`https://vertx.io/docs/vertx-core/java`).

Summary

In this chapter, the reader was introduced to the concept of microservices and how they can be used to create a reactive system. We discussed the significance of the size of an application and how this can affect your decision to convert it into a microservice. You also learned how existing microservice frameworks support message-driven architecture and had a chance to use one of them – the Vert.x toolkit – in practice.

In the next chapter, we are going to explore the **Java Microbenchmark Harness (JMH)** project, which allows you to measure code performance and other parameters. We will define what JMH is, how to create and run a benchmark, what the benchmark parameters are, and the supporting IDE plugins.

Quiz

1. Select all the correct statements:

    a. Microservices allow more refined scaling.
    b. Microservice deployment is easier than a monolith.
    c. Microservices facilitate choosing technologies that are best suited for the job.
    d. Microservice programming is much easier than any other application programming.

2. Can a microservice be bigger than some of the monolith applications?
3. How can microservices talk to each other?
4. Name two frameworks that are created in support of microservices.
5. What is the main building block of microservices in Vert.x?
6. What is the difference between the `send` and `publish` event bus message in Vert.x?
7. How does the `send` method of an event bus decide which receiver to send a message in Vert.x?
8. Can Vert.x verticles be clustered?
9. Where can you find out more information about Vert.x?

17
Java Microbenchmark Harness

In this chapter, the reader will be introduced to a **Java Microbenchmark Harness (JMH)** project that allows for measuring various code performance characteristics. If performance is an important issue for your application, this tool can help you to identify bottlenecks with precision—up to the method level. Using it, the reader will be able to not only measure the average execution time of the code and other performance values (such as throughput, for example) but to do it in a controlled manner—with or without the JVM optimizations, warmup runs, and so on.

In addition to theoretical knowledge, the reader will have a chance to run JMH using practical demo examples and recommendations.

The following topics will be covered in this chapter:

- What is JMH?
- Creating a JMH benchmark
- Running the benchmark
- Using the IDE plugin
- JMH benchmark parameters
- JMH usage examples

What is JMH?

According to the dictionary, a **benchmark** is *a standard or point of reference against which things may be compared or assessed*. In programming, it is the way to compare the performance of applications, or just methods. The **micro preface** is focused on the latter—the smaller code fragments rather than an application as a whole. The JMH is a framework for measuring the performance of a single method.

That may appear to be very useful. Can we not just run a method a thousand or a hundred thousand times in a loop, measure how long it took, and then calculate the average of the method performance? We can. The problem is that JVM is a much more complicated program than just a code-executing machine. It has optimization algorithms focused on making the application code run as fast as possible.

For example, let's look at the following class:

```java
class SomeClass {
    public int someMethod(int m, int s) {
        int res = 0;
        for(int i = 0; i < m; i++){
            int n = i * i;
            if (n != 0 && n % s == 0) {
                res =+ n;
            }
        }
        return res;
    }
}
```

We filled the `someMethod()` method with code that does not make much sense but keeps the method busy. To test the performance of this method, it is tempting to copy the code into some test method and run it in a loop:

```java
public void testCode() {
    StopWatch stopWatch = new StopWatch();
    stopWatch.start();
    int xN = 100_000;
    int m = 1000;
    for(int x = 0; i < xN; x++) {
        int res = 0;
        for(int i = 0; i < m; i++){
            int n = i * i;
            if (n != 0 && n % 250_000 == 0) {
                res += n;
            }
        }
    }
    System.out.println("Average time = " +
                        (stopWatch.getTime() / xN /m) + "ms");
}
```

However, JVM will see that the `res` result is never used and qualify the calculations as **dead code** (the code section that is never executed). So, why bother executing this code at all?

You may be surprised to see that the significant complication or simplification of the algorithm does not affect the performance. That is because, in every case, the code is not actually executed.

You may change the test method and pretend that the result is used by returning it:

```
public int testCode() {
    StopWatch stopWatch = new StopWatch();
    stopWatch.start();
    int xN = 100_000;
    int m = 1000;
    int res = 0;
    for(int x = 0; i < xN; x++) {
        for(int i = 0; i < m; i++){
            int n = i * i;
            if (n != 0 && n % 250_000 == 0) {
                res += n;
            }
        }
    }
    System.out.println("Average time = " +
                        (stopWatch.getTime() / xN / m) + "ms");
    return res;
}
```

This may convince JVM to execute the code every time, but it is not guaranteed. The JVM may notice that the input into the calculations does not change and this algorithm produces the same result every run. Since the code is based on constant input, this optimization is called **constant folding**. The result of this optimization is that this code may be executed only once and the same result is assumed for every run, without actually executing the code.

In practice though, the benchmark is often built around a method, not a block of code. For example, the test code may look as follows:

```
public void testCode() {
    StopWatch stopWatch = new StopWatch();
    stopWatch.start();
    int xN = 100_000;
    int m = 1000;
    SomeClass someClass = new SomeClass();
    for(int x = 0; i < xN; x++) {
        someClass.someMethod(m, 250_000);
    }
    System.out.println("Average time = " +
                        (stopWatch.getTime() / xN / m) + "ms");
}
```

But even this code is susceptible to the same JVM optimization we have just described.

The JMH was created to help to avoid this, and similar pitfalls. In the *JMH Usage examples* section, we will show you how to use JMH to work around the dead code and constants folding optimization, using the @State annotation and the Blackhole object.

Besides, JMH allows for measuring not only average execution time but also throughput and other performance characteristics.

Creating a JMH benchmark

To start using JMH, the following dependencies have to be added to the pom.xml file:

```
<dependency>
    <groupId>org.openjdk.jmh</groupId>
    <artifactId>jmh-core</artifactId>
    <version>1.21</version>
</dependency>
<dependency>
    <groupId>org.openjdk.jmh</groupId>
    <artifactId>jmh-generator-annprocess</artifactId>
    <version>1.21</version>
</dependency>
```

The name of the second .jar file, annprocess, provides a hint that JMH uses annotations. If you guessed so, you were correct. Here is an example of a benchmark created for testing the performance of an algorithm:

```
public class BenchmarkDemo {
    public static void main(String... args) throws Exception{
        org.openjdk.jmh.Main.main(args);
    }
    @Benchmark
    public void testTheMethod() {
        int res = 0;
        for(int i = 0; i < 1000; i++){
            int n = i * i;
            if (n != 0 && n % 250_000 == 0) {
                res += n;
            }
        }
    }
}
```

Please notice the `@Benchmark` annotation. It tells the framework that this method performance has to be measured. If you run the preceding `main()` method, you will see an output similar to the following:

```
# Run progress: 40.00% complete, ETA 00:15:49
# Fork: 2 of 5
# Warmup Iteration   1: ≈ 10⁻⁹ s/op
# Warmup Iteration   2: ≈ 10⁻⁹ s/op
# Warmup Iteration   3: ≈ 10⁻⁹ s/op
# Warmup Iteration   4: ≈ 10⁻⁹ s/op
# Warmup Iteration   5: ≈ 10⁻⁹ s/op
Iteration   1: ≈ 10⁻⁹ s/op
Iteration   2: ≈ 10⁻⁹ s/op
Iteration   3: ≈ 10⁻⁹ s/op
Iteration   4: ≈ 10⁻⁹ s/op
Iteration   5: ≈ 10⁻⁹ s/op
```

This is only one segment of an extensive output that includes multiple iterations under different conditions with the goal being to avoid or offset the JVM optimization. It also takes into account the difference between running the code once and running it multiple times. In the latter case, the JVM starts using the just-in-time compiler, which compiles the often-used bytecodes code into native binary code and does not even read the bytecodes. The warmup cycles serve this purpose—the code is executed without measuring its performance as a dry run that *warms up* the JVM.

There are also ways to tell the JVM which method to compile and use as binary directly, which method to compile every time, and to provide similar instructions to disable certain optimization. We will talk about this shortly.

Let's now see how to run the benchmark.

Running the benchmark

As you have probably guessed, one way to run a benchmark is just to execute the `main()` method. It can be done using `java` command directly or using the IDE. We talked about it in `Chapter 1`, *Getting Started with Java 12*. Yet there is an easier and more convenient way to run a benchmark: by using an IDE plugin.

Using an IDE plugin

All major Java supporting IDEs have such a plugin. We will demonstrate how to use the plugin for IntelliJ installed on a macOS computer, but it is equally applicable to Windows systems.

Here are the steps to follow:

1. To start installing the plugin, press the *command* key and comma (,) together, or just click the wrench symbol (with the hover text **Preferences**) in the top horizontal menu:

2. It will open a window with the following menu in the left pane:

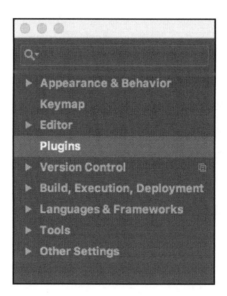

3. Select **Plugins**, as shown in the preceding screenshot, and observe the window with the following top horizontal menu:

4. Select **Marketplace**, type *JMH* in the **Search plugins in marketplace** input field ,and press *Enter*. If you have an internet connection, it will show you a **JMH plugin** symbol, similar to the one shown in this screenshot:

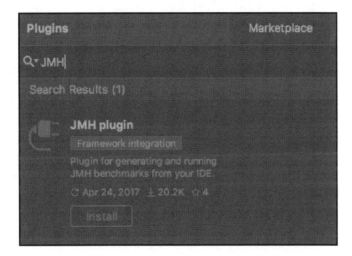

5. Click the **Install** button and then, after it turns into **Restart IDE**, click it again:

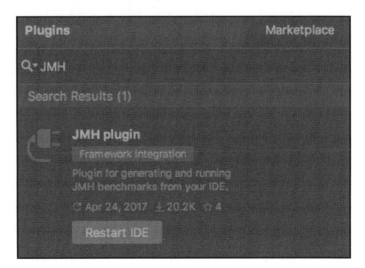

6. After the IDE restarts, the plugin is ready to be used. Now you can not only run the `main()` method but you can also pick and choose which of the benchmark methods to execute if you have several methods with the `@Benchmark` annotation. To do this, select **Run...** from the **Run** drop-down menu:

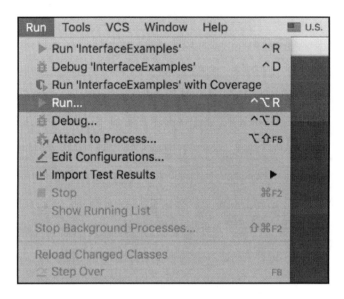

7. It will bring up a window with a selection of methods you can run:

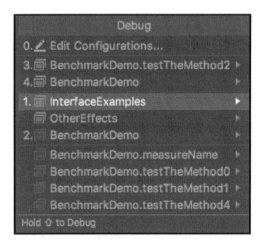

8. Select the one you would like to run and it will be executed. After you have run a method at least once, you can just right-click on it and execute it from the pop-up menu:

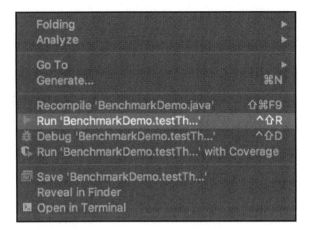

9. You can also use the shortcuts shown to the right of each menu item.

Now let's review the parameters that can be passed to the benchmark.

JMH benchmark parameters

There are many benchmark parameters that allow for fine-tuning the measurements for the particular needs of the task at hand. We are going to present only the major ones.

Mode

The first set of parameters defines the performance aspect (mode) the particular benchmark has to measure:

- `Mode.AverageTime`: Measures the average execution time
- `Mode.Throughput`: Measures the throughput by calling the benchmark method in an iteration

- `Mode.SampleTime`: Samples the execution time, instead of averaging it; allows us to infer the distributions, percentiles, and so on
- `Mode.SingleShotTime`: Measures the single method invocation time; allows for the testing of a cold startup without calling the benchmark method continuously

These parameters can be specified in the annotation `@BenchmarkMode`. For example:

```
@BenchmarkMode(Mode.AverageTime)
```

It is possible to combine several modes:

```
@BenchmarkMode({Mode.Throughput, Mode.AverageTime, Mode.SampleTime,
Mode.SingleShotTime}
```

It is also possible to request all of them:

```
@BenchmarkMode(Mode.All)
```

The described parameters and all the parameters we are going to discuss later in this chapter can be set at the method and/or class level. The method-level set value overrides the class-level value.

Output time unit

The unit of time used for presenting the results can be specified using the `@OutputTimeUnit` annotation:

```
@OutputTimeUnit(TimeUnit.NANOSECONDS)
```

The possible time units come from the `java.util.concurrent.TimeUnit` enum.

Iterations

Another group of parameters defines the iterations used for the warmups and measurements. For example:

```
@Warmup(iterations = 5, time = 100, timeUnit = TimeUnit.MILLISECONDS)
@Measurement(iterations = 5, time = 100, timeUnit = TimeUnit.MILLISECONDS)
```

Forking

While running several tests, the `@Fork` annotation allows you to set each test to be run in a separate process. For example:

```
@Fork(10)
```

The passed-in parameter value indicates how many times the JVM can to be forked into independent processes. The default value is –1. Without it, the test's performance can be mixed, if you use several classes implementing the same interface in tests and they affect each other.

The `warmups` parameter is another one which can be set to indicate how many times the benchmark has to execute without collecting measurements:

```
@Fork(value = 10, warmups = 5)
```

It also allows you to add Java options to the `java` command line. For example:

```
@Fork(value = 10, jvmArgs = {"-Xms2G", "-Xmx2G"})
```

The full list of JMH parameters and examples of how to use them can be found in the `openjdk` project (`http://hg.openjdk.java.net/code-tools/jmh/file/tip/jmh-samples/src/main/java/org/openjdk/jmh/samples`). For example, we did not mention `@Group`, `@GroupThreads`, `@Measurement`, `@Setup`, `@Threads`, `@Timeout`, `@TearDown`, or `@Warmup`.

JMH usage examples

Let's now run a few tests and compare them. First, we run the following test method:

```
@Benchmark
@BenchmarkMode(Mode.All)
@OutputTimeUnit(TimeUnit.NANOSECONDS)
public void testTheMethod0() {
    int res = 0;
    for(int i = 0; i < 1000; i++){
        int n = i * i;
        if (n != 0 && n % 250_000 == 0) {
            res += n;
        }
    }
}
```

As you can see, we have requested to measure all the performance characteristics and to use nanoseconds while presenting the results. On our system, the test execution took around 20 minutes and the final results summary looked like this:

```
Benchmark                                                 Mode     Cnt       Score        Error  Units
BenchmarkDemo.testTheMethod0                              thrpt     25       0.001 ±      0.001  ops/ns
BenchmarkDemo.testTheMethod0                              avgt      25    1041.096 ±     18.841  ns/op
BenchmarkDemo.testTheMethod0                              sample 7406325   1107.536 ±      1.555  ns/op
BenchmarkDemo.testTheMethod0:testTheMethod0·p0.00        sample             929.000             ns/op
BenchmarkDemo.testTheMethod0:testTheMethod0·p0.50        sample            1020.000             ns/op
BenchmarkDemo.testTheMethod0:testTheMethod0·p0.90        sample            1210.000             ns/op
BenchmarkDemo.testTheMethod0:testTheMethod0·p0.95        sample            1230.000             ns/op
BenchmarkDemo.testTheMethod0:testTheMethod0·p0.99        sample            1860.000             ns/op
BenchmarkDemo.testTheMethod0:testTheMethod0·p0.999       sample           18496.000             ns/op
BenchmarkDemo.testTheMethod0:testTheMethod0·p0.9999      sample           41792.000             ns/op
BenchmarkDemo.testTheMethod0:testTheMethod0·p1.00        sample          582656.000             ns/op
BenchmarkDemo.testTheMethod0                              ss         5   29973.000 ± 10585.906  ns/op
```

Let's now change the test as follows:

```java
@Benchmark
@BenchmarkMode(Mode.All)
@OutputTimeUnit(TimeUnit.NANOSECONDS)
public void testTheMethod1() {
    SomeClass someClass = new SomeClass();
    int i = 1000;
    int s = 250_000;
    someClass.someMethod(i, s);
}
```

If we run the `testTheMethod1()` now, the results will be slightly different:

```
Benchmark                                                 Mode     Cnt       Score        Error  Units
BenchmarkDemo.testTheMethod1                              thrpt     25       0.001 ±      0.001  ops/ns
BenchmarkDemo.testTheMethod1                              avgt      25    1037.961 ±     22.561  ns/op
BenchmarkDemo.testTheMethod1                              sample 7531674   1091.872 ±      2.760  ns/op
BenchmarkDemo.testTheMethod1:testTheMethod1·p0.00        sample             898.000             ns/op
BenchmarkDemo.testTheMethod1:testTheMethod1·p0.50        sample             991.000             ns/op
BenchmarkDemo.testTheMethod1:testTheMethod1·p0.90        sample            1200.000             ns/op
BenchmarkDemo.testTheMethod1:testTheMethod1·p0.95        sample            1228.000             ns/op
BenchmarkDemo.testTheMethod1:testTheMethod1·p0.99        sample            1860.000             ns/op
BenchmarkDemo.testTheMethod1:testTheMethod1·p0.999       sample           18720.000             ns/op
BenchmarkDemo.testTheMethod1:testTheMethod1·p0.9999      sample           40256.000             ns/op
BenchmarkDemo.testTheMethod1:testTheMethod1·p1.00        sample         3874816.000             ns/op
BenchmarkDemo.testTheMethod1                              ss         5  303431.200 ± 185625.725  ns/op
```

The results are mostly different around sampling and single-shot running. You can play with these methods and change the forking and number of warmups.

Using the @State annotation

This JMH feature allows you to hide the source of the data from the JVM, thus preventing dead code optimization. You can add a class as the source of the input data as follows:

```
@State(Scope.Thread)
public static class TestState {
    public int m = 1000;
    public int s = 250_000;
}

@Benchmark
@BenchmarkMode(Mode.All)
@OutputTimeUnit(TimeUnit.NANOSECONDS)
public int testTheMethod3(TestState state) {
    SomeClass someClass = new SomeClass();
    return someClass.someMethod(state.m, state.s);
}
```

The `Scope` value is used for sharing data between tests. In our case, with only one test using the `TestCase` class object, we do not have a need for sharing. Otherwise, the value can be set to `Scope.Group` or `Scope.Benchmark`, which means we could add setters to the `TestState` class and read/modify it in other tests.

When we ran this version of the test, we got the following results:

Benchmark	Mode	Cnt	Score	Error	Units
BenchmarkDemo.testTheMethod3	thrpt	25	$\approx 10^{-3}$		ops/ns
BenchmarkDemo.testTheMethod3	avgt	25	3064.479 ±	58.388	ns/op
BenchmarkDemo.testTheMethod3	sample	5893288	3342.995 ±	7.652	ns/op
BenchmarkDemo.testTheMethod3:testTheMethod3·p0.00	sample		2624.000		ns/op
BenchmarkDemo.testTheMethod3:testTheMethod3·p0.50	sample		3072.000		ns/op
BenchmarkDemo.testTheMethod3:testTheMethod3·p0.90	sample		3656.000		ns/op
BenchmarkDemo.testTheMethod3:testTheMethod3·p0.95	sample		4312.000		ns/op
BenchmarkDemo.testTheMethod3:testTheMethod3·p0.99	sample		6000.000		ns/op
BenchmarkDemo.testTheMethod3:testTheMethod3·p0.999	sample		29120.000		ns/op
BenchmarkDemo.testTheMethod3:testTheMethod3·p0.9999	sample		73856.000		ns/op
BenchmarkDemo.testTheMethod3:testTheMethod3·p1.00	sample		6324224.000		ns/op
BenchmarkDemo.testTheMethod3	ss	5	365093.400 ±	256801.341	ns/op

The data has changed again. Notice that the average time for execution has increased three-fold which indicates that more JVM optimization was not applied.

Using the Blackhole object

This JMH feature allows for simulating the results usage, thus preventing the JVM from folding constants optimization:

```
@Benchmark
@BenchmarkMode(Mode.All)
@OutputTimeUnit(TimeUnit.NANOSECONDS)
public void testTheMethod4(TestState state, Blackhole blackhole) {
    SomeClass someClass = new SomeClass();
    blackhole.consume(someClass.someMethod(state.m, state.s));
}
```

As you can see, we have just added a parameter `Blackhole` object and called the `consume()` method on it, thus pretending that the result of the tested method is used.

When we ran this version of the test, we got the following results:

Benchmark	Mode	Cnt	Score	Error	Units
BenchmarkDemo.testTheMethod4	thrpt	25	≈ 10^{-3}		ops/ns
BenchmarkDemo.testTheMethod4	avgt	25	3253.271 ±	82.781	ns/op
BenchmarkDemo.testTheMethod4	sample	6538933	3398.419 ±	5.033	ns/op
BenchmarkDemo.testTheMethod4:testTheMethod4·p0.00	sample		2724.000		ns/op
BenchmarkDemo.testTheMethod4:testTheMethod4·p0.50	sample		3080.000		ns/op
BenchmarkDemo.testTheMethod4:testTheMethod4·p0.90	sample		3668.000		ns/op
BenchmarkDemo.testTheMethod4:testTheMethod4·p0.95	sample		4328.000		ns/op
BenchmarkDemo.testTheMethod4:testTheMethod4·p0.99	sample		5624.000		ns/op
BenchmarkDemo.testTheMethod4:testTheMethod4·p0.999	sample		29728.000		ns/op
BenchmarkDemo.testTheMethod4:testTheMethod4·p0.9999	sample		66573.645		ns/op
BenchmarkDemo.testTheMethod4:testTheMethod4·p1.00	sample		4964352.000		ns/op
BenchmarkDemo.testTheMethod4	ss	5	296304.600 ±	114666.024	ns/op

This time, the results look not that different. Apparently, the constant folding optimization was neutralized even before the `Blackhole` usage was added.

Using the @CompilerControl annotation

Another way to tune up the benchmark is to tell the compiler to compile, inline (or not), and exclude (or not) a particular method from the code. For example, consider the following class:

```
class SomeClass{
    public int oneMethod(int m, int s) {
        int res = 0;
        for(int i = 0; i < m; i++){
            int n = i * i;
```

```
            if (n != 0 && n % s == 0) {
                res = anotherMethod(res, n);
            }
        }
        return res;
    }

    @CompilerControl(CompilerControl.Mode.EXCLUDE)
    private int anotherMethod(int res, int n){
        return res +=n;
    }

}
```

Assuming we are interested in how the method `anotherMethod()` compilation/inlining affects the performance, we can set the `CompilerControl` mode on it to the following:

- `Mode.INLINE`: To force this method inlining
- `Mode.DONT_INLINE`: To avoid this method inlining
- `Mode.EXCLUDE`: To avoid this method compiling

Using the @Param annotation

Sometimes, it is necessary to run the same benchmark for a different set of input data. In such a case, the `@Param` annotation is very useful.

`@Param` is a standard Java annotation used by various frameworks, for example, JUnit. It identifies an array of parameter values. The test with the `@Param` annotation will be run as many times as there are values in the array. Each test execution picks up a different value from the array.

Here is an example:

```
@State(Scope.Benchmark)
public static class TestState1 {
    @Param({"100", "1000", "10000"})
    public int m;
    public int s = 250_000;
}

@Benchmark
@BenchmarkMode(Mode.All)
@OutputTimeUnit(TimeUnit.NANOSECONDS)
public void testTheMethod6(TestState1 state, Blackhole blackhole) {
```

```
        SomeClass someClass = new SomeClass();
        blackhole.consume(someClass.someMethod(state.m, state.s));
    }
```

The testTheMethod6() benchmark is going to be used with each of the listed values of the parameter m.

A word of caution

The described harness takes away most of the worries of the programmer who measures the performance. And yet, it is virtually impossible to cover all the cases of JVM optimization, profile sharing, and similar aspects of the JVM implementation, especially if we take into account that JVM code evolves and differs from one implementation to another. The authors of JMH acknowledge this fact by printing the following warning along with the test results:

```
REMEMBER: The numbers below are just data. To gain reusable insights, you need to follow up on
why the numbers are the way they are. Use profilers (see -prof, -lprof), design factorial
experiments, perform baseline and negative tests that provide experimental control, make sure
the benchmarking environment is safe on JVM/OS/HW level, ask for reviews from the domain experts.
Do not assume the numbers tell you what you want them to tell.
```

The description of the profilers and their usage can be found in the openjdk project (http://hg.openjdk.java.net/code-tools/jmh/file/tip/jmh-samples/src/main/java/org/openjdk/jmh/samples). Among the same samples, you will encounter the description of the code generated by JMH, based on the annotations.

If you would like to get really deep into the details of your code execution and testing, there is no better way to do it than to study the generated code. It describes all the steps and decisions JMH makes in order to run the requested benchmark. You can find the generated code in the target/generated-sources/annotations.

The scope of this book does not allow for going into too many details on how to read it, but it is not very difficult, especially if you start with a simple case of testing one method only. We wish you all the best in this endeavor.

Summary

In this chapter, the reader has learned about the JMH tool and was able to use it for specific practical cases similar to those they encountered while programming their applications. The reader has learned how to create and run a benchmark, how to set the benchmark parameters, and how to install IDE plugins if needed. We also have provided practical recommendations and references for further reading.

In the next chapter, readers will be introduced to the useful practices of designing and writing application code. We will talk about Java idioms, their implementation and usage, and provide recommendations for implementing `equals()`, `hashCode()`, `compareTo()`, and `clone()` methods. We will also discuss the difference between the usage of the `StringBuffer` and `StringBuilder` classes, how to catch exceptions, best design practices, and other proven programming practices.

Quiz

1. Select all the correct statements:

    a. JMH is useless since it runs methods outside the production context.
    b. JMH is able to work around some of the JVM optimizations.
    c. JMH allows for measuring not only average performance time but other performance characteristics too.
    d. JMH can be used to measure the performance of small applications too.

2. Name two steps necessary to start using JMH.
3. Name four ways JMH can be run.
4. Name two modes (performance characteristics) that can be used (measured) with JMH.
5. Name two of time units that can be used to present JMH test results.
6. How can the data (results, state) be shared between JMH benchmarks?
7. How do you tell JMH to run the benchmark for the parameter with the enumerated list of values?

8. How can the compilation of a method be forced or turned off?
9. How can the JVM's constant folding optimization be turned off?
10. How can Java command options be provided programmatically for running the particular benchmark?

Best Practices for Writing High-Quality Code

18

When programmers talk to each other, they often use jargon that cannot be understood by non-programmers, or vaguely understood by the programmers of different programming languages. But those who use the same programming language understand each other just fine. Sometimes it may also depend on how knowledgeable a programmer is. A novice may not understand what an experienced programmer is talking about, while a seasoned colleague nods and responds in kind.

In this chapter, readers will be introduced to some Java programming jargon—the Java idioms that describe certain features, functionality, design solutions, and so on. The reader will also learn the most popular and useful practices of designing and writing application code. By the end of this chapter, the reader will have a solid understanding of what other Java programmers are talking about while discussing their design decisions and the functionalities they use.

The following topics will be covered in this chapter:

- Java idioms, their implementation, and their usage
- The `equals()`, `hashCode()`, `compareTo()`, and `clone()` methods
- The `StringBuffer` and `StringBuilder` classes
- `try`, `catch`, and `finally` clauses

- Best design practices
- Code is written for people
- Testing—the shortest path to quality code

Java idioms, their implementation, and their usage

In addition to serving the means of communication among professionals, programming idioms are also proven programming solutions and common practices that are not directly derived from the language specification, but born out of the programming experience. In this section, we are going to discuss the ones that are used most often. You can find and study the full list of idioms in the official Java documentation (https://docs.oracle.com/javase/tutorial).

The equals() and hashCode() methods

The default implementation of the equals() and hashCode() methods in the java.lang.Object class looks as follows:

```
public boolean equals(Object obj) {
    return (this == obj);
}
/**
* Whenever it is invoked on the same object more than once during
* an execution of a Java application, the hashCode method
* must consistently return the same integer...
* As far as is reasonably practical, the hashCode method defined
* by class Object returns distinct integers for distinct objects.
*/
@HotSpotIntrinsicCandidate
public native int hashCode();
```

As you can see, the default implementation of the `equals()` method compares only memory references that point to the addresses where the objects are stored. Similarly, as you can see from the comments (quoted from the source code), the `hashCode()` method returns the same integer for the same object and a different integer for different objects. Let's demonstrate it using the `Person` class:

```
public class Person {
    private int age;
    private String firstName, lastName;
    public Person(int age, String firstName, String lastName) {
        this.age = age;
        this.lastName = lastName;
        this.firstName = firstName;
    }
    public int getAge() { return age; }
    public String getFirstName() { return firstName; }
    public String getLastName() { return lastName; }
}
```

Here is an example of how the default `equals()` and `hashCode()` methods behave:

```
Person person1 = new Person(42, "Nick", "Samoylov");
Person person2 = person1;
Person person3 = new Person(42, "Nick", "Samoylov");
System.out.println(person1.equals(person2)); //prints: true
System.out.println(person1.equals(person3)); //prints: false
System.out.println(person1.hashCode());       //prints: 777874839
System.out.println(person2.hashCode());       //prints: 777874839
System.out.println(person3.hashCode());       //prints: 596512129
```

The `person1` and `person2` references and their hash codes are equal because they point to the same object (the same area of the memory, and the same address), while the `person3` reference points to another object.

In practice, though, as we have described in Chapter 6, *Data Structures, Generics, and Popular Utilities*, we would like the equality of the object to be based on the value of all or some of the object properties, so here is a typical implementation of the `equals()` and `hashCode()` methods:

```
@Override
public boolean equals(Object o) {
    if (this == o) return true;
    if (o == null) return false;
    if(!(o instanceof Person)) return false;
    Person person = (Person)o;
    return getAge() == person.getAge() &&
```

```
            Objects.equals(getFirstName(), person.getFirstName()) &&
            Objects.equals(getLastName(), person.getLastName());
}
@Override
public int hashCode() {
    return Objects.hash(getAge(), getFirstName(), getLastName());
}
```

It used to be more involved, but using `java.util.Objects` utilities makes it much easier, especially if you notice that the method `Objects.equals()` method handles `null` too.

We have added the described implementation of `equals()` and `hashCode()` methods to the `Person1` class and have executed the same comparisons:

```
Person1 person1 = new Person1(42, "Nick", "Samoylov");
Person1 person2 = person1;
Person1 person3 = new Person1(42, "Nick", "Samoylov");
System.out.println(person1.equals(person2)); //prints: true
System.out.println(person1.equals(person3)); //prints: true
System.out.println(person1.hashCode());      //prints: 2115012528
System.out.println(person2.hashCode());      //prints: 2115012528
System.out.println(person3.hashCode());      //prints: 2115012528
```

As you can see, the change we have made not only makes the same objects equal but makes equal two different objects with the same values of the properties too. Furthermore, the hash code value is now based on the values of the same properties as well.

In `Chapter 6`, *Data Structures, Generics, and Popular Utilities*, we explained why it is important to implement the `hasCode()` method while implementing the `equals()` method.

 It is very important that exactly the same set of properties is used for establishing equality in the `equals()` method and for the hash calculation in the `hashCode()` method.

Having the `@Override` annotation in front of these methods assures that they really override the default implementation in the `Object` class. Otherwise, a typo in the method name may create the illusion that the new implementation is used when in fact it is not. Debugging such cases has proved much more difficult and costly than just adding the `@Override` annotation, which generates an error if the method does not override anything.

The compareTo() method

In Chapter 6, *Data Structures, Generics, and Popular Utilities*, we used the compareTo() method (the only method of the Comparable interface) extensively and pointed out that the order that is established based on this method (its implementation by the elements of a collection) is called a **natural order**.

To demonstrate it, we created the Person2 class:

```
public class Person2 implements Comparable<Person2> {
    private int age;
    private String firstName, lastName;
    public Person2(int age, String firstName, String lastName) {
        this.age = age;
        this.lastName = lastName;
        this.firstName = firstName;
    }
    public int getAge() { return age; }
    public String getFirstName() { return firstName; }
    public String getLastName() { return lastName; }
    @Override
    public int compareTo(Person2 p) {
        int result = Objects.compare(getFirstName(),
                    p.getFirstName(), Comparator.naturalOrder());
        if (result != 0) {
            return result;
        }
        result = Objects.compare(getLastName(),
                    p.getLastName(), Comparator.naturalOrder());
        if (result != 0) {
            return result;
        }
        return Objects.compare(age, p.getAge(),
                                    Comparator.naturalOrder());
    }
    @Override
    public String toString() {
        return firstName + " " + lastName + ", " + age;
    }
}
```

Then we composed a list of `Person2` objects and sorted it:

```
Person2 p1 = new Person2(15, "Zoe", "Adams");
Person2 p2 = new Person2(25, "Nick", "Brook");
Person2 p3 = new Person2(42, "Nick", "Samoylov");
Person2 p4 = new Person2(50, "Ada", "Valentino");
Person2 p6 = new Person2(50, "Bob", "Avalon");
Person2 p5 = new Person2(10, "Zoe", "Adams");
List<Person2> list = new ArrayList<>(List.of(p5, p2, p6, p1, p4, p3));
Collections.sort(list);
list.stream().forEach(System.out::println);
```

The result looks as follows:

```
Ada Valentino, 50
Bob Avalon, 50
Nick Brook, 25
Nick Samoylov, 42
Zoe Adams, 10
Zoe Adams, 15
```

There are three things worth noting:

- According to the `Comparable` interface, the `compareTo()` method must return a negative integer, zero, or a positive integer as the object is less than, equal to, or greater than another object. In our implementation, we returned the result immediately if the values of the same property of two objects were different. We know already that this object is *bigger* or *smaller* no matter what the other properties are. But the sequence, in which you compare the properties of two objects, has an effect on the final result. It defines the precedence in which the property value affects the order.

- We have put the result of `List.of()` into a `new ArrayList()` object. We did so because, as we have mentioned already in Chapter 6, *Data Structures, Generics, and Popular Utilities*, the collection created by a factory method `of()` is unmodifiable. No elements can be added or removed from it and the order of the elements cannot be changed either, while we need to sort the created collection. We used the `of()` method, only because it is more convenient and provides a shorter notation.

- Finally, using `java.util.Objects` for properties comparison makes the implementation much easier and more reliable than custom coding.

While implementing the `compareTo()` method, it is important to make sure that the following rules are not violated:

- `obj1.compareTo(obj2)` returns the same value as `obj2.compareTo(obj1)` only when the returned value is 0.
- If the returned value is not 0, `obj1.compareTo(obj2)` has the opposite sign of `obj2.compareTo(obj1)`.
- If `obj1.compareTo(obj2) > 0` and `obj2.compareTo(obj3) > 0` then `obj1.compareTo(obj3) > 0`.
- If `obj1.compareTo(obj2) < 0` and `obj2.compareTo(obj3) < 0` then `obj1.compareTo(obj3) < 0`.
- If `obj1.compareTo(obj2) == 0` then `obj2.compareTo(obj3)` and `obj1.compareTo(obj3) > 0` have the same sign.
- Both `obj1.compareTo(obj2)` and `obj2.compareTo(obj1)` throw the same exceptions, if any.

It is also recommended, but not always required, that if `obj1.equals(obj2)` then `obj1.compareTo(obj2) == 0` and, at the same time, if `obj1.compareTo(obj2) == 0` then `obj1.equals(obj2)`.

The clone() method

The `clone()` method implementation in the `java.lang.Object` class looks like this:

```
@HotSpotIntrinsicCandidate
protected native Object clone() throws CloneNotSupportedException;
```

The comment states the following:

```
/**
 * Creates and returns a copy of this object.  The precise meaning
 * of "copy" may depend on the class of the object.
 ***
```

The default result of this method returns a copy of the object fields as is, which is fine if the values are of primitive types. However, if an object property holds a reference to another object, only the reference itself will be copied, not the referred object itself. That is why such a copy is called a **shallow copy**. To get a **deep copy**, one has to override the clone() method and clone each of the object properties that refers an object.

In any case, to be able to clone an object, it has to implement the Cloneable interface and make sure that all the objects along the inheritance tree (and the properties that are objects) implement the Cloneable interface too (except the java.lang.Object class). The Cloneable interface is just a marker interface that tells the compiler that the programmer made a conscious decision to allow this object to be cloned (whether because the shallow copy was good enough or because the clone() method was overridden). An attempt to call clone() on an object that does not implement the Cloneable interface will result in a CloneNotSupportedException.

It looks complex already, but in practice, there are even more pitfalls. For example, let's say that the Person class has an address property of type Address class. The shallow copy p2 of the Person object p1 will refer the same object of Address so that p1.address == p2.address. Here is an example. The Address class looks as follows:

```java
class Address {
    private String street, city;
    public Address(String street, String city) {
        this.street = street;
        this.city = city;
    }
    public void setStreet(String street) { this.street = street; }
    public String getStreet() { return street; }
    public String getCity() { return city; }
}
```

The Person3 class uses it like this:

```java
class Person3 implements Cloneable{
    private int age;
    private Address address;
    private String firstName, lastName;

    public Person3(int age, String firstName,
                        String lastName, Address address) {
        this.age = age;
        this.address = address;
        this.lastName = lastName;
        this.firstName = firstName;
    }
}
```

```
    public int getAge() { return age; }
    public Address getAddress() { return address; }
    public String getLastName() { return lastName; }
    public String getFirstName() { return firstName; }
    @Override
    public Person3 clone() throws CloneNotSupportedException{
        return (Person3) super.clone();
    }
}
```

Notice that the method clone does a shallow copy because it does not clone the address property. Here is the result of using such a clone() method implementation:

```
Person3 p1 = new Person3(42, "Nick", "Samoylov",
                            new Address("25 Main Street", "Denver"));
Person3 p2 = p1.clone();
System.out.println(p1.getAge() == p2.getAge());                      // true
System.out.println(p1.getLastName() == p2.getLastName());            // true
System.out.println(p1.getLastName().equals(p2.getLastName()));  // true
System.out.println(p1.getAddress() == p2.getAddress());             // true
System.out.println(p2.getAddress().getStreet());  //prints: 25 Main Street
p1.getAddress().setStreet("42 Dead End");
System.out.println(p2.getAddress().getStreet());  //prints: 42 Dead End
```

As you can see, after the cloning is complete, the change made to the address property of the source object is reflected in the same property of the clone. That isn't very intuitive, is it? While cloning we expected independent copy, didn't we?

To avoid sharing the Address object, one needs to clone it explicitly too. In order to do it, one has to make the Address object cloneable, as follows:

```
public class Address implements Cloneable{
    private String street, city;
    public Address(String street, String city) {
        this.street = street;
        this.city = city;
    }
    public void setStreet(String street) { this.street = street; }
    public String getStreet() { return street; }
    public String getCity() { return city; }
    @Override
    public Address clone() throws CloneNotSupportedException {
        return (Address)super.clone();
    }
}
```

With that implementation in place, we can now add the `address` property cloning:

```
class Person4 implements Cloneable{
    private int age;
    private Address address;
    private String firstName, lastName;
    public Person4(int age, String firstName,
                            String lastName, Address address) {
        this.age = age;
        this.address = address;
        this.lastName = lastName;
        this.firstName = firstName;
    }
    public int getAge() { return age; }
    public Address getAddress() { return address; }
    public String getLastName() { return lastName; }
    public String getFirstName() { return firstName; }
    @Override
    public Person4 clone() throws CloneNotSupportedException{
        Person4 cl = (Person4) super.clone();
        cl.address = this.address.clone();
        return cl;
    }
}
```

Now, if we run the same test, the results are going to be as we expected them originally:

```
Person4 p1 = new Person4(42, "Nick", "Samoylov",
        new Address("25 Main Street", "Denver"));
Person4 p2 = p1.clone();
System.out.println(p1.getAge() == p2.getAge());              // true
System.out.println(p1.getLastName() == p2.getLastName());       // true
System.out.println(p1.getLastName().equals(p2.getLastName())); // true
System.out.println(p1.getAddress() == p2.getAddress());         // false
System.out.println(p2.getAddress().getStreet()); //prints: 25 Main Street
p1.getAddress().setStreet("42 Dead End");
System.out.println(p2.getAddress().getStreet()); //prints: 25 Main Street
```

So, if the application expects all the properties to be deeply copied, all the objects involved have to be cloneable. That is fine as long as none of the related objects, whether a property in the current object or in the parent class (and their properties and parents), do not acquire a new object property without making them cloneable and are cloned explicitly in the `clone()` method of the container object. This last statement is complex. And the reason for its complexity is the underlying complexity of the cloning process. That is why programmers often stay away from making objects cloneable.

Instead, they prefer to clone the object manually, if need be. For example:

```
Person4 p1 = new Person4(42, "Nick", "Samoylov",
                         new Address("25 Main Street", "Denver"));
Address address = new Address(p1.getAddress().getStreet(),
                                      p1.getAddress().getCity());
Person4 p2 = new Person4(p1.getAge(), p1.getFirstName(),
                                      p1.getLastName(), address);
System.out.println(p1.getAge() == p2.getAge());           // true
System.out.println(p1.getLastName() == p2.getLastName());       // true
System.out.println(p1.getLastName().equals(p2.getLastName())); // true
System.out.println(p1.getAddress() == p2.getAddress());         // false
System.out.println(p2.getAddress().getStreet()); //prints: 25 Main Street
p1.getAddress().setStreet("42 Dead End");
System.out.println(p2.getAddress().getStreet()); //prints: 25 Main Street
```

This approach still requires code changes if another property is added to any related object. However, it provides more control over the result and has less chance of unexpected consequences.

Fortunately, the `clone()` method is not used very often. In fact, you may never encounter a need to use it.

The StringBuffer and StringBuilder classes

We have talked about the difference between the `StringBuffer` and `StringBuilder` classes in `Chapter 6`, *Data Structures, Generics, and Popular Utilities*. We are not going to repeat it here. Instead, we will just mention that, in a single-threaded process (which is the vast majority of cases), the `StringBuilder` class is a preferred choice because it is faster.

Try, catch, and finally clauses

This book contains `Chapter 4`, *Exception Handling*, dedicated to the usage of `try`, `catch`, and `finally` clauses, so we are not going to repeat it here. We would like just to repeat again that using a try-with-resources statement is a much-preferred way to release resources (traditionally done in a `finally` block). Deferring to the library makes the code simpler and more reliable.

Best design practices

The term *best* is often subjective and context dependent. That is why we would like to disclose that the following recommendations are based on the vast majority of cases in mainstream programming. However, they should not be followed blindly and unconditionally because there are cases when some of these practices in some contexts are useless or even wrong. Before following them, try to understand the motivation behind them and use it as your guide for your decisions. For example, size matters. If the application is not going to grow beyond a few thousand lines of code, a simple monolith with laundry-list style code is good enough. But if there are complicated pockets of code and several people working on it, breaking the code into specialized pieces would be beneficial for code understanding, maintenance, and even scaling, if one particular code area requires more resources than others.

We will start with higher-level design decisions in no particular order.

Identifying loosely coupled functional areas

These design decisions can be made very early on, based just on the general understanding of the main parts of the future system, their functionality, and the data they produce and exchange. There are several benefits of doing this:

- An identification of the structure of the future system that has bearings on the further design steps and implementation
- Specialization and deeper analysis of parts
- Parallel development of parts
- A better understanding of data flow

Breaking the functional area into traditional tiers

With each functional area in place, there can be specializations based on the technical aspects and technologies used. The traditional separation of technical specialization is:

- The frontend (user graphic or web interface)
- The middle tier with extensive business logic
- The backend (data storage or data source)

The benefits of doing this include the following:

- Deployment and scaling by tiers
- Programmer specialization based on their expertise
- Parallel development of parts

Coding to an interface

The specialized parts, based on the decisions described in the previous two subsections, have to be described in an interface that hides the implementation details. The benefits of such a design lie in the foundations of object-oriented programming and are described in detail in Chapter 2, *Java Object-Oriented Programming (OOP)*, so we are not going to repeat it here.

Using factories

We talked about this in Chapter 2, *Java Object-Oriented Programming (OOP)*, too. An interface, by definition, does not and cannot describe the constructor of a class that implements the interface. Using factories allows you to close this gap and expose just an interface to a client.

Preferring composition over inheritance

Originally, object-oriented programming was focused on inheritance as the way to share the common functionality between objects. Inheritance is even one of the four object-oriented programming principles as we have described them in Chapter 2, *Java Object-Oriented Programming (OOP)*. In practice, however, this method of functionality sharing creates too much of the dependency between classes included in the same inheritance line. The evolution of application functionality is often unpredictable, and some of the classes in the inheritance chain start to acquire functionality unrelated to the original purpose of the class chain. We can argue that there are design solutions that allow us not to do it, and keep the original classes intact. But, in practice, such things happen all the time, and the subclasses may suddenly change behavior just because they acquired new functionality through inheritance. We cannot choose our parents, can we? Besides, it breaks the encapsulation this way, while encapsulation is another foundation principle of OOP.

Composition, on the other hand, allows us to choose and control which functionality of the class to use and which to ignore. It also allows the object to stay light and not be burdened by the inheritance. Such a design is more flexible, more extensible, and more predictable.

Using libraries

Throughout the book, we mentioned many times that using the **Java Class Library** (**JCL**) and external (to the **Java Development Kit** (**JDK**) Java libraries makes programming much easier and produces a code of higher quality. There is even a dedicated chapter, Chapter 7, *Java Standard and External Libraries*, which contains an overview of the most popular Java libraries. People who create libraries invest a lot of time and effort, so you should take advantage of them any time you can.

In Chapter 13, *Functional Programming*, we described standard functional interfaces that reside in the java.util.function package of JCL. That is another way to take advantage of a library—by using the set of well-known and shared interfaces, instead of defining your own ones.

This last statement is a good segue to the next topic of this chapter about writing code that is easily understood by other people.

Code is written for people

The first decades of programming required writing machine commands so that electronic devices could execute them. Not only was it a tedious and error-prone endeavor, but it also required you to write the instructions in a manner that yielded the best performance possible because the computers were slow and did not do much code optimization, if at all.

Since then, we have made a lot of progress in both hardware and programming. The modern compiler went a long way towards making the submitted code work as fast as possible, even when a programmer did not think about it. We talked about it with specific examples in the previous chapter, Chapter 17, *Java Microbenchmark Harness*.

It allowed programmers to write more lines of code without thinking much about the optimization. But tradition and many books about programming continued to call for it, and some programmers still worry about their code performance, more so than the results it produces. It is easier to follow the tradition than to break away from it. That is why programmers tend to pay more attention to the way they write code than to the business they automate, although a good code that implements an incorrect business logic is useless.

However, back to the topic. With modern JVM, the need for code optimization by a programmer is not as pressing as it used to be. Nowadays, a programmer has to pay attention mostly to a big picture, to avoid structural mistakes that lead to poor code performance and to code that is used multiple times. The latter becomes less pressing as the JVM becomes more sophisticated, observing the code in real time and just returning the results (without execution) when the same code block is called several times with the same input.

That leaves us with the only conclusion possible: while writing code, one has to make sure it is easy to read and understand for a human, not for a computer. Those who have worked in the industry for some time have been puzzled over the code they have themselves written a few years before. One improves the code-writing style via clarity and the transparency of the intent.

We can discuss the need for comments until the cows are back in the barn. We definitely do not need comments that literally echo what the code does. For example:

```
//Initialize variable
int i = 0;
```

The comments that explain the intent are much more valuable:

```
// In case of x > 0, we are looking for a matching group
// because we would like to reuse the data from the account.
// If we find the matching group, we either cancel it and clone,
// or add the x value to the group, or bail out.
// If we do not find the matching group,
// we create a new group using data of the matched group.
```

The commented code can be very complex. Good comments explain the intent and provide guidance that helps us to understand the code. Yet programmers often do not bother to write comments. The argument against writing comments typically includes two statements:

- Comments have to be maintained and evolve along with the code, otherwise, they may become misleading, but there is no tool that can prompt the programmer to adjust the comments along with changing the code. Thus, comments are dangerous.
- The code itself has to be written so (including name selection for variables and methods) that no extra explanation is needed.

Both statements are true, but it is also true that comments can be very helpful, especially those that capture the intent. Besides, such comments tend to require fewer adjustments because the code intent doesn't change often, if ever.

Testing is the shortest path to quality code

The last best practice we will discuss is this statement: *testing is not an overhead or a burden; it is the programmer's guide to success*. The only question is when to write the test.

There is a compelling argument that requires writing a test before any line of code is written. If you can do it, that is great. We are not going to try and talk you out of it. But if you do not do it, try to start writing a test after you have written one, or all lines of code, you had been tasked to write.

In practice, many experienced programmers find it helpful to start writing testing code after some of the new functionality is implemented, because that is when the programmer understands better how the new code fits into the existing context. They may even try and hard-code some values to see how well the new code is integrated with the code that calls the new method. After making sure the new code is well integrated, the programmer can continue implementing and tuning it all the time testing the new implementation against the requirements in the context of the calling code.

One important qualification has to be added: while writing the test, it is better if the input data and the test criteria are set not by you, but by a person who assigned you the task or the tester. Setting the test according to the results the code produces is a well-known programmer's trap. Objective self-assessment is not easy, if possible at all.

Summary

In this chapter, we have discussed the Java idioms that a mainstream programmer encounters daily. We have also discussed the best design practices and related recommendations, including code-writing style and testing.

In this chapter, readers have learned about the most popular Java idioms related to certain features, functionalities, and design solutions. The idioms were demonstrated with practical examples, and readers have learned how to incorporate them into their code and the professional language they use to communicate with other programmers.

In the next chapter, we will introduce the reader to four projects that add new features to Java: Panama, Valhalla, Amber, and Loom. We hope it will help the reader to follow the Java development and envision the roadmap of future releases.

Quiz

1. Select all the correct statements:
    a. Idioms can be used to communicate the code intent.
    b. Idioms can be used to explain what the code does.
    c. Idioms can be misused and obscure the topic of conversation.
    d. Idioms should be avoided in order to express the idea clearly.

2. Is it necessary to implement `hasCode()` every time `equals()` is implemented?
3. If `obj1.compareTo(obj2)` returns a negative value, what does it mean?
4. Does the deep copy notion apply to a primitive value during cloning?
5. Which is faster, `StringBuffer` or `StringBuilder`?
6. What are the benefits of coding to an interface?
7. What are the benefits of using composition versus the inheritance?
8. What is the advantage of using libraries versus writing your own code?
9. Who is the target audience of your code?
10. Is testing required?

Java - Getting New Features

19

In this chapter, the reader will learn about the current most significant projects that will add new features to Java and enhance it in other aspects. After reading this chapter, the reader will understand how to follow Java development and will envision the road map of future Java releases. If so desired, the reader can become a JDK source contributor too.

The following topics will be covered in this chapter:

- Java continues to evolve
- Panama project
- Valhalla project
- Amber project
- Loom project
- Skara project

Java continues to evolve

This is the best news for any Java developer: Java is actively supported and continues to be enhanced to stay abreast with the latest demands of the industry. It means that whatever you hear about other languages and newest technologies, you will get the best features and functionality added to Java soon. And with the new release schedule—every half year—you can be assured that the new additions will be released as soon as they prove to be useful and practical.

While thinking about designing a new application or new functionality to be added to the existing one, it is important to know how Java may be enhanced in the near future. Such knowledge can help you to design the new code in such a manner that it will be easier to accommodate new Java functions and make your application simpler and more powerful. To follow all the **JDK Enhancement Proposals (JEP)** could be impractical for a mainstream programmer because one has to follow too many different threads of discussion and development. By contrast, staying on top of one of the Java enhancement projects in the area of your interests is much easier. You could even try to contribute to one such project as an expert in a particular area, or just as an interested party.

In this chapter, we are going to review the five most important, in our opinion, Java enhancement projects:

- **Project Panama**: Focused on the interoperability with non-Java libraries
- **Project Valhalla**: Conceived around the introduction of a new value type and related generics enhancements
- **Project Amber**: Aims to bring features that can make writing Java code more readable and concise and target specific use cases such as data class, pattern match, raw string literals, concise method bodies, and lambda enhancements, to name the most significant sub-projects
- **Project Loom**: Addresses the creation of lightweight threads called **fibers** and makes asynchronous coding easier

Panama project

Throughout the book, we advised using various Java libraries— the standard **Java Class Library (JCL)** and external Java libraries that help to improve code quality and make the development time shorter. But there are also non-Java external libraries that may be needed for your application. Such a need has increased recently with the growing demand for using machine learning algorithms for data processing. The porting of these algorithms to Java does not always keep up with the latest achievements in the area of recognizing faces, classifying human actions in videos, and tracking camera movements, for example.

The existing mechanism of utilizing the libraries written in different languages is **Java Native Interface (JNI)**, **Java Native Access (JNA)**, and **Java Native Runtime (JNR)**. Despite all these facilities, accessing the native code (the code in other languages compiled for the particular platform) is not as easy as using a Java library. Besides, it limits the **Java Virtual Machine (JVM)** code optimization and often requires writing code in C.

The **Panama** project (`https://openjdk.java.net/projects/panama`) is set to address these issues, including the support of C++ functionality. The authors use the term **foreign libraries**. This term includes all the libraries in other languages. The idea behind the new approach is to translate the native headers into the corresponding Java interfaces using a tool called **jextract**. The generated interfaces allow accessing the native methods and data structures directly, without writing C code.

Not surprisingly, the supporting classes are planned to be stored in the `java.foreign` package.

At the time of writing (March 2019), the early-access builds of Panama are based on an incomplete version of Java 13 and intended for expert users. It is expected to reduce the amount of work for creating Java bindings for native libraries by 90% and produce code that performs four to five times faster than JNI at least.

Valhalla project

The motivation for the **Valhalla project** (`https://openjdk.java.net/projects/valhalla`) came from the fact that, since Java was first introduced almost 25 years ago, the hardware has changed and the decisions made at that time would have a different outcome today. For example, the operation of getting a value from memory and an arithmetic operation incurred roughly the same cost in terms of the performance time. Nowadays, the situation has changed. The memory access is from 200 to 1,000 times longer than an arithmetic operation. This means that an operation that involves primitive types is much cheaper than the operation based on their wrapping types.

When we do something with two primitive types, we grab values and use them in an operation. When we do the same operation with wrapper types, we first use the reference to access the object (which is now much longer—relative to the operation itself—than 20 years ago), and only then can we grab the value. That is why the Valhalla project attempts to introduce for a reference type a new **value** type, which provides access to a value without using the reference—the same way a primitive type is available by value.

It will also save on memory consumption and the efficiency of wrapping arrays. Each element will be now represented by a value, not by a reference.

Such a solution logically leads to the question of generics. Today, generics can be used only for a wrapping type. We can write `List<Integer>`, but we cannot write `List<int>`. And that is what the Valhalla project is poised to address too. It is going to *extend generic types to support the specialization of generic classes and interfaces over primitive types*. The extension will allow using a primitive type in generics too.

Amber project

The **Amber project** (https://openjdk.java.net/projects/amber) is focused on small Java syntax enhancements that would make it more expressive, concise, and simpler. These improvements are going to increase the Java programmers' productivity and make their code-writing more enjoyable.

Two Java features created by the Amber project have been delivered already and we talked about them:

- The type holder var (see Chapter 1, *Getting Started with Java 12*) has been available to use since Java 10.
- The local-variable syntax for lambda parameters (see Chapter 13, *Functional Programming*) was added to Java 11.
- The less-verbose switch statement (see Chapter 1, *Getting Started with Java 12*) was introduced as a preview feature with Java 12.

Other new features are going to be released with the future Java version. We will look closer only at five of them in the following subsections:

- Data class
- Pattern match
- Raw string literals
- Concise method bodies
- Lambda leftovers

Data class

There are classes that carry data only. Their purpose is to keep several values together and nothing else. For example:

```
public class Person {
    public int age;
    public String firstName, lastName;

    public Person(int age, String firstName, String lastName) {
        this.age = age;
        this.lastName = lastName;
        this.firstName = firstName;
    }
}
```

They may also include the standard set of the `equals()`, `hashCode()`, and `toString()` methods. If that is the case, why bother and write the implementation for these methods? They can be automatically generated - the same way your IDE can do it today. That is the idea behind the new entity called **data class** that can be defined as simply as follows:

```
record Person(int age, String firstName, String lastName) {}
```

The rest will be assumed present by default.

But then, as Brian Goetz wrote (`https://cr.openjdk.java.net/~briangoetz/amber/datum.html`), the questions start coming:

> *"Are they extensible? Are the fields mutable? Can I control the behavior of the generated methods or the accessibility of the fields? Can I have additional fields and constructors?"*

> *- Brian Goetz*

That is where the current state of this idea is—in the middle of the attempt to limit the scope and still provide a value to the language.

Stay tuned.

Pattern match

From time to time, almost every programmer encounters the need to switch to different processing of the value depending on its type. For example:

```
SomeClass someObj = new SomeClass();
Object value = someOtherObject.getValue("someKey");
if (value instanceof BigDecimal) {
    BigDecimal v = (BigDecimal) value;
    BigDecimal vAbs = v.abs();
    ...
} else if (value instanceof Boolean) {
    Boolean v = (Boolean)value;
    boolean v1 = v.booleanValue();
    ...
} else if (value instanceof String) {
    String v = (String) value;
    String s = v.substring(3);
    ...
}
...
```

While writing such a code, you get bored pretty quickly. And that is what pattern matching is going to fix. After the feature is implemented, the preceding code example can be changed to look as follows:

```
SomeClass someObj = new SomeClass();
Object value = someOtherObject.getValue("someKey");
if (value instanceof BigDecimal v) {
    BigDecimal vAbs = v.abs();
    ...
} else if (value instanceof Boolean v) {
    boolean v1 = v.booleanValue();
    ...
} else if (value instanceof String v) {
    String s = v.substring(3);
    ...
}
...
```

Nice, isn't it? It will also support the inlined version, such as the following one:

```
if (value instanceof String v && v.length() > 4) {
    String s = v.substring(3);
    ...
}
```

This new syntax will first be allowed in an `if` statement and later added to a `switch` statement too.

Raw string literals

Once in a while, you may wish to have an output indented, so it will look something like this, for example:

```
The result:
    - the symbol A was not found;
    - the type of the object was not Integer either.
```

To achieve it, the code looks as follows:

```
String s = "The result:\n" +
        "    - the symbol A was not found;\n" +
        "    - the type of the object was not Integer either.";
System.out.println(s);
```

After adding the new *raw string literal*, the same code can be changed to look like this:

```
String s = `The result:
                - the symbol A was not found;
                - the type of the object was not Integer either.
           `;
System.out.println(s);
```

This way, the code looks much less cluttered and easier to write. It will also be possible to align the raw string literal against left margin using the `align()` method, set an indent value using the `indent(int n)` method, and the value of the indent after alignment using the `align(int indent)` method.

Similarly, putting the string inside the symbols (`` ` ``) will allow us to avoid using the escape indicator backslash (\). For example, while executing a command, the current code may contain this line:

```
Runtime.getRuntime().exec("\"C:\\Program Files\\foo\" bar");
```

With a raw string literal in place, the same line can be changed to the following:

```
Runtime.getRuntime().exec(`"C:\Program Files\foo" bar`);
```

Again, it is easier to write and to read.

Concise method bodies

The idea of this feature was expired by the lambda expressions syntax, which can be very compact. For example:

```
Function<String, Integer> f = s -> s.length();
```

Or, using method reference, it can be expressed even shorter:

```
Function<String, Integer> f = String::length;
```

The logical extension of this approach was this: why not apply the same short-hand style to the standard getters? Look at this method:

```
String getFirstName() { return firstName; }
```

It can easily be shortened to the following form:

```
String getFirstName() -> firstName;
```

Or, consider the case when the method uses another method:

```
int getNameLength(String name) { return name.length(); }
```

It can be shortened by the method reference too, as follows:

```
int getNameLength(String name) = String::length;
```

But, as of this writing (March 2019), this proposal is still in the early stages, and many things can be changed in the final release.

Lambda leftovers

The Amber project plans three additions to the lambda expressions syntax:

- Shadowing local variable
- Netter disambiguation of functional expressions
- Using an underscore to indicate a not-used parameter

Using an underscore instead of a parameter name

In many other programming languages, an underscore (_) in a lambda expression denotes an unnamed parameter. After Java 9 made it illegal to use an underscore as an identifier, the Amber project plans to use it for a lambda parameter in the cases when this parameter is not actually needed for the current implementation. For example, look at this function:

```
BiFunction<Integer, String, String> f = (i, s) -> s.substring(3);
```

The parameter (i) is not used in the function body, but we still provide the identifier as a placeholder.

With the new addition, it will be possible to replace it with the underscore, thus avoiding using an identifier and indicating that the parameter is never used:

```
BiFunction<Integer, String, String> f = (_, s) -> s.substring(3);
```

This way, it is more difficult to miss the fact that one input value is not used.

Shadowing a local variable

Currently, it is not possible to give a parameter of a lambda expression the same name as is used as an identifier in the local context. For example:

```
int i = 42;
//some other code
BiFunction<Integer, String, String> f = (i, s) -> s.substring(3); //error
```

In future releases, such name reuse will be possible.

Better disambiguation of functional expressions

As of this writing, it is possible to have a method overloaded as follows:

```
void method(Function<String, String> fss){
    //do something
}
void method(Predicate<String> predicate){
    //do something
}
```

But, it is possible to use it only by defining the type of the passed-in function explicitly:

```
Predicate<String> pred = s -> s.contains("a");
method(pred);
```

An attempt to use it with the inlined lambda expression will fail:

```
method(s -> s.contains("a"));    // compilation error
```

The compiler complains because of an ambiguity it cannot resolve because both functions have one input parameter of the same type and are different only when it comes to the return type.

The Amber project may address it, but the final decision is not made yet because it depends on the effect this proposal has on the compiler implementation.

Loom project

The **Loom project** (https://openjdk.java.net/projects/loom) may be the most significant of the projects listed in this chapter that can give Java a power boost. From the very early days almost 25 years ago, Java provided a relatively simple multi-threading model with a well-defined synchronization mechanism. We described it in Chapter 8, *Multithreading and Concurrent Processing*. This simplicity, as well as the overall simplicity and security of Java, was one of the main factors of Java's success. Java servlets allowed the processing of many concurrent requests and were at the foundation of Java-based HTTP servers.

The thread in Java is based on the OS kernel thread, though, which is a general-purpose thread. But the kernel OS thread was designed to perform many different system tasks too. It makes such a thread too heavy (requiring too many resources) for the business needs of a particular application. The actual business operations necessary to satisfy a request received by an application, typically do not require all the thread capability. This means that the current thread-model limits the application power. To estimate how strong the limitation is, it is enough to observe that an HTTP server today can handle more than a million concurrent open sockets, while the JVM cannot handle more than a few thousand.

That was the motivation for introducing asynchronous processing, using the threads minimally and introducing lightweight processing workers instead. We talked about it in Chapter 15, *Reactive Programming* and Chapter 16, *Microservices*. The asynchronous processing model works very well, but its programming is not as simple as in other Java programming. It also requires a significant effort to integrate with legacy code based on the threads and even more effort to migrate the legacy code to adopt the new model.

Adding such a complexity made Java not as easy to learn as it used to be, and the Loom project is set to bring the simplicity of Java concurrent processing back into use by making it more lightweight.

The project plans to add to Java a new class, Fiber, in support of a lightweight thread construct, managed by the JVM. Fibers take much fewer resources. They also have almost no or very little overhead of context switching, the procedure necessary when a thread is suspended and another thread has to start or continue its own job that was suspended because of the CPU time-share or similar. The context switching of the current threads is one of the main reasons for performance limitation.

To give you an idea of how light the fibers are, compared to the threads, the Loom developers Ron Pressler and Alan Bateman provided the following numbers (`http://cr.openjdk.java.net/~rpressler/loom/JVMLS2018.pdf`):

- **Thread**:
    - Typically 1 MB reserved for stack + 16 KB of kernel data structures
    - ~2,300 bytes per started thread, includes **Virtual Machine (VM)** metadata
- **Fiber**:
    - Continuation stack: hundreds of bytes to KBs
    - 200-240 bytes per fiber in the current prototype

As you can see, we can hope there will a significant improvement in the performance of concurrent processing.

The term **continuation** is not new. It was used before *fibers*. It denotes *a sequence of instructions that execute sequentially, and may suspend itself*. Another part of the concurrent processor is a **scheduler** that *assigns continuations to CPU cores, replacing a paused one with another that's ready to run, and ensuring that a continuation that is ready to resume will eventually be assigned to a CPU core*. A current thread model also has a continuation and a scheduler, even if they are not always exposed as APIs. The Loom project intends to separate the continuation and the scheduler and to implement Java fibers on top of them. The existing `ForkJoinPool` will probably serve as the fiber.

You can read more about the motivation and goals of the Loom project in the project proposal (`https://cr.openjdk.java.net/~rpressler/loom/Loom-Proposal.html`), which is a relatively easy and very instructive read for any Java developer.

Skara project

The **Skara project** (`http://openjdk.java.net/projects/skara`) is not adding new features to Java. It is focused on improving access to the Java source code of JDK.

To access the source code today, you need to download it from a Mercurial repository and compile it manually. The Skara project's goal is to move the source code to Git because Git is now the most popular source repository, and many programmers use it already. The source code of the examples in this book, as you know, is stored on GitHub too.

The current results of the Skara project you can see in GitHub already (`https://github.com/Project-Skara/jdk`). Well, it still uses a mirror of the JDK Mercurial repository. But, in the future, it will become more independent.

Summary

In this chapter, the reader has learned about the current most significant projects that enhance the JDK. We hope that you were able to understand how to follow Java development and have envisioned the road map of future Java releases. There are many more on-going projects (`https://openjdk.java.net/projects`) that you can look at too. We also hope that you got sufficiently excited by the prospect to become a productive JDK source contributor and active community member. Welcome!

Assessments

Chapter 1 – Getting Started with Java 12

1. c) Java Development Kit
2. b) Java Class Library
3. d) Java Standard Edition
4. b) Integrated Development Environment
5. a) Project building, b) Project configuration, c) Project documentation
6. a) boolean, b) numeric
7. a) long, c) short, d) byte
8. d) Value representation
9. a) \ \ , b) 2_0 , c) 2__0f , d) \f
10. a) % , c) & , d) ->
11. a) 0
12. b) false, false
13. d) 4
14. c) Compilation error
15. b) 2
16. a), c), d)
17. d) 20 -1
18. c) The x value is within the 11 range
19. c) result = 32
20. a) A variable can be declared, b) A variable can be assigned
21. b) A selection statement, d) An increment statement

Chapter 2 – Java Object-Oriented Programming (OOP)

1. a), d)
2. b), c), d)
3. a), b), c)
4. a), c), d)
5. d)
6. c), d)
7. a), b)
8. b), d)
9. d)
10. b)
11. a), c)
12. b), c), d)
13. a), b)
14. b), c)
15. b), c), d)
16. b), c)
17. c)
18. a), b), c)
19. b), c), d)
20. a), c)
21. a), c), d)

Chapter 3 – Java Fundamentals

1. a), d)
2. c), d)
3. a), b), d)
4. a), c), d)
5. a), c)
6. a), b), d)

7. a), b), c), d)
8. c), d)
9. d)
10. c)
11. b)
12. c)

Chapter 4 – Exception Handling

1. a), b), c)
2. b)
3. c)
4. a), b), c), d)
5. a)
6. a), c)
7. d)

Chapter 5 – Strings, Input/Output, and Files

1. b)
2. c)
3. b)
4. a)
5. d)
6. a), c), d)
7. c)
8. d)
9. a), b), c)
10. c), d) (notice the usage of the `mkdir()` method, instead of `mkdirs()`)

Chapter 6 – Data Structures, Generics, and Popular Utilities

1. d)
2. b), d)
3. a), b), c), d)
4. a), b), c), d)
5. a), b), d)
6. a), b), c)
7. c)
8. a), b), c), d)
9. b), d)
10. b)
11. b), c)
12. a)
13. c)
14. d)
15. b)
16. c)
17. a)
18. b)
19. c)

Chapter 7 – Java Standard and External Libraries

1. a), b), c)
2. a), b), d)
3. b), c)
4. b), d)
5. a), c)
6. a), b), c), d)
7. b), c), d)

8. b), c)
9. b)
10. c), d)
11. a), c)
12. b), d)
13. a), d)
14. b), c), d)
15. a), b), d)
16. b), d)

Chapter 8 – Multithreading and Concurrent Processing

1. a), c), d)
2. b), c), d)
3. a)
4. a), c), d)
5. b), c), d)
6. a), b), c), d)
7. c), d)
8. a), b), c)
9. b), c)
10. b), c), d)
11. a), b), c)
12. b), c)
13. b), c)

Chapter 9 – JVM Structure and Garbage Collection

1. b), d)
2. c)

3. d)
4. b), c)
5. a), d)
6. c)
7. a), b), c), d)
8. a), c), d)
9. b), d)
10. a), b), c), d)
11. a)
12. a), b), c)
13. a), c)
14. a), c), d)
15. b), d)

Chapter 10 – Managing Data in a Database

1. c)
2. a), d)
3. b), c), d)
4. a), b), c), d)
5. a), b), c)
6. a), d)
7. a), b), c)
8. a), c)
9. a), c), d)
10. a), b)
11. a), d)
12. a), b), d)
13. a), b), c)

Chapter 11 – Network Programming

1. The correct answer may include FTP, SMTP, HTTP, HTTPS, WebSocket, SSH, Telnet, LDAP, DNS, or some other protocols
2. The correct answer may include UDP, TCP, SCTP, DCCP, or some other protocols
3. `java.net.http`
4. UDP
5. Yes
6. `java.net`
7. Transmission Control Protocol
8. They are synonyms
9. The TCP session is identified by IP address and port of the source and IP address and port of the destination
10. `ServerSocket` can be used without the client running. It just listens on the specified port
11. UDP
12. TCP
13. The correct answers may include HTTP, HTTPS, Telnet, FTP, or SMTP
14. a), c), d)
15. They are synonyms
16. They are synonyms
17. `/something/something?par=42`
18. The correct answer may include binary format, header compression, multiplexing, or push capability
19. `java.net.http.HttpClient`
20. `java.net.http.WebSocket`
21. No difference
22. `java.util.concurrent.CompletableFuture`

Chapter 12 – Java GUI Programming

1. Stage
2. Node

3. Application
4. `void start(Stage pm)`
5. `static void launch(String... args)`
6. `--module-path` **and** `--add-modules`
7. `void stop()`
8. `WebView`
9. `Media`, `MediaPlayer`, `MediaView`
10. `--add-exports`
11. Any five from the following
    list: `Blend`, `Bloom`, `BoxBlur`, `ColorAdjust`, `DisplacementMap`, `DropShadow`, `Glow`, `InnerShadow`, `Lighting`, `MotionBlur`, `PerspectiveTransform`, `Reflection`, `ShadowTone`, **and** `SepiaTone`

Chapter 13 – Functional Programming

1. c)
2. a), d)
3. One
4. `void`
5. One
6. `boolean`
7. None
8. `T`
9. One
10. `R`
11. The enclosing context
12. `Location::methodName`

Chapter 14 – Java Standard Streams

1. a), b)
2. `of()`, without parameters, produces an empty stream
3. `java.util.Set`

4. 135
5. 42
6. 2121
7. No, but it extends the functional interface `Consumer` and can be passed around as such
8. None
9. 3
10. 1.5
11. 42,X,a
12. Compilation error, because `peek()` cannot return anything
13. 2
14. An alternative `Optional` object
15. `a`
16. One
17. Any of `filter()`, `map()`, and `flatMap()`
18. Any of `distinct()`, `limit()`, `sorted()`, `reduce()`, and `collect()`

Chapter 15 – Reactive Programming

1. a), b), c)
2. Yes
3. Non-Blocking Input/Output
4. No
5. Reactive eXtension
6. `java.util.concurrent`
7. a), d)
8. The blocking operator name starts with **blocking**
9. A hot observable emits values at its own pace. A cold observable emits the next value after the previous one has reached the Terminal operator
10. The observable stops emitting values and the pipeline stops operating
11. a), c), d)
12. For example, any two from the following: `buffer()`, `flatMap()`, `groupBy()`, `map()`, `scan()`, and `window()`

13. For example, any two from the
    following: `debounce()`, `distinct()`, `elementAt(long
    n)`, `filter()`, `firstElement()`, `ignoreElements()`, `lastElement()`,
    `sample()`, `skip()`, and `take()`

14. Drop excessive values, take the latest, use buffer

15. `subscribeOn()`, `observeOn()`, `fromFuture()`

Chapter 16 – Microservices

1. a), c)
2. Yes
3. The same way the traditional applications do; plus, they often have their own
   way to communicate (using an event bus, for example)
4. Any two from the list: Akka, Dropwizard, Jodd, Lightbend Lagom, Ninja, Spotify
   Apollo, and Vert.x.
5. A class that implements the interface `io.vertx.core.Verticle`
6. `Send` only sends to one receiver registered with the address; `publish` sends
   messages to all receivers registered with the same address
7. It uses a round-robin algorithm
8. Yes
9. `https://vertx.io/`

Chapter 17 – Java Microbenchmark Harness

1. b), c), d)
2. Add a dependency on JMH to the project (or classpath, if run manually) and add
   the annotation `@Benchmark` to the method you would like to test for
   performance
3. As the main method using a Java command with an explicitly named main class,
   as the main method using a java command with an executable `.jar` file, and
   using an IDE running as the main method or using a plugin and running an
   individual method
4. Any two of the following: `Mode.AverageTime`, `Mode.Throughput`,
   `Mode.SampleTime`, and `Mode.SingleShotTime`

5. Any two of the following: `TimeUnit.NANOSECONDS`, `TimeUnit.MICROSECONDS`, `TimeUnit.MILLISECONDS`, `TimeUnit.SECONDS`, `TimeUnit.MINUTES`, `TimeUnit.HOURS`, and `TimeUnit.DAYS`
6. Using an object of a class with the annotation `@State`
7. Using the annotation `@Param` in front of the `state` property
8. Using the annotation `@CompilerConrol`
9. Using a parameter of the type `Blackhole` that consumes the produced result
10. Using the annotation `@Fork`

Chapter 18 – Best Practices for Writing High-Quality Code

1. a), b), c)
2. Generally, it is recommended but not required. It is required for certain situations for example, an when object of the class is going to be placed and searched inside a hash-based data structure
3. `obj1` is less than `obj2`
4. No
5. `StringBuilder`
6. Allowing the implementation to change without changing the client code
7. More control over the code evolution and code flexibility for accommodating a change
8. More reliable code, quicker to write, less testing, easier for other people to understand
9. Other programmers that are going to maintain your code and you some time later
10. No, but it is very helpful to you

Other Books You May Enjoy

If you enjoyed this book, you may be interested in these other books by Packt:

Java 11 Cookbook - Second Edition
Nick Samoylov

ISBN: 9781789132359

- Set up JDK and understand what's new in the JDK 11 installation
- Implement object-oriented designs using classes and interfaces
- Manage operating system processes
- Create a modular application with clear dependencies
- Build graphical user interfaces using JavaFX
- Use the new HTTP Client API
- Explore the new diagnostic features in Java 11
- Discover how to use the new JShell REPL tool

Java 11 and 12 - New Features
Mala Gupta

ISBN: 9781789133271

- Study type interference and how to work with the var type
- Understand Class-Data Sharing, its benefits, and limitations
- Discover platform options to reduce your application's launch time
- Improve application performance by switching garbage collectors
- Get up to date with the new Java release cadence
- Define and assess decision criteria for migrating to a new version of Java

Leave a review - let other readers know what you think

Please share your thoughts on this book with others by leaving a review on the site that you bought it from. If you purchased the book from Amazon, please leave us an honest review on this book's Amazon page. This is vital so that other potential readers can see and use your unbiased opinion to make purchasing decisions, we can understand what our customers think about our products, and our authors can see your feedback on the title that they have worked with Packt to create. It will only take a few minutes of your time, but is valuable to other potential customers, our authors, and Packt. Thank you!

Index

PrintStream class 189, 190
process
    about 323
    versus thread 279
Program Counter (PC) registers 328
programming 628
pull model 543

Q

queue 216

R

reactive 539
Reactive Extension (RX) 539
Reactive Manifesto
    reference 540
reactive programming
    about 538, 539
    elastic 541
    message-driven 542
    resilient 541
    responsive 540
Reactive Streams
    about 539, 542, 543
    reference 540
reactive system, of microservices
    EventBus message receiver 587, 588
    EventBus message senders 589, 591, 592
    HTTP server 583, 584, 586
reactive system
    demonstration 592, 593
reactive systems 538
reactor design pattern 538
Reactor
    reference 540
Reader class
    subclasses 192
reference type
    about 125
    as method parameter 131, 132
requestor 533
reserved keywords 136
restricted keywords 137
Reverse Address Resolution Protocol (RARP) 365
run loop 538

Runnable implementation
    versus thread extension 283, 284
runtime data areas, JVM memory
    shared areas 331
    unshared areas 331
runtime exceptions 157
Rx programming 539
rxfied Vert.x API 581
RxJava 2 API
    reference 546
RxJava 2.2.7 544
RxJava
    about 539, 544, 545
    Disposable object 553, 554
    multithreading 568, 569, 570, 572, 573
    observable types 546
    observable, creating 555, 557
    operators 557
    reference 540

S

scheduler 643
Secure Shell (SSH) 366
SELECT statement
    about 348, 349
    execute(String sql) method 352
    executeQuery(String sql) method 354
    executeUpdate(String sql) method 356
SequenceInputStream class 185
sequential processing
    considerations 530
sequential streams
    versus parallel streams 534, 535
Service Provider Interface (SPI) 543
Service-Oriented Architecture (SOA) 578
Set interface
    about 227
    initializing 218, 220
set interfaces 215
shallow copy 92, 622
Simple Mail Transfer Protocol (SMTP) 366
Simple Network Management Protocol (SNMP)
    378
Skara project
    about 643